T0183235

LIBRARY
Astronomy and Astrophysics Library

Series Editors

Martin A. Barstow, Department of Physics and Astronomy, University of Leicester, Leicester, UK

Andreas Burkert, University Observatory Munich, Munich, Germany

Athena Coustenis, LESIA, Paris-Meudon Observatory, Meudon, France

Roberto Gilmozzi, European Southern Observatory (ESO), Garching, Germany

Georges Meynet, Geneva Observatory, Versoix, Switzerland

Shin Mineshige, Department of Astronomy, Kyoto University, Kyoto, Japan

Ian Robson, The UK Astronomy Technology Centre, Edinburgh, UK

Peter Schneider, Argelander-Institut für Astronomie, Bonn, Germany

Steven N. Shore, Dipartimento di Fisica "Enrico Fermi", Università di Pisa, Pisa, Italy

Virginia Trimble, Department of Physics & Astronomy, University of California, Irvine, CA, USA

Derek Ward-Thompson, School of Physical Sciences and Computing, University of Central Lancashire, Preston, UK

More information about this series at http://www.springer.com/series/848

Hannu E. J. Koskinen • Emilia K. J. Kilpua

Physics of Earth's Radiation Belts

Theory and Observations

Springer

Hannu E. J. Koskinen
Department of Physics
University of Helsinki
Helsinki, Finland

Emilia K. J. Kilpua
Department of Physics
University of Helsinki
Helsinki, Finland

ISSN 0941-7834   ISSN 2196-9698   (electronic)
Astronomy and Astrophysics Library
ISBN 978-3-030-82169-2   ISBN 978-3-030-82167-8   (eBook)
https://doi.org/10.1007/978-3-030-82167-8

© The Editor(s) (if applicable) and The Author(s) 2022. This book is an open access publication.
Open Access This book is licensed under the terms of the Creative Commons Attribution 4.0 International License (http://creativecommons.org/licenses/by/4.0/), which permits use, sharing, adaptation, distribution and reproduction in any medium or format, as long as you give appropriate credit to the original author(s) and the source, provide a link to the Creative Commons license and indicate if changes were made.
The images or other third party material in this book are included in the book's Creative Commons license, unless indicated otherwise in a credit line to the material. If material is not included in the book's Creative Commons license and your intended use is not permitted by statutory regulation or exceeds the permitted use, you will need to obtain permission directly from the copyright holder.
The use of general descriptive names, registered names, trademarks, service marks, etc. in this publication does not imply, even in the absence of a specific statement, that such names are exempt from the relevant protective laws and regulations and therefore free for general use.
The publisher, the authors, and the editors are safe to assume that the advice and information in this book are believed to be true and accurate at the date of publication. Neither the publisher nor the authors or the editors give a warranty, expressed or implied, with respect to the material contained herein or for any errors or omissions that may have been made. The publisher remains neutral with regard to jurisdictional claims in published maps and institutional affiliations.

Cover illustration: Artist's view of the Earth's magnetosphere with emphasis of the Van Allen's Belts. Credit Jim Wilkie.

This Springer imprint is published by the registered company Springer Nature Switzerland AG.
The registered company address is: Gewerbestrasse 11, 6330 Cham, Switzerland

Foreword

The last decade has been good for studying radiation belt physics. A primary reason for this is NASA's *Van Allen Probes* mission, a pair of satellites bristling with a robust assortment of sensors carefully designed to assess the physical properties of relativistic electrons and the other particles and electromagnetic fields responsible for electron dynamics in near-Earth space. Collecting data for just over seven years, these twin satellites provide us with arguably the biggest and best data set humanity as ever had for disentangling the physics of the radiation belts. The data from the *Van Allen Probes* mission is augmented by that from several other missions that also surveyed this same region of outer space, namely the *Time-History of Events and Macroscale Interactions During Substorms* (THEMIS) mission, the *Magnetospheric Multiscale Mission*, the *Arase* mission, geosynchronous orbiting spacecraft, and low-Earth orbiting spacecraft, in particular several CubeSat missions. There has also been an extensive long-duration high-altitude balloon program in recent years focused on energetic electron physics, in particular the *Balloon Array for Radiation-belt Relativistic Electron Losses* (BARREL) campaign, for which 40 such payloads were launched from the ice sheets of Antarctica. The analysis of all of this data was enhanced through the use of theoretical advancements and improved numerical tools, including sophisticated suites of coupled models. The end result has been hundreds, perhaps thousands, of new studies about the radiation belts, written and published in the peer-reviewed disciplinary journals over the last decade, yielding a substantially new understanding of the energetic particle environment encircling our planet. The need of an updated holistic view on this topic is, therefore, critical.

As the Editor in Chief of one of those disciplinary journals in which many of these studies were published—I was EiC of the Journal of Geophysical Research–Space Physics for six years, from the beginning of 2014 through 2019—I am familiar with the development of our new thinking about the radiation belts. I was regularly amazed at the quantity of research articles produced by radiation belt physicists. We solicited manuscripts for several special sections on this topic during my EiC term, and each time I thought this would be the last, as surely the research community was running out of new findings on the subject. But no; each of these special sections was a huge success, with dozens of high-quality studies resulting in large hundreds of citations—one measure of their impact on the direction of the research field—over the next few years. The physics of Earth's radiation belts was definitely among the "hot topics" of space physics during my EiC term.

This book, *Physics of the Radiation Belts—Theory and Observations* by Drs. Hannu Koskinen and Emilia Kilpua of the University of Helsinki, offers an excellent distillation of those numerous new studies. They expertly blend the latest findings with our long-standing theories, developed over many decades, explaining the dominant physical processes governing relativistic charge particles in near-Earth space. While there will always be new studies published with additional contributions to our understanding, with some appearing right after publication of this book, it is useful to periodically assemble the collected knowledge of the research community in a single volume. The synthesis compiled herein of all of the many original research contributions in this field over the past decade is reason alone to read it.

Some might ask the question of why topical books like this one are needed anymore. In the age of the internet, with seemingly all possible information we could desire just a few clicks away, the concept of a book may seem antiquated. I disagree. I not only vehemently oppose this viewpoint but also think that modern technology—and the ease with which facts can be recalled to our electronic devices—leads to a greater need for long-form compilations of our knowledge. I like to think of it as "deep learning," analogous "deep work," deftly described by Cal Newport's book of that title. Deep learning is the process of minimizing distractions and letting our minds focus on engaging with a single topic for an extended interval. Books are more than a collection of many details but offer an integrative synthesis of the subject, bringing together disparate and seemingly disconnected facets the matter to compose a collective conceptual view that is greater than any one of the interleaved components. This is not possible in the short-form writing found among the brief descriptions of the issue available across the internet. New books are as essential today as they ever were.

This book adroitly covers each of the important topics of its chosen subject matter, examining each of the pieces of the radiation belt puzzle before bringing all of these together. By providing several chapters of introductory plasma physics, the book clearly defines the equations of motion governing why these fast-moving electrically charged particles behave the way they do. Because they are flying at relativistic speeds, they don't spend much time at any one location, the forces are mere nudges on their trajectories. It takes a persistent nudging to change their flight path, and this is most effectively accomplished through their interaction with electric and magnetic waves in space. The book devotes two full chapters to waves, a necessary inclusion to fully describe their properties. That is, the book systematically and robustly covers each of the principal topics of radiation belt physics; this aspect makes it a worthwhile reference text for anyone in the field. It doesn't stop there, however. The content of the last two chapters are an equally compelling reason to read the book, weaving those earlier sections into a comprehensive tapestry of the relative importance of those processes on the observed structure and dynamics of the radiation belts. Space physics, as a field, is moving towards a systems-level approach to geospace science, and this book makes the case that a systems-level approach is needed for Earth's radiation belts.

This is not the only new book on the radiation belts. There have been several book-by-committee compilations on space physics in recent years, including some on energetic particles in near-Earth space. The "chapters" of those books, however, are independently written review articles. The unique contribution of this work is that it is not an aggregated collection written by many different authors but a unified story, building towards a synergistic conclusion.

Speaking of story, I greatly enjoy a good novel. A narrative that develops over the course of hundreds of pages, taking hours to read, is one that carries me away along the author's carefully designed route, immersing me in a different realm among new characters with difficult problems that they address through creative solutions. Novels require many pages because there is a need to fully construct a setting, reveal personality traits of the characters, and explore relationships between them that are interwoven into a storyline. The world-building process revealed throughout a good novel captivates the reader, compelling the continuation from one page to the next, as the reader anticipates the progression towards a final climatic scene.

So it is with *Physics of the Radiation Belts*. There are many specific scientific concepts needed to fully understand Earth's radiation belts, and this book takes the reader on a well-planned journey through these topics. By the end, the reader is rewarded with a view of the radiation belts that incorporates all of those scientific threads into a comprehensive comparative analysis. The result is a beautiful gestalt of the radiation belts, tailor-made for deep learning about a hot topic of space physics.

Michael W. Liemohn

Department of Climate and Space Sciences
and Engineering
University of Michigan
Ann Arbor, MI USA
March 2021

Preface

The discovery of James A. Van Allen and his team in 1958 that the Earth is surrounded by belts of intense corpuscular radiation trapped in the quasi-dipolar terrestrial magnetic field can be considered as the birth of magnetospheric physics as we understand it today. An authoritative account of how everything happened was given by Van Allen (1983) himself in the monograph *Origins of Magnetospheric Physics*.

At the time of the first artificial satellites, the progress in magnetospheric physics was extremely rapid. The inner radiation belt was found using Geiger–Müller tubes onboard *Explorer* I and III in February and March 1958, and the two-belt structure was confirmed by the unsuccessful Moon probe *Pioneer* III in December 1958. While *Pioneer* III did not achieve the escape velocity and fell back to Earth from an altitude of more than 100,000 km, it crossed the outer belt twice and contributed valuable observations of space radiation. Soon thereafter Thomas Gold (1959) introduced the term "magnetosphere" to describe the (non-spherical) domain where the Earth's magnetic field determines the motion of charged particles. Later, the designation "Van Allen radiation belts" became common to credit Van Allen's pioneering role.

Already three months before *Explorer* I, the second satellite of the Soviet Union, *Sputnik* 2, had carried two Geiger–Müller tubes of Sergei Nikolaevich Vernov. Due to several reasons, including a limited amount of data and the tight secrecy around the Soviet space program, Vernov and his collaborators were not able to interpret the fluctuations in the counting rates of the instrument having been due to trapped radiation before the publication of *Explorer* and *Pioneer* observations (see, e.g., Baker and Panasyuk 2017).

From the very beginning it was clear that understanding, monitoring, and forecasting the rapid temporal evolution of the radiation belts were critical to both civilian and military space activities. At the end of 2020, more than 3300 active satellites were in orbit and several hundred are being launched annually. Thus, the knowledge and understanding of the radiation belts is more important than ever. The energetic corpuscular radiation is a common reason for satellite malfunctions in Earth orbit and an obvious risk to the health of astronauts. In fact, every now and then, satellite operators "rediscover" the radiation belts with unwelcome consequences. Due to the strong temporal variations of the intensity and spectrum of the Earth's radiation environment, in particular during geomagnetic storms, monitoring and forecasting the radiation environment is a key element in space weather services. Radiation belts also offer a unique natural plasma laboratory to study fundamental plasma physical processes and phenomena, including wave–particle interactions and acceleration of charged particles to relativistic energies.

The radiation belts have now been investigated for more than six decades, but many details of the underlying physical processes remain enigmatic and new surprises are found with increasingly detailed observations. Remarkable scientific progress was taking place at time of writing this volume, owing to the highly successful *Van Allen Probes*, also known as *Radiation Belt Storm Probes* (RBSP), of NASA, which were launched in 2012 and deactivated in 2019. The mission consisted of two satellites crossing through the heart of the outer radiation belt with unprecedented instrumentation for this particular purpose. The authors of this book were amazed and perplexed by the complexity of new observations and the consequent development of new modeling and theoretical approaches to match with the widening and deepening view on this important and intriguing domain of near-Earth space. We are convinced that this was the right time to write a modern textbook-style monograph combining the theoretical foundations with new data in a form accessible to students in space physics and engineering as well as young scientists already active in, or moving to, this exciting field of research.

We emphasize that there is a large and rapidly increasing amount of scientific publications on radiation belts. As this volume is meant to be a *textbook*, not a comprehensive review of the past and present literature, we have tried to be selective with citations and included mainly references that we think are necessary to follow the presentation. However, to credit the many recent contributions, the list of references has become longer than is customary in textbooks. It is evident that future studies will bring new light to the radiation belt phenomena and make parts of the content of our book obsolete, even erroneous.

Of the earlier literature, we want to highlight the classic monographs *The Adiabatic Motion of Charged Particles* (Northrop 1963), *Dynamics of Geomagnetically Trapped Radiation* (Roederer 1970) and its thoroughly revised edition *Dynamics of Magnetically Trapped Particles* (Roederer and Zhang 2014), *Particle Diffusion in the Radiation Belts* (Schulz and Lanzerotti 1974), and *Quantitative Aspects of Magnetospheric Physics* (Lyons and Williams 1984). Of the more recent sources, particularly recommendable reading are the articles in the compilation *Waves, Particles, and Storms in Geospace* edited by Balasis et al. (2016). A comprehensive summary of the recent advances of understanding the radiation belts from the space weather viewpoint is the review article by Baker et al. (2018).

On the Style and Content of the Book

Our aim has been to write a book that is accessible to readers with various backgrounds, in particular graduate students and young scientists as well as magnetospheric researchers and space engineers, who feel that they need more understanding of the physical foundations of radiation belt phenomena. While we assume that the reader has some familiarity with basic plasma physics and the Earth's plasma environment, we briefly review the central concepts in the first four chapters, simultaneously introducing the notations and conventions used later in the book. For understandable reasons, we follow the presentation and notations of the mongraph *Physics of Space Storms—From the Solar Surface to the Earth* (Koskinen 2011), which is also our main reference to more thorough discussions of basic space plasma physics not included in the present volume. A careful reader may notice that we have corrected a number of errors and typos in that book.

We wish to strongly emphasize the close ties between theory and observations. Thus, we have included several examples from present and past observations and their current interpretation. We remind that new and more comprehensive observations will, every now and then, invalidate earlier conclusions, which of course is the purpose of scientific research. As authors of a textbook, we try to avoid taking side among competing ideas in the current scientific debate.

We begin with a brief description of the magnetic and plasma environment of the radiation belts and magnetospheric dynamics in Chap. 1. The basics of single-particle motion in a magnetic field and the adiabatic invariants, with the focus on the quasi-dipolar field of the inner magnetosphere, are reviewed in Chap. 2. The chapter is concluded with an introduction to drift shell splitting and magnetopause shadowing. The basic concepts of plasma physics and the most important velocity space distribution functions in the inner magnetosphere are discussed in Chap. 3. Because the phase space density as a function of adiabatic invariants has due to the improved observations become an important—but not always quite well understood—tool in analysis of radiation belt data, the chapter concludes with a presentation of the procedure how the phase space density can be obtained from particle observations and its limitations.

Wave–particle interactions are the most important processes in transport, acceleration, and loss of radiation belt particles. There is no unique best approach to treat this complex in the most logical way in a textbook. We have selected a strategy where we first introduce in Chap. 4 the inner magnetospheric plasma wave phenomena in general, yet keeping the focus on wave modes relevant to the topic of the book. Thereafter, we discuss the drivers of the waves in Chap. 5 and the effects of the waves to particle populations in Chap. 6. We want, however, to emphasize that the growth and attenuation of the waves through particle acceleration/scattering are intimately tied to each other. Thus, these three chapters should be studied together.

Chapter 7 is dedicated to the structure and evolution of the electron belts, which became a primary focus during the *Van Allen Probes* era. Here, we also discuss the effects of different solar wind drivers of magnetospheric dynamics on radiation belts as well as the effects of energetic electron precipitation on the atmosphere.

In the end of the book, Appendix A reviews some basic concepts of electromagnetic fields and waves. Appendix B contains a brief historical reference to the spacecraft, the observations of which we have used in our presentation. We also open the acronyms of the names of the satellites in the Appendix, where they are easier to find than inside the main text.

We have deliberately left out two important topics suggested to us by some of our colleagues. We do not explicitly deal with technological consequences of corpuscular space radiation. There are several extensive compilations of articles on technological and health risks posed by high-energy particles, from radiation belts to cosmic rays, penetrating through the magnetosphere and practically any feasible shielding of components and systems. Recommendable reading is *Space Weather— Physics and Effects* (Bothmer and Daglis 2007).

Another wide research topic is physics of the radiation belts around other magnetized planets, in particular Jupiter and Saturn. While the basic physics is the same, the physical environments of the high-energy radiation belts around the giant planets are very different from the terrestrial magnetosphere. The colocation of radiation belts with moons and rings, which act as sources and sinks of heavy neutrals and ions, makes the interparticle collisions and wave–particle interactions much more complicated than in the Earth's radiation belts. Furthermore, the large-scale plasma dynamics of the fast-rotating massive magnetospheres is different. A proper treatment of these issues would require a textbook of its own.

We and several of our colleagues have tested the basic space plasma physics material included in Chaps. 1–6 in classroom practice over a period of more than 30 years. In our minds, extensive problem solving is an essential part of learning physics. However, we decided not to include exercise problems in this volume. If the book is used as course material, as we hope, the instructor can ask the students to derive some of the theoretical results that have been skipped in the text, to read and summarize seminal papers, to try to explain some peculiarities in the data presentations, or to plot various quantities as functions of their variables. Today, most observational data are available in various web-servers, which makes it possible for the more advanced students to train their skills in scientific data analysis and interpretation either using readily available tools or writing their own scripts to illustrate the data.

Helsinki, Finland
Helsinki, Finland
June, 2021

Hannu E. J. Koskinen
Emilia K. J. Kilpua

Acknowledgments

A textbook is always based on long-time experiences and innumerable discussions with several colleagues, all of whom are impossible to properly acknowledge afterwards. Hannu Koskinen's gratitude extends to the 1980s and to the entire team of the first Swedish scientific satellite project *Viking*. This phase affected his career and understanding of wave–particle interactions more than anything else. He is particularly indebted to the discussions with Mats André during his years in Sweden. Another contact from the 1980s deserving special thanks is Bob Lysak whose insights to the ULF waves have turned out to be invaluable to the writing of this volume. For Emilia Kilpua's career, her postdoctoral years 2005–2008 at the University of California, Berkeley, where she analyzed solar eruptions in interplanetary space using multi-spacecraft observations under the guidance of Janet Luhmann and Stuart Bale, were most instructive. Emilia also expresses her gratitude to discussions held in the Young Centre for Advanced Study (CAS) at the Norwegian Academy of Science and Letters Fellowship network led by Hilde Tyssøy.

For the topic of this book, two early workshops of the International Space Science Institute (ISSI) in Bern 1996–1997 on transport across the boundaries of the magnetosphere and on source and loss processes of magnetospheric plasma were of significant importance. Hannu wants to express his gratitude to all participants of the workshops. In particular, discussions with Larry Lyons at these meetings and during the writing of the consequent ISSI volumes were most interesting and instructive. He also wants to thank warmly the collaborators in the EU FP7 space weather forecasting project SPACECAST 2011–2014. The deep insights to the radiation belt modeling of the project leader Richard Horne made a permanent impression.

The need for a coherent textbook-type monograph on physical foundations of radiation belts started to become evident during the flood of articles discussing *Van Allen Probes* observations. A key event was again an ISSI workshop in 2016, this time on physical foundations of space weather, in which both of us participated. Throughout the years, several ISSI workshops and teams have been most influential to our understanding of solar-terrestrial Physics. The friendly support from the ISSI staff deserves a special recognition.

We started sketching this book in spring 2018. In order to educate ourselves, we kicked-off a local Radiation Belt Journal Club. We appreciate very much the contributions of the past and present members of the club. Adnane, Harriet, Lucile, Maxime D., Maxime G., Mikko, Milla, Sanni, Stepan, Thiago, and Yann, discussions with you have not only been utmost important to our book but also great fun!

One of the best, if not the best, methods of learning physics is to teach it to the students. Our approach how to present space plasma physics has evolved over a period of more than three decades of lectures and supervision of undergraduate and doctoral students. Thank you all who have participated in our lectures. You cannot imagine how much this has meant to us.

The list of our co-authors in radiation belt articles and several other colleagues who have directly or indirectly contributed to the contents of the book is long: Timo Asikainen, Dan Baker, Bernie Blake, Daniel Boscher, Seth Claudepierre, Stepan Dubyagin, Jim Fennell, Rainer Friedel, Natalia Ganushkina, Harriet George, Sarah Glauert, Daniel Heyndericks, Heli Hietala, Richard Horne, Allison Jaynes, Liisa Juusola, Milla Kalliokoski, Sri Kanekal, Antti Kero, Solène Lejosne, Mike Liemohn, Vincent Maget, Nigel Meredith, Paul O'Brien, Adnane Osmane, Minna Palmroth, David Pitchford, Tuija Pulkkinen, Graig Rodger, Angelica Sicard, Jim Slavin, Harlan Spence, Tero Raita, Geoff Reeves, Jean-Francois Ripoll, Juan Rodriguez, Kazuo Takahashi, Naoko Takahashi, Lucile Turc, Drew Turner, and Rami Vainio. We are deeply indebted to all of you.

We wish to express our special gratitude to Maxime Grandin, Adnane Osmane, Noora Partamies, Yann Pfau-Kempf, and Lucile Turc, who read and commented on the manuscript in various phases of the writing.

We also wish to thank the Faculty of Science at the University of Helsinki and the Finnish Centre of Excellence in Research of Sustainable Space for providing an outstanding environment for our work on radiation belts. The financial support from the Finnish Society of Sciences and Letters and the Magnus Ehrnrooth Foundation to cover the Open Access fee is greatly appreciated.

Last but not least, we are grateful to the efficient support of Ramon Khanna and the editorial team at Springer Nature in the production of this volume. We appreciate very highly the approach of Springer Nature toward open access publishing.

Contents

About the Authors

Hannu E. J. Koskinen and Emilia K. J. Kilpua are professors in space physics in the Faculty of Science at the University of Helsinki.

Hannu learned about particle dynamics in the magnetic field during a course on cosmical electrodynamics at the University of Helsinki in 1979. He moved to Uppsala in 1981 and worked more than 6 years at the Uppsala Division of the Swedish Institute of Space Physics as a member of the team that built the low-frequency wave instrument for the first Swedish magnetospheric satellite *Viking*, launched in February 1986. He received his Ph.D. degree from the University of Uppsala in 1985 under the supervision of Rolf Boström. During those years, he became influenced by the ideas of Hannes Alfvén, including the importance of the guiding center approximation. Hannu returned to Finland after the highly successful *Viking* mission in 1987 and joined the emerging space research activities at the Finnish Meteorological Institute (FMI). In 1997, he was appointed as a professor in space physics in the Department of Physics at the University of Helsinki, a position shared with the FMI. During 2014–2017, he served as the director of the Department of Physics. He retired from his professorship in 2018 as professor emeritus. He has taught several courses on space physics, classical mechanics, and classical electrodynamics over three decades and written textbooks in all these fields in Finnish as well as an English textbook *Introduction to Plasma Physics* with Emilia Kilpua, published by the University of Helsinki student organization Limes r.y. In 2011, Springer/Praxis published his textbook *Physics of Space Storms – From the Solar Surface to the Earth*, which is one of the main reference works in the present volume. Hannu has been a co-investigator in a dozen spacecraft instrument projects investigating the plasma environments of Earth, Mars, Venus, and comet Churyumov–Gerasimenko. He has held several positions in the European Space Agency, including the membership of Solar System Working Group 1993–1996, a national delegate position in the Science Programme Committee 2002–2016, and the Programme Board of Space Situational Awareness 2010–2016. He acted as the chair of the latter in 2011–2014. Hannu Koskinen is a member of the Finnish Society of Sciences and Letters, the Finnish Academy of Sciences and Letters, Academia Europaea, and the International Academy of Astronautics.

Emilia became involved in space physics at the end of 1990s when she selected solar activity and its consequences in the magnetosphere as the topic of her MSc and Ph.D. studies in the Department of Physics at the University of Helsinki under the supervision of Hannu Koskinen. She received her Ph.D. degree in 2005 after which she began a 3-year postdoctoral period at the Space Sciences Laboratory of the University of California, Berkeley. There she analyzed solar eruptions in interplanetary space using recent multi-spacecraft observations under the guidance of Janet Luhmann and Stuart Bale. During 2009–2015, she was an Academy Research Fellow of the Academy of Finland. In 2015, she was appointed as tenure-track associate professor in space physics in the Department of Physics at the University of Helsinki, and she was promoted to full professor in 2020. When the first observations from Van Allen Probes became available in 2012, Emilia took the lead in the local space physics research group to investigate how different solar wind drivers of the magnetospheric dynamics influence the radiation belts. She has taught several courses on space physics at various levels from undergraduate to Ph.D. students and introductory electromagnetism to first-year students. She is the lead author of the above-mentioned textbook *Introduction to Plasma Physics*. Emilia has supervised several doctoral students and postdocs in the fields of radiation belts and solar activity. She has a prestigious European Research Council Consolidator Grant for studies of Solar Magnetic Flux Ropes and Their Magnetosheaths 2017–2022 and she is a group leader in the Finnish Centre of Excellence in Research of Sustainable Space 2018–2023. Emilia is member of the Finnish Academy of Sciences and Letters.

Chapter 1
Radiation Belts and Their Environment

The Van Allen radiation belts of high-energy electrons and ions, mostly protons, are embedded in the Earth's inner magnetosphere where the geomagnetic field is close to that of a magnetic dipole. Understanding of the belts requires a thorough knowledge of the inner magnetosphere and its dynamics, the coupling of the solar wind to the magnetosphere, and wave–particle interactions in different temporal and spatial scales. In this introductory chapter we briefly describe the basic structure of the inner magnetosphere, its different plasma regions and the basics of magnetospheric activity.

1.1   The Overall View to the Belts

The discovery of radiation belts dates back to the dawn of the space age when the knowledge of the physical properties of the magnetosphere was still in its infancy. In February 1958 the first U.S. satellite *Explorer* I[1] carried a Geiger–Müller instrument that was designed to measure cosmic radiation. It indeed did so until the spacecraft reached the altitude of about 700 km when the instrument mysteriously fell silent. The observations from *Explorer* III confirmed *Explorer* I observations only a month later. In their seminal paper James Van Allen and his co-workers (Van Allen et al. 1958) suggested that the instrument was saturated due to high-intensity corpuscular radiation trapped in the Earth's magnetic field.

   In December 1958 *Pioneer* III ventured further into space and understanding of the basic structure of inner and outer radiation belts started to evolve. It soon became clear that a population of multi-MeV protons, up to 1–2 GeV, dominates the ion radiation at equatorial geocentric distances of about $1.1 - 3\,R_E$ ($R_E \simeq 6370$ km

---

[1] The spacecraft mentioned in the text are briefly introduced and their acronyms deciphered in Appendix B.

© The Author(s) 2022
H. E. J. Koskinen, E. K. J. Kilpua, *Physics of Earth's Radiation Belts*,
Astronomy and Astrophysics Library, https://doi.org/10.1007/978-3-030-82167-8_1

Fig. 1.1 A sketch of showing the inner and outer electron belts and a slot region in between embedded in the dipolar magnetic field of the Earth. The inner belt is within $2\,R_E$ form the center of the Earth. The figure illustrates the structure when the outer belt is split to two spatially distinct domains as observed by the *Van Allen Probes*. (Image credits: NASA's Goddard Space Flight Center and Grant Stevens, Rob Barnes and Sasha Ukhorskiy of the Applied Physics Laboratory of the Johns Hopkins University)

is the radius of Earth).[2] The high-energy electrons exhibit a two-belt structure with a *slot region* in between (Fig. 1.1). The inner electron belt is partially co-located with the proton belt at equatorial distances of about $1.1 - 2\,R_E$. The outer belt is beyond about $3\,R_E$ extending to distances of $7 - 10\,R_E$ with electron energies from tens of keV to several MeV. Sometimes the outer belt exhibits two or even three spatially distinct parts. As the proton mass is $931\,\text{MeV}\,\text{c}^{-2}$ and the electron mass $511\,\text{keV}\,\text{c}^{-2}$, the highest-energy inner belt protons and the outer belt electrons are relativistic moving at almost the speed of light.

Since the early space age, the radiation belts have been investigated using a large number of satellites.[3] The observations now cover more than five solar cycles and have revealed the extremely complex and highly variable structure of the belts.

---

[2] When giving an altitude in terms of Earth radius, we always refer to geocentric distance.

[3] A brief introduction to satellites cited in the book is given in Appendix B.

Based on these observations and theoretical reasoning great number of different
numerical models of radiation belts have been constructed not only for scientific
purposes but also to meet the needs of spacecraft engineers and space mission
planners. As our focus is on the physical processes, we will not go into the details
of these models. An interested reader can find the models with their descriptions
at several web-sites, e.g., the Community Coordinated Modeling Center (CCMC)[4]
and the Space Environment Information System (SPENVIS)[5] It is evident that the
observations during the *Van Allen Probes* era—many of which are discussed in this
book and the subsequent modeling efforts will lead to important revisions and
refinements of the models.

Although the fluxes of the highest-energy particles and their energy densities are
considerably lower than those of the background plasma in the inner magnetosphere,
they are of a significant concern due to their space weather effects, both posing
risks to spacecraft and humans in orbit and affecting the upper atmosphere through
energetic electron and proton precipitation. The energization of radiation belt
particles is an interesting fundamental plasma physical process and much emphasis
has been placed on understanding the dynamics of relativistic and ultra-relativistic
populations.

The inner belt, in particular the proton population, is relatively stable, whereas
the outer electron belt is in continuous change. The high-energy electron fluxes can
change several orders of magnitude within minutes: the outer belt may suddenly
become almost completely depleted of, or get abruptly filled with, relativistic
electrons. Most activity occurs in "the heart of the outer belt", at equatorial distances
of about $4 - 5\,R_E$. While the *Van Allen Probes* mission has shown that there is
an almost impenetrable inner edge of the outer belt ultra-relativistic ($\gtrsim 4$ MeV)
electrons at an equatorial distance of 2.8 R_E, there have been a few observed events
when the slot region was filled with ultra-relativistic electrons and the electrons
remained trapped in the region up to several months.

The highly variable configuration and complex dynamics of the outer belt owe
to the continuous changes in the plasma and geomagnetic field conditions driven by
variable properties of the solar wind caused, in particular, by coronal mass ejections,
stream interaction regions, and fast solar wind flows carrying Alfvénic fluctuations.
Locally the kinetic response to particle injections from nightside magnetosphere
affect the thermodynamic properties of the radiation belt electrons.

The radiation belts overlap with different plasma domains of the inner magnetosphere: the ring current, the plasmasphere and the plasma sheet, whose properties
and locations vary in time. In particular the boundary of the plasmasphere, moving
between equatorial distances of 3 and 5 R_E as a response to the solar wind driving,
is a critical region to the dynamics of the outer radiation belt.

The inner magnetospheric plasma exhibits complex wave activity transferring
energy and momentum between different plasma populations. The waves are known

[4] https://ccmc.gsfc.nasa.gov/models/.

[5] https://www.spenvis.oma.be.

to scatter and energize the electrons depending on the particle energy, wave amplitude and the direction of wave propagation. While much of elementary space plasma theory has been developed under the approximation of linear perturbations, in the case of observed large-amplitude waves nonlinear effects need to be considered. Furthermore, the plasma and magnetic environment of the belts is not spatially symmetric, but varies as function of local time sector and geomagnetic latitude, and of course, temporally.

1.2   Earth's Magnetic Environment

In the first approximation the Earth's magnetic field is that of a magnetic dipole. The dipole axis is tilted $11°$ from the direction of the Earth's rotation axis. The current circuit giving rise to the magnetic field is located in the liquid core about 1200–3400 km from the center of the planet. The current system is asymmetric displacing the dipole moment from the center, which together with inhomogeneous distribution of magnetic matter above the core gives rise to large deviations from the dipole field on the surface. The pure dipole field on the surface would be $30\,\mu T$ at the dipole equator and $60\,\mu T$ at the poles. However, the actual surface field exceeds $66\,\mu T$ in the region between Australia and Antarctica and is weakest, about $22\,\mu T$, in a region called *South Atlantic Anomaly* (SAA). The magnetic poles migrate slowly, and the SAA has during the past decades moved slowly from Africa toward South America being presently deepest in Paraguay. The SAA has a specific practical interest, as the inner radiation belt reaches down to low Earth orbiting (LEO) satellites at altitudes of 700–800 km above the anomaly.

1.2.1   The Dipole Field

Knowledge of the charged particle motion in the dipole field is essential in studies of radiation belts. In the main radiation belt domain at geocentric distances 2–7 R_E the dipole field is a good first approximation for the quiet state of the magnetic field. In reality, the dipole field is an idealization where the source current is assumed to be confined into a point at the origin. The source of planetary and stellar dipoles is a finite, actually a large, current system within the celestial body. Such fields, including the Terrestrial magnetic field, are customarily represented as a multipole expansion: dipole, quadrupole, octupole, etc. When moving away from the source, the higher multipoles vanish faster than the dipole making the dipole field a good starting point to consider the motion of charged particles in radiation belts. In the dipole field charged particles behave adiabatically as long as their gyro radii are smaller than the gradient scale length of the field (Chap. 2) and their orbits are not disturbed by collisions or time-varying electromagnetic field.

For the geomagnetic field it is customary to define the spherical coordinates in a special way. The *dipole moment* ($\mathbf{m}_E$) is in the origin and points approximately toward geographic *south*, tilted $11°$ as mentioned above. Similar to the geographic coordinates the latitude (λ) is zero at the dipole equator and increases toward the north, whereas the latitudes in the southern hemisphere are negative. The longitude (ϕ) increases toward the east from a given reference longitude. In magnetospheric physics the longitude is often given as the *magnetic local time* (MLT). In the dipole approximation MLT is determined by the flare angle between two planes: the dipole meridional plane containing the subsolar point on the Earth's surface, and the dipole meridional plane which contains a given point on the surface, i.e., the local dipole meridian. Magnetic noon (MLT = 12 h) points toward the Sun, midnight (MLT = 24 h) anti-sunward. Magnetic dawn (MLT = 6 h) is approximately in the direction of the Earth's orbit around the Sun.[6] The abbreviation h (for hour) is often dropped and fractional MLTs are given by decimals instead of minutes and seconds.

The SI-unit of m_E is $\mathrm{A\,m^2}$. In the radiation belt context it is convenient to replace m_E by $k_0 = \mu_0 m_E / 4\pi$, which is also customarily called dipole moment. The strength of the terrestrial dipole moment varies slowly. For our discussion a sufficiently accurate approximation is

$$
\begin{aligned}
m_E &= 8 \times 10^{22}\,\mathrm{A\,m^2} \\
k_0 &= 8 \times 10^{15}\,\mathrm{Wb\,m} \quad (\mathrm{SI : Wb = T\,m^2}) \\
&= 8 \times 10^{25}\,\mathrm{G\,cm^3} \quad (\text{Gaussian units,} \quad 1\,\mathrm{G} = 10^{-4}\,\mathrm{T}) \\
&= 0.3\,\mathrm{G}\,R_E^3 \quad\quad (R_E \simeq 6370\,\mathrm{km})
\end{aligned}
$$

The last expression is convenient in practice because the dipole field on the surface of the Earth (at $1\,R_E$) varies in the range 0.3–0.6 G.

Outside its source, the dipole field is a curl-free potential field $\mathbf{B} = -\nabla\Psi$, where the scalar potential is given by

$$
\Psi = -\mathbf{k}_0 \cdot \nabla \frac{1}{r} = -k_0 \frac{\sin\lambda}{r^2}, \tag{1.1}
$$

yielding

$$
\mathbf{B} = \frac{1}{r^3}[3(\mathbf{k}_0 \cdot \mathbf{e}_r)\mathbf{e}_r - \mathbf{k}_0]. \tag{1.2}
$$

---

[6] The definition of MLT in non-dipolar coordinate systems is more complicated but the main directions are approximately the same.

The components of the magnetic field are

$$B_r = -\frac{2k_0}{r^3}\sin\lambda$$

$$B_\lambda = \frac{k_0}{r^3}\cos\lambda \tag{1.3}$$

$$B_\phi = 0$$

and its magnitude is

$$B = \frac{k_0}{r^3}(1 + 3\sin^2\lambda)^{1/2}. \tag{1.4}$$

The equation of a magnetic field line is

$$r = r_0\cos^2\lambda, \tag{1.5}$$

where r_0 is the distance where the field line crosses the equator. The length element of the magnetic field line element is

$$ds = (dr^2 + r^2 d\lambda^2)^{1/2} = r_0\cos\lambda(1 + 3\sin^2\lambda)^{1/2}d\lambda. \tag{1.6}$$

This can be integrated in a closed form, yielding the length of the dipole field line S_d as a function of r_0

$$S_d \approx 2.7603\, r_0. \tag{1.7}$$

The curvature radius $R_C = |d^2\mathbf{r}/ds^2|^{-1}$ of the magnetic field is an important parameter for the motion of charged particles. For the dipole field the *radius of curvature* is

$$R_C(\lambda) = \frac{r_0}{3}\cos\lambda\frac{(1 + 3\sin^2\lambda)^{3/2}}{2 - \cos^2\lambda}. \tag{1.8}$$

Any dipole field line is determined by its (constant) longitude ϕ_0 and the distance where the field line crosses the dipole equator. This distance is often given in terms of the *L-parameter*

$$L = r_0/R_E. \tag{1.9}$$

The parameter was introduced in the early days of *Explorer* data analysis by Carl E. McIlwain to organize the observations in magnetic field-related coordinates. Consequently, L is known as *McIlwain's L-parameter*.

For a given L the corresponding field line reaches the surface of the Earth at the (dipole) latitude

$$\lambda_e = \arccos \frac{1}{\sqrt{L}} . \qquad (1.10)$$

For example, $L = 2$ (the inner belt) intersects the surface at $\lambda_e = 45°$, $L = 4$ (the heart of the outer belt) at $\lambda_e = 60°$ and $L = 6.6$ (the *geostationary orbit*)[7] at $\lambda_e = 67.1°$.

The dipole field line length in (1.7) was calculated from the dipole itself. Now we can calculate also the dipole field line length from a point on the surface to the surface on the opposite hemisphere to be

$$S_e \approx (2.7755 \times L - 2.1747) \, R_E , \qquad (1.11)$$

which is a good approximation when $L \gtrsim 2$.

The field magnitude along a given field line as a function of latitude is

$$B(\lambda) = [B_r(\lambda)^2 + B_\lambda(\lambda)^2]^{1/2} = \frac{k_0}{r_0^3} \frac{(1 + 3\sin^2\lambda)^{1/2}}{\cos^6\lambda} . \qquad (1.12)$$

For the Earth

$$\frac{k_0}{r_0^3} = \frac{0.3}{L^3} \, G = \frac{3 \times 10^{-5}}{L^3} \, T . \qquad (1.13)$$

At the magnetic equator on the surface of the Earth, the dipole field is $0.3\,G$ ($30\,\mu T$), at the poles $0.6\,G$ ($60\,\mu T$).

The actual geomagnetic field has considerable deviations from the dipolar field because the dipole is not quite in the center of the Earth, the source is not a point, and the electric conductivity of the Earth is not uniform. The geomagnetic field is described by the *International Geomagnetic Reference Field* (IGRF) model, which is regularly updated to reflect the slow secular variations of the field, i.e., changes in timescales of years or longer (Fig. 1.2).

1.2.2 Deviations from the Dipole Field due to Magnetospheric Current Systems

The Earth's *magnetosphere* is the region where the near-Earth magnetic field controls the motion of charged particles. It is formed by the interaction between the

---

[7] The geostationary distance is an altitude where a satellite on equatorial plane moves around the Earth in 24 h. The orbit is called *Geostationary Earth Orbit* (GEO) or *geosynchronous orbit*.

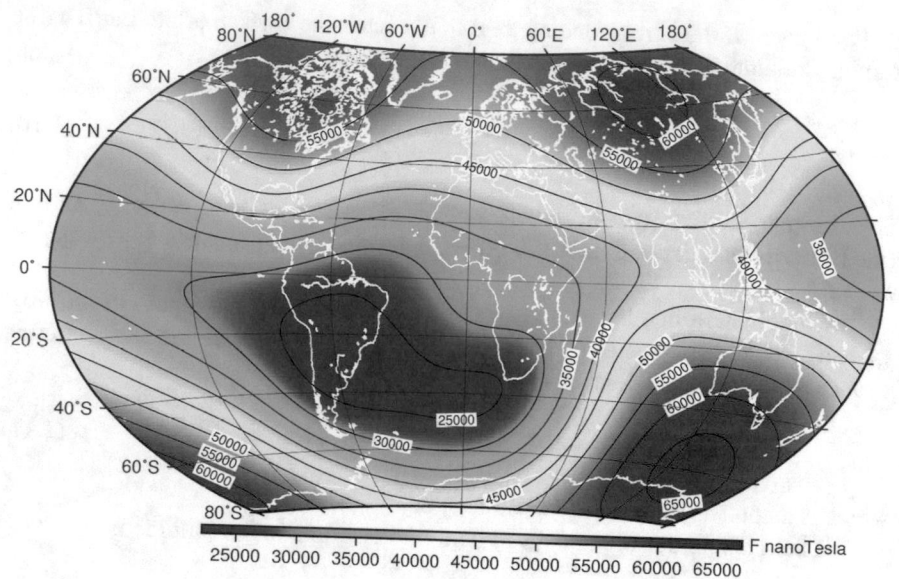

Fig. 1.2 The magnetic field magnitude on the surface of the Earth according to the 13th generation IGRF model released in December 2019. The South Atlantic Anomaly is the deep blue region extending from the southern tip of Africa to South America. The model is available at National Centers for Environmental Information (NCEI, https://www.ncei.noaa.gov)

geodipole and the solar wind. The deformation of the field, caused by the variable solar wind pressure, sets up time-dependent magnetospheric current systems that dominate deviations from the dipole field in the outer radiation belt and beyond.

The solar wind plasma cannot easily penetrate to the Earth's magnetic field and the outer magnetosphere is essentially a cavity around which the solar wind flows. The cavity is bounded by a flow discontinuity called the *magnetopause*. The shape and location of the magnetopause is determined by the balance between the solar wind dynamic plasma pressure and the magnetospheric magnetic field pressure. The nose, or apex, of the magnetopause is, under average solar wind conditions, at the distance of about $10\,R_E$ from the center of the Earth but can be pushed to the vicinity of the geostationary distance $(6.6\,R_E)$ during periods of large solar wind pressure, which has important consequences to the dynamics of the outer radiation belt. In the dayside the dipole field is compressed toward the Earth, whereas in the nightside the field is stretched to form a long *magnetotail*. The deviations from the curl-free dipole field correspond to electric current systems according to Ampère's law $\mathbf{J} = \nabla \times \mathbf{B}/\mu_0$.

In the frame of reference of the Earth the solar wind is supersonic, or actually super-magnetosonic, exceeding the local magnetosonic speed $v_{ms} = \sqrt{v_s + v_A}$, where v_s is the sound speed, $v_A = B/\sqrt{\mu_0 \rho_m}$ the Alfvén speed and ρ_m the mass density of the solar wind. Because fluid-scale perturbations cannot propagate faster than v_{ms}, this leads to a formation of a collisionless shock front, called the *bow shock*, upstream of the magnetosphere. Under typical solar wind conditions the apex of the shock in the solar direction is about 3 R_E upstream of the magnetopause. The shock converts a considerable fraction of solar wind kinetic energy to heat and electromagnetic energy. The irregular shocked flow region between the bow shock and the magnetopause is called the *magnetosheath*.

The current system on the dayside magnetopause shielding the Earth's magnetic field from the solar wind is known as the *Chapman–Ferraro current*, recognizing the early attempt of Chapman and Ferraro (1931) to explain how magnetic storms would be driven by corpuscular radiation from the Sun. In the first approximation the Chapman–Ferraro current density $\mathbf{J}_{CF}$ can be expressed as

$$\mathbf{J}_{CF} = \frac{\mathbf{B}_{MS}}{B_{MS}^2} \times \nabla P_{dyn}, \tag{1.14}$$

where $\mathbf{B}_{MS}$ is the magnetospheric magnetic field and P_{dyn} the dynamic pressure of the solar wind. Because the *interplanetary magnetic field* (IMF) at the Earth's orbit is only a few nanoteslas, the magnetopause current must shield the magnetospheric field to almost zero just outside the current layer. Consequently, the magnetic field immediately inside the magnetopause doubles: about one half comes from the Earth's dipole and the second half from the magnetopause current.

The Chapman–Ferraro model describes a teardrop-like closed magnetosphere that is compressed in the dayside and stretched in the nightside, but not very far. Since the 1960s spacecraft observations have shown that the nightside magnetosphere, the *magnetotail*, is very long, extending far beyond the orbit of the Moon. This requires a mechanism to transfer energy from the solar wind into the magnetosphere to keep up the current system that sustains the tail-like configuration.

Figure 1.3 is a sketch of the magnetosphere with the main large-scale magnetospheric current systems. The overwhelming fraction of the magnetospheric volume consists of *tail lobes*, connected magnetically to the polar caps in the ionized upper atmosphere, known as the *ionosphere*. The polar caps are bounded by *auroral ovals*. Consequently, in the northern lobe the magnetic field points toward the Earth, in the southern away from the Earth. To maintain the lobe structure, there must be a current sheet between the lobes where the current points from dawn to dusk. This *cross-tail current* is embedded within the *plasma sheet* (Sect. 1.3.1) and closes around the tail lobes forming the nightside part of the the *magnetopause current*.

The cusp-like configurations of weak magnetic field above the polar regions known as *polar cusps* do not connect magnetically to magnetic poles, but instead to the southern and northern auroral ovals at noon, because the entire magnetic flux enclosed by the ovals is connected to the tail lobes. Tailward of the cusps the Chapman–Ferraro current and the tail magnetopause current smoothly merge with each other. Figure 1.3 also illustrates the westward flowing *ring current* (RC) and the *magnetic field-aligned currents* (FAC) connecting the magnetospheric currents to the horizontal *ionospheric currents* in auroral regions at an altitude of about 100 km.

The magnetospheric current systems can have significant temporal variations, which makes the mathematical description of the magnetic field complicated. A

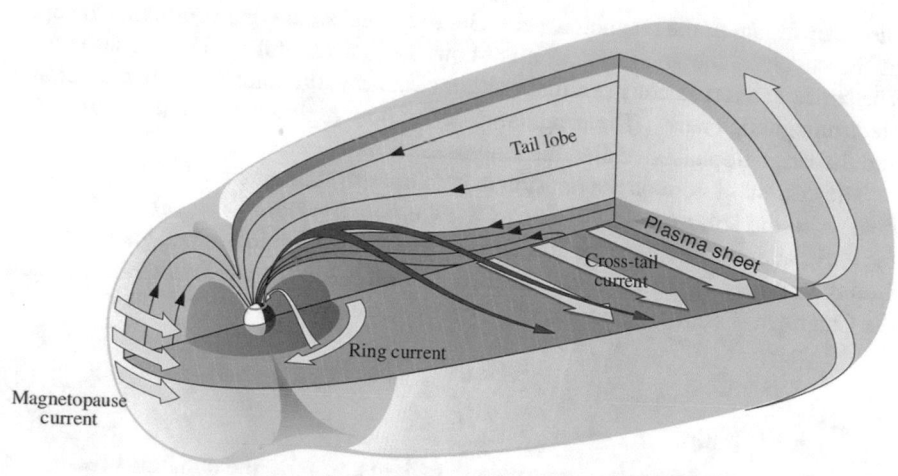

Fig. 1.3 The magnetosphere and the large scale magnetospheric current systems. (Figure courtesy T. Mäkinen, from Koskinen 2011, reprinted by permission from SpringerNature)

common approach is to apply some of the various models developed by Nikolai Tsyganenko (for a review, see Tsyganenko 2013).[8] Particularly popular in radiation belt studies is the model known as TS04 (Tsyganenko and Sitnov 2005).

For illustrative purposes simpler models are sometimes useful. For example, the early time-independent model of Mead (1964) reduces in the magnetic equatorial (r, ϕ) plane to

$$ B(r, \phi) = B_E \left(\frac{R_E}{r} \right)^3 \left[1 + \frac{b_1}{B_E} \left(\frac{r}{R_E} \right)^3 - \frac{b_2}{B_E} \left(\frac{r}{R_E} \right)^4 \cos \phi \right], \qquad (1.15) $$

where we have adopted the notation of Roederer and Zhang (2014). Here B_E is the equatorial dipole field on the surface of the Earth (approximately $30.4 \, \mu T = 30,400 \, \text{nT}$) and ϕ is the longitude east of midnight. The $\cos \phi$ term describes the azimuthal asymmetry due to the dayside compression and nightside stretching of the field. The coefficients b_1 and b_2 depend on the distance of the subsolar point of the magnetopause R_s (in units of R_E), which, in turn, depends on the upstream solar wind pressure

$$ b_1 = 25 \left(\frac{10}{R_s} \right)^3 \text{nT} $$

$$ b_2 = 2.1 \left(\frac{10}{R_s} \right)^4 \text{nT} . \qquad (1.16) $$

---

[8] Tsyganenko models are available at Community Coordinated Modeling Center: https://ccmc.gsfc.nasa.gov/models/.

This model is fairly accurate during quiet and moderately disturbed times at geocentric distances 1.5–7 R_E.

1.2.3 Geomagnetic Activity Indices

The intensity and variations of magnetospheric and ionospheric current systems are traditionally described in terms of *geomagnetic activity indices* (Mayaud 1980), which are available at the International Service of Geomagnetic Indices webpages maintained by the University of Strasbourg.[9] The indices are calculated from ground-based magnetometer measurements. The large number of useful indices illustrates the great variability of geomagnetic activity; sometimes the effects are stronger at high latitudes, sometimes at low, sometimes the background current systems are strong already before the main perturbation, etc. As different indices describe different features of magnetospheric currents, there is no one-to-one correspondence between them. The choice of a particular index depends on physical processes being investigated. Here we briefly introduce the most widely used indices for global storm levels, *Dst* and *Kp*, and for the activity at auroral latitudes, *AE*, which will be used later when discussing the relation of radiation belt dynamics with evolving geomagnetic activity.

The *Dst* index aims at measuring the intensity of the ring current. It is calculated once an hour as a weighted average of the deviation from the quiet level of the horizontal magnetic field component (*H*) measured at four low-latitude stations distributed around the globe. *Geomagnetic storms* (also known as *magnetospheric storms* or *magnetic storms*) are defined as periods of strongly negative *Dst* index, signalling enhanced westward the ring current. The more negative the *Dst* index is, the stronger is the storm. There is no canonical lower threshold for the magnetic perturbation beyond which the state of the magnetosphere is to be called a storm and identification of weak storms is often ambiguous. In this book we call storms with *Dst* from –50 to –100 nT *moderate*, from –100 to –200 nT *intense*, and those with *Dst* < −200 nT *big*. A similar 1-min index derived from a partly different set of six low-latitude stations (*SYM–H*) is also in use.

A sensitive ground-based magnetometer reacts to all magnetospheric current systems and, thus, *Dst* has contributions from other currents in addition to the ring current. These include the magnetopause and cross-tail currents, as well as induced currents in the ground due to rapid temporal changes of ionospheric currents. Large solar wind pressure pushes the magnetopause closer to the Earth forcing the magnetopause current to increase to be able to shield a locally stronger geomagnetic field from the solar wind. The effect is strongest on the dayside where

---

[9] http://isgi.unistra.fr/.

the magnetopause current flows in the direction opposite to the ring current. The *pressure corrected Dst* index can be defined as

$$Dst^* = Dst - b\sqrt{P_{dyn}} + c \,, \tag{1.17}$$

where P_{dyn} is the solar wind dynamic pressure and b and c are empirical parameters, whose exact values depend on the used statistical analysis methods, e.g., $b = 7.26\,\mathrm{nT\,nPa}^{-1/2}$ and $c = 11\,\mathrm{nT}$ as determined by O'Brien and McPherron (2000).

The contribution from the dawn-to-dusk directed tail current to the Dst index is more difficult to estimate. During strong activity the cross-tail current intensifies and moves closer to the Earth, enhancing the nightside contribution to Dst. The estimates of this effect on Dst vary in the range 25–50% (e.g., Turner et al. 2000; Alexeev et al. 1996). Furthermore, fast temporal changes in the ionospheric currents induce strong localized currents in the ground, which may contribute up to 25% to the Dst index (Langel and Estes 1985; Häkkinen et al. 2002).

Another widely used index is the planetary K index, Kp. Each magnetic observatory has its own K index and Kp is an average of K indices from 13 mid-latitude stations. It is a quasi-logarithmic range index expressed in a scale of one-thirds: 0, 0+, 1−, 1, 1+, . . . , 8+, 9−, 9. Kp is based on mid-latitude observations and thus more sensitive to high-latitude auroral current systems and to substorm activity than the Dst index. Kp is a 3-h index and does not reflect rapid changes in the magnetospheric currents.

The fastest variations in the current systems take place at auroral latitudes. To describe the strength of the auroral currents the *auroral electrojet indices* (AE) are commonly used. The standard AE index is calculated from 11 or 12 magnetometer stations located under the average auroral oval in the northern hemisphere. It is derived from the magnetic north component at each station by determining the envelope of the largest negative deviation from the quiet time background, called the AL index, and the largest positive deviation, called the AU index. The AE index itself is $AE = AU - AL$ (all in nT). Thus AL is the measure of the strongest westward current in the auroral oval, AU is the measure of the strongest eastward current, and AE characterizes the total electrojet activity. AE, AU, AL are typically given with 1-min time resolution.

As the *auroral electrojets* flow at the altitude of about 100 km, their magnetic deviations on the ground are much larger than those caused by the ring current. For example, during typical substorm activations AE is in the range 200–400 nT and can during strong storms exceed 2000 nT, whereas the equatorial Dst perturbations exceed −200 nT only during the strongest storms.

1.3 Magnetospheric Particles and Plasmas

The magnetosphere is a vast domain with a wide range of relevant physical parameters. The energies, temperatures and densities vary by several orders of magnitude and change also significantly as response to variable solar wind conditions. The inner magnetosphere consists of three main particle domains; the cold and relatively dense *plasmasphere*, the more energetic *ring current* and the high-energy *radiation belts*. They *are not* spatially distinct regions, but partially overlap and their mutual interactions are critical to the physics of radiation belts. The plasma sheet in the outer magnetosphere acts as the source of suprathermal particles that are injected into the inner magnetosphere during periods of magnetospheric activity.

This introductory discussion remains at a very general level. We introduce the details of individual particle motion in Chap. 2 and the basic plasma concepts in Chap. 3.

1.3.1 Outer Magnetosphere

The outer magnetosphere can be considered to begin at distances of about 7–8 R_E where the nightside magnetic field becomes increasingly stretched. Table 1.1 summarizes typical plasma parameters in the mid-tail region, at about $X = -20 R_E$ from the Earth. Here X is the Earth-centered coordinate along the Earth–Sun line, positive toward the Sun. The tail lobes are almost empty, particle number densities being of the order of $0.01 \, \text{cm}^{-3}$. The central plasma sheet where the cross-tail current is embedded (Fig. 1.3) is, in turn, a region of hot high-density plasma. It is surrounded by the plasma sheet boundary layer with density and temperature intermediate to values in the central plasma sheet and tail lobes. The field lines of the boundary layer connect to the poleward edge of the auroral oval. The actual numbers differ considerably from the typical values under changing solar wind conditions and, in particular, during strong magnetospheric disturbances.

Table 1.1 also includes typical parameters in the magnetosheath at the same X-coordinate. The magnetosheath consists of solar wind plasma that has been compressed and heated by the Earth's bow shock. It has higher density and lower temperature than observed in the outer magnetosphere. Typical densities of the unperturbed solar wind at 1 AU extend from about $3 \, \text{cm}^{-3}$ in the fast ($\sim 750 \, \text{km s}^{-1}$) to about $10 \, \text{cm}^{-3}$ in the slow ($\sim 350 \, \text{km s}^{-1}$) solar wind, again with large deviations. Table 1.1 shows that, while the magnetic field magnitude is rather similar in

Table 1.1 Typical values of plasma parameters in the mid-tail. *Plasma beta* (β) is the ratio between kinetic and magnetic pressures (Eq. 3.28)

	Magneto-sheath	Tail lobe	Plasma sheet boundary	Central plasma sheet
$n \, (\text{cm}^{-3})$	8	0.01	0.1	0.3
$T_i \, (\text{eV})$	150	300	1000	4200
$T_e \, (\text{eV})$	25	50	150	600
$B \, (\text{nT})$	15	20	20	10
β	2.5	$3 \cdot 10^{-3}$	0.1	6

all regions shown, plasma beta (the ratio between the kinetic and magnetic field pressures), is a useful parameter to distinguish between different regions.

1.3.2   Inner Magnetosphere

The inner magnetosphere is the region where the magnetic field is quasi-dipolar. It is populated by different spatially overlapping particle species with different origins and widely different energies: the ring current, the radiation belts and the plasmasphere. The ring current and radiation belts consist mainly of trapped particles in the quasi-dipolar field drifting due to magnetic field gradient and curvature effects around the Earth, whereas the motion and spatial extent of plasmaspheric plasma is mostly influenced by the corotation and convection electric fields (Chap. 2).

The *ring current* arises from the azimuthal drift of energetic charged particles around the Earth; positively charged particles drifting toward the west and electrons toward the east. Basically all drifting particles contribute to the ring current. The drift currents are proportional to the energy density of the particles and the main ring current carriers are positive ions in the energy range 10–200 keV, whose fluxes are much larger than those of the higher-energy radiation belt particles. The ring current flows at geocentric distances 3–8 R_E, and peaks at about 3–4 R_E. At the earthward edge of the ring current the negative pressure gradient introduces a local eastward diamagnetic current, but the net current remains westward.

During magnetospheric activity the role of the ionosphere as the plasma source of ring current enhances, increasing the relative abundance of oxygen (O^+) and helium (He^+) ions in the magnetosphere (to be discussed in Sect. 6.3.1). As a result a significant fraction of ring current can at times be carried by oxygen ions of atmospheric origin. The heavy-ion content furthermore modifies the properties of plasma waves in the inner magnetosphere, which has consequences on the wave–particle interactions with the radiation belt electrons, as will be discussed from Chap. 4 onward.

The *plasmasphere* is the innermost part of the magnetosphere. It consists of cold ($\sim$1 eV) and dense ($\gtrsim 10^3$ cm^{-3}) plasma of ionospheric origin. The existence of the plasmasphere was already known before the spaceflight era based on the propagation characteristics of lightning-generated and man-made very low-frequency (VLF) waves. The plasmasphere has a relatively clear outer edge, the *plasmapause*, where the proton density drops several orders of magnitude. The location and structure of the plasmapause vary considerably as a function of magnetic activity (Fig. 1.4). During magnetospheric quiescence the density decreases smoothly at distances from 4–6 R_E, whereas during strong activity the plasmapause is steeper and pushed closer to the Earth.

Fig. 1.4 Plasma density in the night sector organized by the activity index Kp. $Kp < 1+$ corresponds to a very quiet magnetosphere, whereas $Kp = 4 - 5$ indicates a significant activity level, although not yet a big magnetic storm. The L-shell is defined in Sect. 2.6. It corresponds to the magnetic field lines of a given L-parameter. (Adapted from Chappell (1972), reprinted by permission from American Geophysical Union)

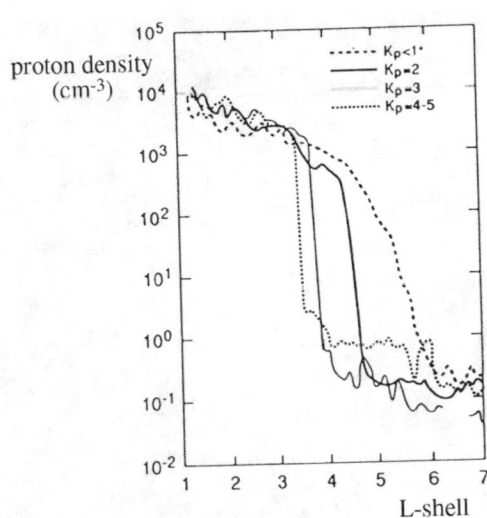

The location of the plasmapause is determined by the interplay between the sunward convection of plasma sheet particles and the plasmaspheric plasma corotating with the Earth. In Sect. 2.3 we add the convective and corotational electric fields to the guiding center motion of charged particles and find that an outward bulge called *plasmaspheric plume* develops on the duskside around 18 h magnetic local time (MLT). Plasmaspheric plumes are most common and pronounced during geomagnetic storms and substorms, but they can exist also during quiet conditions (e.g., Moldwin et al. 2016, and references therein). During geomagnetic storms the plume can expand out to geostationary orbit and bend toward earlier MLT.

Figure 1.5 shows global observations of the plasmasphere taken by the EUV instrument onboard the IMAGE satellite before and after a moderate geomagnetic storm in June 2000. Before the storm the plasmasphere was more or less symmetric. After the storm the plasmasphere was significantly eroded leaving a plume extending from the dusk toward the dayside magnetopause. When traversing the plume, the trapped radiation belt electrons, otherwise outside the plasmapause, encounter a colder and higher-density plasma with plasma wave environment similar to the plasmasphere proper. Consequently, the influence of the plasmasphere on radiation belt particles extends beyond its nominal boundary depicted in Fig. 1.4.

The plasma parameters in the plasmasphere, in the plume and at the plasmapause are critical to the generation and propagation of plasma waves that, in turn, interact with the energetic particles in the ring current and radiation belts. Thus, the coldest and the hottest components of the inner magnetosphere are intimately coupled to each other through wave–particle interactions.

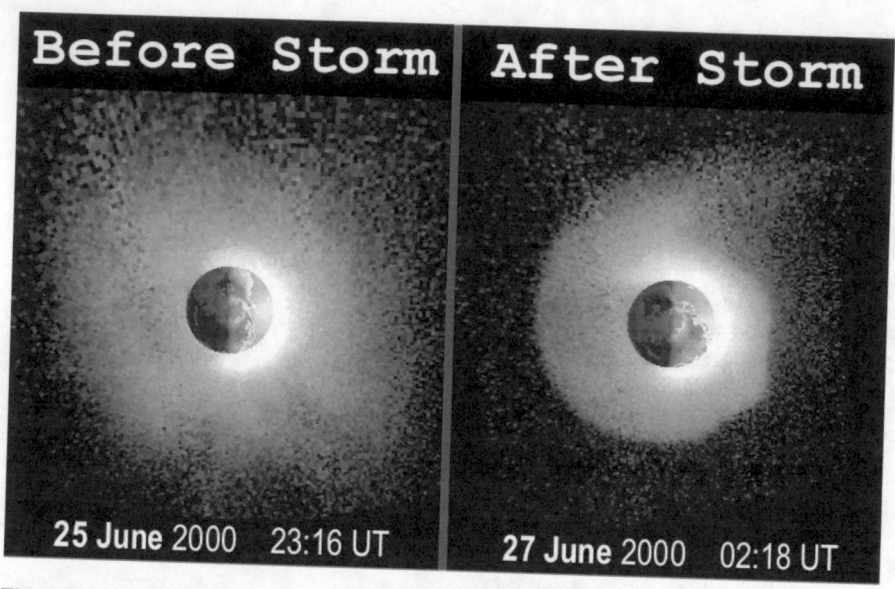

Fig. 1.5 Plasmapheric plume and plasmaspheric erosion as observed by the IMAGE EUV instrument. The picture is taken from above the northern hemisphere and the Sun is to the right. (Figure courtesy: Jerry Goldstein, Southwest Research Institute, for more information of this particular storm see Goldstein et al. 2004)

1.3.3 Cosmic Rays

In addition to ion and electron radiation belts another important component of corpuscular radiation in the near-Earth space consists of *cosmic rays*. The kinetic energies of a large fraction of cosmic ray particles are so large that the geomagnetic field cannot trap them. Instead, the particles traverse through the Earth's magnetosphere without much deflection of their trajectories. Some of them hit the atmosphere interacting with nuclei of atmospheric atoms and molecules causing showers of elementary particles being possible to detect on ground. Those with highest energies can penetrate all the way to the ground.

The spectrum of cosmic ray ions at energies below about 10^{15} eV per nucleon in the near-Earth space has three main components:

- *Galactic cosmic rays* (GCR), whose spectrum peaks at energies above 100 MeV per nucleon, are most likely accelerated by supernova remnant shock waves in our galaxy.
- *Solar cosmic rays* (SCR) are accelerated by coronal and interplanetary shocks related to solar eruptions. Their energies are mostly below 100 MeV per nucleon and a fraction of them can become trapped in the inner radiation belt.

- *Anomalous cosmic rays* (ACR) are ions of solar origin captured and accelerated by the heliospheric termination shock, where the supersonic solar wind becomes subsonic before encountering the interstellar plasma, or in the heliosheath outside the heliopause. Some of the ions are injected back toward the Sun. Near the Earth the ACR spectrum peaks at about 10 MeV per nucleon and thus the particles can become trapped in the geomagnetic field.

Although the galactic cosmic rays cannot directly be trapped into the radiation belts, they contribute indirectly to the inner belt composition through the *Cosmic Ray Albedo Neutron Decay* (GRAND) mechanism. The cosmic ray bombardment of the atmosphere produces neutrons that move in all directions. Although the average neutron lifetime is 14 min 38 s, during which a multi-MeV neutron either hits the Earth or escapes far away from the magnetosphere, a small fraction of them decay to protons while still in the inner magnetosphere and may become trapped in the inner radiation belt (to be discussed in Sect. 6.3.3).

Below about 10 GeV GCR and ACR fluxes are modulated by the 11- and 22-year solar cycles, so they provide quasi-stationary background radiation in the timescales of radiation belt observations. The arrivals of SCRs are, in turn, transient phenomena related to solar flares and coronal mass ejections.

The cosmic ray electrons also have galactic and solar components. Furthermore, the magnetosphere of Jupiter accelerates high-energy electrons escaping to the interplanetary space. These *Jovian electrons* can be observed near the Earth at intervals of about 13 months when the Earth and Jupiter are connected by the IMF.

Supernova shock waves are the most likely sources of the accelerated GCR electrons, whereas in the acceleration of SCR and Jovian electrons also other mechanisms besides shock acceleration are important, in particular inductive electric fields associated with magnetic reconnection in solar flares and the Jovian magnetosphere.

The acceleration and identity of the observed very highest-energy cosmic rays up to about 3×10^{20} eV remain enigmatic. It should not be possible to observe protons with energies higher than 6×10^{19} eV, known as the *Greisen–Zatsepin–Kuzmin cut-off*, unless they are accelerated not too far from the observing site. Above the cut-off the interaction of protons with the blue-shifted cosmic microwave background produces pions that carry away the excessive energy. It is possible that the highest-energy particles are nuclei of heavier elements. This is, for the time being, an open question.

1.4 Magnetospheric Dynamics

Strong solar wind forcing drives *storms* and more intermittent *substorms* in the magnetosphere. Both are critical dynamical elements in the temporal and spatial evolution of the radiation belts. They are primarily caused by various large-scale heliospheric structures such as interplanetary counterparts of *coronal mass ejections*

(CMEs/ICMEs),[10] *stream interaction regions* (SIRs) of slow and fast solar wind flows, and fast solar wind supporting Alfvénic fluctuations (to be discussed more in detail in Sect. 7.3.1). ICMEs are often preceded by interplanetary fast forward *shocks* and turbulent *sheath regions* between the shock and the ejecta, which all create their distinct responses in the magnetosphere and radiation belts. Because fast solar wind streams originate from coronal holes, which can persist over several solar rotations, the slow and fast stream pattern repeats in 27-day intervals and SIRs are often called *co-rotating interaction regions* (CIRs). However, stream interaction region is a physically more descriptive term. SIRs may gradually evolve to become bounded by shocks, but fully developed SIR shocks are only seldom observed sunward of the Earth's orbit. The duration of these large-scale heliospheric structures near the orbit of the Earth varies from a few hours to days. On average, the passage of a sheath region past the Earth takes 8–9 h and the passage of an ICME or SIR about 1 day. The fast streams typically influence the Earth's environment for several days.

1.4.1   Magnetospheric Convection

Magnetospheric plasma is in a continuous large-scale advective motion, which in this context is, somewhat inaccurately, called *magnetospheric convection* (for a thorough introduction, see Kennel 1995). The convection is most directly observable in the polar ionosphere, where the plasma flows from the dayside across the polar cap to the nightside and turns back to the dayside through the morning and evening sector auroral region. The non-resistive *ideal magnetohydrodynamics* (MHD, Sect. 3.2.3) is a fairly accurate description of the large-scale plasma motion above the resistive ionosphere. In ideal MHD the magnetic field lines are electric equipotentials and the electric field **E** and plasma velocity **V** are related to each other through the simple relation

$$\mathbf{E} = -\mathbf{V} \times \mathbf{B} .$$

(1.18)

Consequently, the observable convective motion, or alternatively the electric potential, in the ionosphere can be mapped along the magnetic field lines to plasma motion in the tail lobes and the plasma sheet. As the electric field in the tail plasma sheet points from dawn to dusk and the magnetic field to the north, the convection brings plasma particles from the nightside plasma sheet toward the Earth where a fraction of them become carriers of the ring current and form the source population for the radiation belts.

---

[10] Both acronyms are commonly used. We call the ejection CME when it is observed in the Solar corona and ICME further away in the interplanetary space.

In ideal MHD the plasma and the magnetic field lines are said to be *frozen-in* to each other. This means that two plasma elements that are connected by a magnetic field line remain so when plasma flows from one place to another (the proof of this statement can be found in most plasma physics textbooks, e.g., Koskinen 2011). It is convenient to illustrate the motion with moving field lines, although the magnetic field lines are not physical entities and their motion is just a convenient metaphor. A more physical description is that the magnetic field evolves in space and time such that the plasma elements maintain their magnetic connection.

The convection is sustained by solar wind energy input into the magnetosphere. The input is weakest, but yet finite, when the interplanetary magnetic field (IMF) points toward the north, and is enhanced during southward pointing IMF. If the magnetopause were fully closed, plasma would circulate inside the magnetosphere so that the magnetic flux tubes crossing the polar cap from dayside to nightside would reach to the outer boundary of the magnetosphere where some type of *viscous interaction* with the anti-sunward solar wind flow would be needed to maintain the circulation. This was the mechanism proposed by Axford and Hines (1961) to explain the convection. The classical (collisional) viscosity on the magnetopause is vanishingly small, but finite gyro radius effects and wave–particle interactions give rise to some level of *anomalous viscosity*.[11] It is estimated to provide about 10% of the momentum transfer from the solar wind to the magnetosphere.

The magnetosphere is, however, not fully closed. In the same year, when Axford and Hines presented their viscous interaction model, Dungey (1961) explained the convection in terms of magnetic reconnection. The *Dungey cycle* begins with a violation of the frozen-in condition at the dayside magnetopause current sheet. A magnetic field line in the solar wind is cut and reconnected with a terrestrial field line. Reconnection is most efficient for oppositely directed magnetic fields, as is the case in the dayside equatorial plane when the IMF points southward, but remains finite under other orientations. Subsequent to the dayside reconnection the solar wind flow drags the newly-connected field line to the nightside and the part of the field line that is inside the magnetosphere becomes a tail lobe field line. Consequently, an increasing amount of magnetic flux is piling up in the lobes. At some distance far in the tail the oppositely directed field lines in the northern and southern lobes reconnect again across the cross-tail current layer. At this point the ionospheric end of the field line has reached the auroral oval near local midnight. Now the earthward outflow from the reconnection site in the tail drags the newly-closed field line toward the Earth. The return flow cannot penetrate to the plasmasphere corotating with the Earth and the convective flow must proceed via the dawn and dusk sectors around the Earth to the dayside. In the ionosphere the flow returns toward the dayside along the dawnside and duskside auroral oval. Once approaching the dayside magnetopause, the magnetospheric plasma provides the inflow to the dayside reconnection inside of the magnetopause. Note that the

---

[11] This is one of many examples of the questionable use of word "anomalous". There is nothing anomalous in wave–particle interactions or processes beyond fluid description.

resistive ionosphere breaks the frozen-in condition of ideal MHD and it is not reasonable to use the picture of moving field lines in the atmosphere.

The increase in the tail lobe magnetic flux and strengthening of plasma convection inside the magnetosphere during southward IMF have a strong observational basis. Calculating the east-west component of the motion-induced solar wind electric field ($E = VB_{south}$) incident on the magnetopause and estimating the corresponding potential drop over the magnetosphere, some 10% of the solar wind electric field is estimated to "penetrate" into the magnetosphere as the dawn-to-dusk directed convection electric field. Note that $\mathbf{E} = -\mathbf{V} \times \mathbf{B}$ is not a causal relationship indicating whether it is the electric field that drives the magnetospheric convection, or convection that gives rise to the motion-induced electric field. The ultimate driver of the circulation is the solar wind forcing on the magnetosphere.

The plasma circulation is not as smooth as the above discussion may suggest. If the reconnection rates at the dayside magnetopause and nightside current sheet balance each other, a steady-state convection can, indeed, arise. This is, however, seldom the case since the changes in the driving solar wind and in the magnetospheric response are faster than the magnetospheric circulation timescale of a few hours. Reconnection may cause significant erosion of the dayside magnetospheric magnetic field placing the magnetopause closer to the Earth than a simple pressure balance consideration would indicate. The changing magnetic flux in the tail lobes causes expansion and contraction of the polar caps affecting the size and shape of the auroral ovals.

Furthermore, the convection in the plasma sheet has been found to consist of intermittent high-speed *bursty bulk flows* (BBF) with almost stagnant plasma in between (Angelopoulos et al. 1992, and references therein). It is noteworthy that while BBFs are more frequent during high auroral activity, they also appear during auroral quiescence. BBFs have been estimated to be the primary mechanism of earthward mass and energy transport in regions where they have been observed (Angelopoulos et al. 1994). Thus the high-latitude convection observed in the ionosphere corresponds to an average of the BBFs and slower background flows in the outer magnetosphere.

1.4.2 Geomagnetic Storms

Strong perturbations of the geomagnetic field known as *geomagnetic* (or *magnetic*) *storms* have been known since the nineteenth century. Because we look at the storms in this book mostly from the magnetospheric viewpoint, we call them also *magnetospheric storms*. As illustrated in Fig. 1.6, the storms are periods of most dynamic evolution of radiation belts. They often, but not always, commence with a significant positive deviation in the horizontal component of the magnetic field (H) measured on the ground (Fig. 1.7), called *storm sudden commencement* (SSC). An SSC is a signature of an ICME-driven shock and the associated pressure pulse arriving at the Earth's magnetopause. SSCs are also observed during pressure pulses

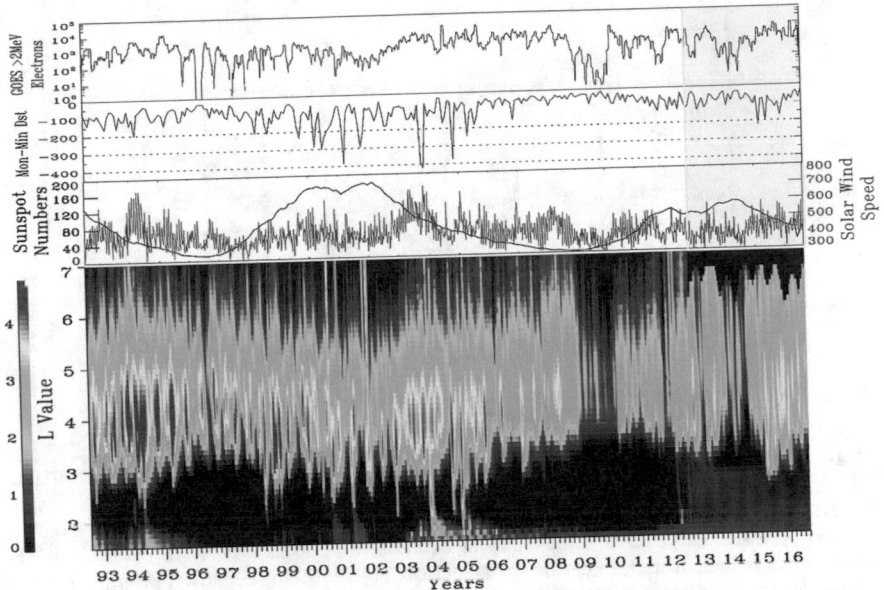

Fig. 1.6 Outer radiation belt response to solar and magnetospheric activity from the SAMPEX satellite and *Van Allen Probes* observations over a period of more than two solar cycles. The uppermost panel shows 27-day window-averaged relativistic (>2 MeV) electron fluxes at geostationary orbit, the second panel the monthly minimum of the *Dst* index, and the third panel the yearly window-averaged sunspot number (black) and weekly window-averaged solar wind speed (red). The spectrogram in the lowest panel is a composite of 27-day window-averaged SAMPEX observations of relativistic (~2 MeV) electron fluxes until September 2012 and *Van Allen Probes* REPT observations of (~2.1 MeV) electron fluxes after 5 September 2012. The shift from SAMPEX to *Van Allen Probes* is visible in the change of sensitivity to particle flux in the slot region (From Li et al. 2017, Creative Commons Attribution-NonCommercial-NoDerivs License)

related to SIRs or to ICMEs that are not sufficiently fast to drive a shock in the solar wind but still disturb and pile-up the solar wind ahead of them. If the solar wind parameters are known, the pressure effect can be removed from the *Dst* index as discussed in Sect. 1.2.3.

Storms in the magnetosphere can also be driven by low-speed ICMEs and SIRs without a significant pressure pulse. SIR-driven storms occur if the field fluctuations have sufficiently long periods of strong enough southward magnetic field to sustain global convention electric field to enhance the ring current. Thus there are storms without a clear SSC signature in the *Dst* index. On the other hand, a shock wave hitting the magnetopause is not always followed by a geomagnetic storm, in particular, if the IMF points dominantly toward the north during the following solar wind structure. In such cases the positive deviation in the magnetograms is called a *sudden impulse* (SI), after which the *Dst* index returns close to its background level with small temporal variations only. If the dynamic pressure remains at enhanced level, *Dst* can maintain positive deviation for some period.

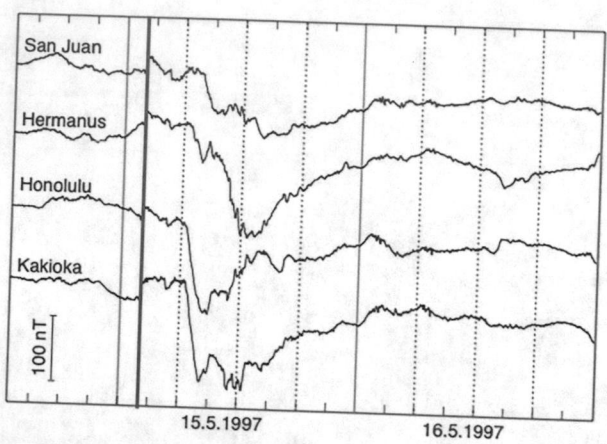

Fig. 1.7 The horizontal component (*H*) of the magnetic field measured at four low-latitude stations during a magnetic storm on 15 May 1997. An ICME-driven solar wind shock hit the magnetosphere on 15 May at about 02 UT causing the storm sudden commencement which is indicated by a sudden positive jump of the *H* component at all stations (thick blue line). The main phase of the storm started after 06 UT as indicated by the strong negative deviation in the *H* component. The solid vertical lines give the UT midnight and the tick-marks on the horizontal axis are given for each 3 h. (Figure courtesy: L. Häkkinen, adapted from Koskinen 2011, reprinted by permission from SpringerNature)

After the SSC an *initial phase* of the storm begins. It is characterized by a positive deviation of *Dst*, typically a few tens of nT. The initial phase is caused by a combination of predominantly northward IMF and high dynamic pressure. The phase can have very different durations depending on the type and structure of the solar wind driver. It can be very brief if the storm is driven by an ICME with a southward magnetic field following immediately a sheath with predominantly southward magnetic field. In such a case the storm *main phase*, which is a period characterized by a rapid decrease of the *H* component of the equatorial magnetic field, starts as soon as the energy transfer into the magnetosphere has become strong enough. If the sheath has a predominantly northward IMF, the main phase will not begin until a southward field of the ejecta enhances reconnection on the dayside magnetopause.

If there is no southward IMF either in the sheath or in the ICME, no regular global storm is expected to take place. However, pressure pulses/shocks followed by northward IMF can cause significant consequences to the radiation belt environment, as they can shake and compress the magnetosphere strongly and trigger a sequence of substorms (Sect. 1.4.3).

During the storm main phase, the enhanced energy input from the solar wind leads to energization and increase of the number of ring current carriers in the inner magnetosphere, as the enhanced magnetospheric convection transports an increasing amount of charged particles from the tail to the ring current region. Here substorms, discussed below, have important contribution, as they inject fresh particles from the near-Earth tail. The ring current enhancement is typically asymmetric because not all current carrying ions are on closed drift paths but a significant fraction of them passes the Earth on the evening side and continue toward the dayside magnetopause. This is illustrated in Fig. 1.7 where the Honolulu and Kakioka magnetometers show the steepest main phase development when these stations were in the dusk side of the globe.

When energy input from the solar wind ceases, the energetic ring current ions are lost faster than fresh ones are supplemented from the tail. The *Dst* index starts to return toward the background level. This phase is called the *recovery phase*. It is usually much longer than the main phase, because the dominating loss processes of the ring current carriers: charge exchange with the low-energy neutral atoms of the Earth's exosphere, wave–particle interactions, and Coulomb collisions (Sect. 6.3.2), are slower than the rapid increase of the current during the main phase. As ICMEs last typically 1 day, storms driven by ICMEs trailed by a slow wind tend to have relatively short recovery phases, whereas storms driven by SIRs and ICMEs followed by a fast stream can have much longer recovery phases. This is because Alfvénic fluctuations, i.e., large-amplitude MHD Alfvén waves (Sect. 4.4), in fast streams interacting with the magnetospheric boundary lead to triggering substorms, which inject particles to the inner magnetosphere. This can keep keep the ring current populated with fresh particles up to or longer than a week. The ring current development can also be more complex, often resulting in multi-step enhancement of *Dst* or events where *Dst* does not recover to quiet-time level between relatively closely-spaced intensifications. This typically occurs when both sheath and ICME ejecta carry southward field or when the Earth is impacted by multiple interacting ICMEs.

1.4.3 Substorms

From the radiation belt viewpoint the key significance of *magnetospheric substorms* is their ability to inject fresh particles in the energy range from tens to a few hundred keV from the tail plasma sheet into the inner magnetosphere. After being injected to the quasi-dipolar magnetosphere, charged particles start to drift around the Earth, contributing to the ring current and radiation belt populations. The injections have a twofold role: They provide particles to be accelerated to high energies. Simultaneously the injected electrons and protons drive waves that can lead to both acceleration and loss of radiation belt electrons and ring current carriers.

Magnetospheric substorms result from piling of tail lobe magnetic flux in the near-to-mid-tail region during enhanced convection. The details of the substorm cycle are still debated after more than half a century of research. Observationally it is clear that substorms encompass global configurational changes in the magnetosphere, namely the stretching of the near-Earth nightside magnetic field and related thinning of the plasma sheet during the flux pile-up (substorm *growth phase*), followed by a relatively rapid return of the near-Earth field toward a dipolar shape (*expansion phase*), and a slower return to a quiet-time stretched configuration associated with thickening of the plasma sheet (*recovery phase*). A substorm cycle typically lasts 2–3 h. The strongest activity occurs following the *onset* of the expansion phase: The cross-tail current in the near-Earth tail disrupts and couples to the polar region ionospheric currents through *magnetic field-aligned currents* forming the so-called *substorm current wedge*. This leads to intense precipitation of magnetospheric particles causing the most fascinating auroral displays. During geomagnetic storms the substorm cycle may not be equally well-defined. For example, a new growth phase may begin and the onset of the next expansion may follow soon after the previous expansion phase.

A widely used, though not the only, description of the substorm cycle is the so-called *near-Earth neutral line model* (NENL model, for a review, see Baker et al. 1996). In the model the current sheet is pinched off by a new magnetic reconnection neutral line once enough flux has piled up in the tail. The new neutral line forms somewhere at distances of 8–$30\,R_E$ from the Earth, which is much closer to the Earth than the far-tail neutral line of the Dungey cycle (Sect. 1.4.1). Earthward of the neutral line plasma is pushed rapidly toward the Earth. Tailward of the neutral line plasma flows tailward, and together with the far-tail neutral line, a tailward moving structure called *plasmoid* forms. Sometimes recurrent substorm onsets can create a chain of plasmoids. While it is common to illustrate the plasmoid formation using two-dimensional cartoons in the noon–midnight meridional plane, the three-dimensional evolution of the substorm process in the magnetotail is far more complex. In reality a plasmoid is a magnetic flux rope whose two-dimensional cut looks like a closed loop of magnetic field around a magnetic null point.

As pointed out in Sect. 1.4.1, the plasma flow in the central plasma sheet is not quite smooth and a significant fraction of energy and mass transport takes place as bursty bulk flows (BBFs). The BBFs are thought to be associated with localized reconnection events in the plasma sheet roughly at the same distances from the Earth as the reconnection line of the NENL model. They create small flux tubes called *dipolarizing flux bundles* (DFBs). The name derives from their enhanced northward magnetic field component B_Z corresponding to a more dipole-like state of the geomagnetic field compared to a more stretched configuration. Once created, DFBs surge toward the Earth due to the force caused by magnetic curvature tension in the fluid picture. They are preceded by sharp increases of B_Z called *dipolarization fronts*. DFBs are also associated with large azimuthal electric fields, up to several $mV\,m^{-1}$, which are capable of accelerating charged particles to high energies. Whether the braking of the bursty bulk flows and coalescence of dipolarization fronts closer to the Earth cause the formation of the substorm current wedge, or not, is a controversial issue.

The NENL model has been challenged by the common observation that the auroral substorm activation starts at the most equatorward arc and expands thereafter poleward. Whether the NENL model or some of the competing approaches (for a discussion, see e.g., Koskinen 2011) is the most appropriate substorm description, is not relevant to our discussion of radiation belts. What is essential is that the substorm expansions dipolarize the tail magnetic field configuration having been stretched during the growth phase and inject fresh particles into the inner magnetosphere. The particle injections can be observed as *dispersionless*, meaning that injected particles arrive to the observing spacecraft simultaneously at all energies, or *dispersive* when particles of higher energies arrive before those of lower energies. Because the dispersion arises from energy-dependent gradient and curvature drifts of the particles (Sect. 2.2.2), a dispersionless injection suggests that the acceleration occurs relatively close to the observing spacecraft, whereas dispersive arrival indicates acceleration further away from the observation when the particle distribution has had time to develop dispersion due to energy-dependent drift motion.

Dispersionless substorm injections are typically observed close to the midnight sector at geostationary orbit (6.6 R_E) and beyond, but have been found all the way down to about 4 R_E (Friedel et al. 1996). The injection sites move earthward as the substorm progresses and are also controlled by geomagnetic activity, although the extent of the dispersionless region is unclear, both in local time and radial directions. Neither have the details of acceleration of the injected particles been fully resolved. It has been suggested to be related both to betatron and Fermi acceleration (Sect. 2.4.4) associated with earthward moving dipolarization fronts. Another important aspect of dipolarization fronts for radiation belts is their braking close to Earth, which can launch magnetosonic waves that can effectively interact with radiation belt electrons.

Open Access This chapter is licensed under the terms of the Creative Commons Attribution 4.0 International License (http://creativecommons.org/licenses/by/4.0/), which permits use, sharing, adaptation, distribution and reproduction in any medium or format, as long as you give appropriate credit to the original author(s) and the source, provide a link to the Creative Commons license and indicate if changes were made.

The images or other third party material in this chapter are included in the chapter's Creative Commons license, unless indicated otherwise in a credit line to the material. If material is not included in the chapter's Creative Commons license and your intended use is not permitted by statutory regulation or exceeds the permitted use, you will need to obtain permission directly from the copyright holder.

Chapter 2
Charged Particles in Near-Earth Space

In this chapter we discuss the concepts that govern the motion of charged particles in the geomagnetic field and the principles how they stay trapped in the radiation belts. The basic particle orbit theory can be found in most plasma physics textbooks. We partly follow the presentation in Koskinen (2011). A more detailed discussion can be found in Roederer and Zhang (2014). A classic treatment of adiabatic motion of charged particles is Northrop (1963).

The Lorentz force and Maxwell's equations are summarized in Appendix A.1 where we also introduce the key concepts and notations of basic electrodynamics used in the book.

2.1 Guiding Center Approximation

The equation of motion of a particle with charge q, mass m and velocity $\mathbf{v}$ under the *Lorentz force* due to the electric ($\mathbf{E}$) and magnetic ($\mathbf{B}$) fields is

$$\frac{d\mathbf{p}}{dt} = q\left(\mathbf{E} + \mathbf{v} \times \mathbf{B}\right), \tag{2.1}$$

where $\mathbf{p} = \gamma m \mathbf{v}$ is the relativistic momentum and $\gamma = (1 - v^2/c^2)^{-1/2}$ the *Lorentz factor*. As discussed in Appendix A.1 we do not use terms "rest mass" or "relativistic mass". The mass of an electron is $m_e = 511\,\text{keV}\,c^{-2}$ and of a proton $m_p = 931\,\text{MeV}\,c^{-2}$.

Integration of the equation of motion in realistic magnetic field configurations must be done numerically. Numerical computation of even a fairly large number of particle orbits is not a problem for present day computers but to get a mental picture of particle motion we need an analytically tractable approach. Such is the *guiding center approximation* introduced by Hannes Alfvén in the 1940s (Alfvén 1950).

© The Author(s) 2022
H. E. J. Koskinen, E. K. J. Kilpua, *Physics of Earth's Radiation Belts*,
Astronomy and Astrophysics Library, https://doi.org/10.1007/978-3-030-82167-8_2

Let us start, for simplicity, from non-relativistic particles and consider a homogeneous static magnetic field in the direction of the z-coordinate with zero electric field. The equation of motion of a charged particle

$$m \frac{d\mathbf{v}}{dt} = q(\mathbf{v} \times \mathbf{B}) \tag{2.2}$$

describes a helical orbit with constant speed along the magnetic field and circular motion around the magnetic field line with the *angular frequency*

$$\omega_c = \frac{|q|B}{m}. \tag{2.3}$$

We call ω_c *gyro frequency* (also terms *cyclotron frequency* or *Larmor frequency* are frequently used). In the literature ω_c sometimes includes the sign of the charge q. In this book we write the gyro frequency as a positive quantity $\omega_c = |q|B/m$ and indicate the sign explicitly. The corresponding *oscillation frequencies* $f_{c\alpha} = \omega_{c\alpha}/(2\pi)$ of electrons and protons are

$$f_{ce}(\text{Hz}) \approx 28 \, B(\text{nT})$$

$$f_{cp}(\text{Hz}) \approx 1.5 \times 10^{-2} \, B(\text{nT}) .$$

The period of the gyro motion is

$$\tau_L = \frac{2\pi}{\omega_c} \tag{2.4}$$

and the radius of the circular motion perpendicular to the magnetic field

$$r_L = \frac{v_\perp}{\omega_c} = \frac{m v_\perp}{|q|B}, \tag{2.5}$$

where $v_\perp = \sqrt{v_x^2 + v_y^2}$ is the velocity perpendicular to the magnetic field. r_L is called the *gyro radius* (*cyclotron radius*, *Larmor radius*). Looking along (against) the magnetic field, the particle rotating clockwise (anticlockwise) has a negative charge. In plasma physics this is the convention of *right-handedness*.

This way we have decomposed the motion into two elements: constant speed $v_\parallel$ along the magnetic field and circular velocity $v_\perp$ perpendicular to the field. The sum of these components is a helical motion with the *pitch angle* α defined as

$$\tan \alpha = v_\perp / v_\parallel$$

$$\Rightarrow \alpha = \arcsin(v_\perp/v) = \arccos(v_\parallel/v) . \tag{2.6}$$

Alfvén pointed out that this decomposition is convenient even in temporally and spatially varying fields if the variations are small compared to the gyro motion and that the field does not change much with the particle motion along the magnetic field during one gyro period. This is the *guiding center approximation*. The center of the gyro motion is the *guiding center* (GC) and we call the frame of reference where $v_\parallel = 0$ is the *guiding center system* (GCS).

In the GCS the charge gives rise to a current $I = q/\tau_L$ along its circular path with associated *magnetic moment*

$$\mu = I\pi r_L^2 = \frac{1}{2}\frac{q^2 r_L^2 B}{m} = \frac{1}{2}\frac{m v_\perp^2}{B} = \frac{W_\perp}{B}. \tag{2.7}$$

Here we have introduced "perpendicular energy" $W_\perp$ to refer to the kinetic energy related to the velocity perpendicular to the magnetic field. Similarly, we define "parallel energy" $W_\parallel = (1/2)m v_\parallel^2$. The total energy of the particle is $W = W_\parallel + W_\perp$. With the word "energy" we refer to the kinetic energy, written relativistically (Appendix A.1) as

$$W = mc^2(\gamma - 1). \tag{2.8}$$

The magnetic moment is actually a vector $(q/2)\,\mathbf{r}_L \times \mathbf{v}_\perp$, where the gyro radius is a vector $\mathbf{r}_L$ pointing from the guiding center to the particle. The magnetic moment of both negatively and positively charged particles is opposite to the ambient magnetic field. Thus charged particles tend to weaken the background magnetic field and the plasma consisting of free charges resembles a *diamagnetic* medium. This is a useful concept when we discuss electromagnetic waves in the cold plasma approximation (Chap. 4).

For relativistic particles the gyro frequency and gyro radius are obtained simply replacing m by γm. In the formulas later in the text ω_c refers to the non-relativistic frequency and γ is introduced explicitly, when needed. Of course, when calculating the gyro periods and gyro radii, γ must be included. Furthermore, the magnetic moment of relativistic particles must must be expressed in terms of momentum ($p = \gamma m v$) as

$$\mu = \frac{p_\perp^2}{2mB}. \tag{2.9}$$

Note that here the constant mass m (not γm) is introduced in the denominator to give the familiar magnetic moment at the non-relativistic limit. We will return to the relativistic magnetic moment in Sect. 2.4.1.

2.2 Drift Motion

Adding a background electric field or letting the magnetic field be inhomogeneous, as it always is in the magnetosphere, modifies the path of the charged particles. If the effect remains small enough during one gyro period, the guiding center approximation is a useful tool to describe the motion.

2.2.1 E×B Drift

We start by adding a constant electric field perpendicular to the constant magnetic field. The equation of motion is again straightforward to solve. The guiding center is found to drift perpendicular to both electric and magnetic fields with the velocity

$$\mathbf{v}_E = \frac{\mathbf{E} \times \mathbf{B}}{B^2}. \tag{2.10}$$

This is called the *electric drift* (or E×B drift). The drift velocity is independent of the charge, mass and energy of the particle. All charged particles move with the same velocity and thus the E×B drift does not give rise to electric current. An example of the drift is the magnetospheric convection (Sec. 1.3.1) transporting particles from the tail plasma sheet toward the Earth.

The E×B drift can also be found by making a non-relativistic ($\gamma \approx 1$) Lorentz transformation to the frame co-moving with the guiding center

$$\mathbf{E}' = \mathbf{E} + \mathbf{v} \times \mathbf{B}. \tag{2.11}$$

In this frame $\mathbf{E}' = 0 \Rightarrow \mathbf{E} = -\mathbf{v} \times \mathbf{B}$, from which we find the solution (2.10) for $\mathbf{v}$. Note that a possible electric field component parallel to $\mathbf{B}$ cannot be eliminated by coordinate transformation because $\mathbf{E} \cdot \mathbf{B}$ is Lorentz invariant.

This coordinate transformation is possible for weak enough perpendicular forces $\mathbf{F}_\perp$ resulting in a general expression for the drift velocity

$$\mathbf{v}_D = \frac{\mathbf{F}_\perp \times \mathbf{B}}{q B^2}. \tag{2.12}$$

Here "weak enough" means $F/qB \ll c$. For example, for the E×B drift the ratio E/B must be much smaller than speed of light. Otherwise the guiding center approximation cannot be used.

2.2.2 *Gradient and Curvature Drifts*

The charged particles in the radiation belts move in a nearly dipolar magnetic field of the Earth that is curved and has gradients perpendicular to and along the field. Consider a spatially inhomogeneous magnetic field, assuming that the perpendicular and field-aligned components of the gradients are small within one gyration of the particle

$$|\nabla_\perp B| \ll B/r_L \; ; \; |\nabla_\parallel B| \ll (\omega_c/v_\parallel)B \,.$$

Under these circumstances it is possible to use perturbation approach to solve the equation of motion (Northrop 1963). The validity of these conditions does not depend on the field geometry alone but also on the energy and mass of the particle whose motion is to be calculated.

As only weak spatial inhomogeneities of the field are assumed, we can make Taylor expansion of the external magnetic field around the particle's GC. The details of the calculation can be found in most advanced plasma physics textbooks (e.g., Koskinen 2011). The solution of the equation of motion can be expressed as a sum of the unperturbed gyro motion and a small correction.

We start by neglecting the field line curvature. Keeping the first order terms and averaging over one gyro period the force reduces to

$$\mathbf{F} = -\mu \nabla B \,. \tag{2.13}$$

If the gradient has a magnetic field-aligned component, the force causes acceleration/deceleration of charged particles along the field

$$\frac{d\mathbf{v}_\parallel}{dt} = -\frac{\mu}{m}\nabla_\parallel B = -\frac{\mu}{m}\frac{\partial B(s)}{\partial s}\,\mathbf{b}\,, \tag{2.14}$$

where s is the coordinate and $\mathbf{b}$ the unit vector along the magnetic field.

In the perpendicular direction we find a drift across the magnetic field in the same way as for the zero order drift. The drift must balance the perpendicular force term (2.12) implying

$$\mathbf{v}_G = \frac{\mu}{qB^2}\mathbf{B} \times (\nabla B) = \frac{W_\perp}{qB^3}\mathbf{B} \times (\nabla B)\,. \tag{2.15}$$

This is called the *gradient drift*. The drift is perpendicular to both the magnetic field and its gradient. It is a result of small changes of the gyro radius as the particle gyrates in the inhomogeneous magnetic field (Fig. 2.1).

The gradient drift depends both on the perpendicular energy and the charge of the particle. Because negatively and positively charged particles drift to opposite directions, the drift contributes to the electric current in the plasma.

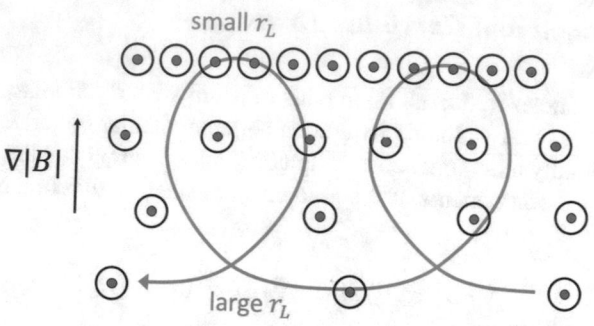

Fig. 2.1 The small variation of the gyro radius due to the magnetic field gradient results in the gradient drift

In a curved magnetic field also the GC motion is curved and the reference frame attached to the GC is not inertial. Let us denote the GC velocity by $\mathbf{V}$. Now $\mathbf{V}_\parallel$ is not exactly equal to $v_\parallel$ because $\mathbf{V}$ is determined at the GC whereas v is the velocity of the charge at the distance of $\mathbf{r}_L$ from the GC. The difference becomes significant for particles whose gyro radii approach the gradient scale length, i.e., near the limit where the GC approximation becomes invalid (for a more detailed discussion, see, Roederer and Zhang 2014).

Let us move to the frame co-moving with the GC. Let the orthogonal basis $\{\mathbf{e}_i\}$ define the coordinate axes and choose $\mathbf{e}_3 \parallel \mathbf{v}_\parallel \parallel \mathbf{B}$. Now $\mathbf{v} = \sum v_i \, \mathbf{e}_i$, and $\{\mathbf{e}_i\}$ rotates when its origin moves with the GC.

$$\frac{d\mathbf{v}}{dt} = \sum_i \left(\frac{dv_i}{dt} \, \mathbf{e}_i + v_i \, \frac{d\mathbf{e}_i}{dt} \right) = \sum_i \left(\frac{dv_i}{dt} \, \mathbf{e}_i + v_i \, (\mathbf{V}_\parallel \cdot \nabla) \, \mathbf{e}_i \right). \tag{2.16}$$

The term $\sum v_i (\mathbf{V}_\parallel \cdot \nabla) \mathbf{e}_i$ is due to the curvature and causes a centrifugal effect. Averaged over one Larmor cycle the curvature force is

$$\mathbf{F}_C = - \left\langle m \sum_i v_i (\mathbf{V}_\parallel \cdot \nabla) \mathbf{e}_i \right\rangle. \tag{2.17}$$

Due to the assumption of weak curvature $(\mathbf{V}_\parallel \cdot \nabla) \, \mathbf{e}_i$ can be approximated constant in every point along the gyro orbit. In the present approximation the perpendicular velocity components v_1 and v_2 oscillate sinusoidally and thus $\langle v_1 \mathbf{e}_1 \rangle = \langle v_2 \mathbf{e}_2 \rangle = 0$. Furthermore, during one gyro period, on average, $v_\parallel \approx V_\parallel$ and we get

$$\mathbf{F}_C = -m V_\parallel^2 (\mathbf{e}_3 \cdot \nabla) \, \mathbf{e}_3. \tag{2.18}$$

A little exercise in differential geometry yields

$$(\mathbf{e}_3 \cdot \nabla)\,\mathbf{e}_3 = \mathbf{R}_C / R_C^2 \,, \tag{2.19}$$

where $\mathbf{R}_C$ is the *radius of curvature* vector, pointing inward. Now

$$\mathbf{F}_C = -m V_\parallel^2 \frac{\mathbf{R}_C}{R_C^2} \,. \tag{2.20}$$

Because $\mathbf{B} = B\mathbf{e}_3$,

$$(\mathbf{e}_3 \cdot \nabla)\mathbf{e}_3 = (\mathbf{B} \cdot \nabla \mathbf{B})/B^2 \tag{2.21}$$

and we can write the resulting *curvature drift* velocity as

$$\mathbf{v}_C = \frac{-m V_\parallel^2}{q B^2} \frac{\mathbf{R}_C \times \mathbf{B}}{R_C^2} = \frac{m V_\parallel^2}{q B^4} \mathbf{B} \times (\mathbf{B} \cdot \nabla)\mathbf{B} \,. \tag{2.22}$$

Now we can again approximate $v_\parallel \approx V_\parallel$ and express the curvature drift in terms of the parallel energy of the particle $W_\parallel = (1/2)m v_\parallel^2$.

If there are no local currents ($\nabla \times \mathbf{B} = 0$), as in the case of a pure dipole field, the curvature drift simplifies to

$$\mathbf{v}_C = \frac{2 W_\parallel}{q B^3} \mathbf{B} \times \nabla B \tag{2.23}$$

and $\mathbf{v}_G$ and $\mathbf{v}_C$ can be combined as

$$\mathbf{v}_{GC} = \frac{W_\perp + 2 W_\parallel}{q B^3} \mathbf{B} \times \nabla B = \frac{W}{q B R_C} (1 + \cos^2 \alpha)\,\mathbf{n} \times \mathbf{t} \,, \tag{2.24}$$

where $\mathbf{t} \parallel \mathbf{B}$ and $\mathbf{n} \parallel \mathbf{R}_C$ are unit vectors. The drift velocities are straightforward to write relativistically replacing m by γm.

As a consequence of the gradient and curvature of the near-Earth geomagnetic field charged high-energy particles drift around the Earth, electrons to the east and positively charged particles to the west, resulting in net westward current. The motion of low-energy particles is dominated by the E×B-drift.

The perturbation approach can be continued to higher orders. The recipe is the same as above. First determine the force due to the higher order perturbation and calculate the drift velocity to balance its effect. Mathematically this procedure results in an *asymptotic* expansion of the guiding center position as a function of time (for a detailed discussion, see, Northrop 1963).

2.3   Drifts in the Magnetospheric Electric Field

In magnetospheric physics a commonly used frame of reference is the *Geocentric Solar Magnetospheric* coordinate system (GSM). In the GSM system the X-axis points toward the Sun and the Earth's dipole axis is in the XZ-plane. Z points approximately northward and Y is opposite to the Earth's orbital motion around the Sun. As the Earth rotates, the XZ-plane flaps about the X-axis such that the dipole remains in the XZ-plane but can be tilted from the Z-direction maximally $34°$, which is the sum of the tilts of the Earth's rotation axis ($\approx 23°$) from the ecliptic plane and of the dipole axis ($\approx 11°$) from the rotation axis.

The large-scale magnetospheric plasma flow in the GSM system gives rise to the electric field $\mathbf{E} = -\mathbf{V} \times \mathbf{B}$. If the magnetic field is time-independent, the electric field is curl-free and can be expressed as the gradient of a scalar potential $\mathbf{E} = -\nabla \varphi$. During rapid changes of the magnetic field, such as geomagnetic storms, also the inductive electric field given by Faraday's law $\partial \mathbf{B}/\partial t = -\nabla \times \mathbf{E}$ must be taken into account.

Let us, for simplicity, consider the motion of relatively low-energy plasma sheet and plasmaspheric particles in the GSM equatorial plane. Assume further that the magnetospheric magnetic field is perpendicular to the equatorial plane pointing upward (Z-direction). Thus the sunward advection in the plasma sheet corresponds to a dawn-to-dusk (Y-direction) pointing electric field $E_0 \mathbf{e}_y$, which we here assume to be constant. Let r be the distance from the center of the Earth and ϕ the angle from the direction of the Sun. Then the electric field, called in this context convection electric field, is given by

$$\mathbf{E}_{conv} = -\nabla(-E_0 r \sin \phi) \tag{2.25}$$

and its potential is

$$\varphi_{conv} = -E_0 r \sin \phi . \tag{2.26}$$

The Earth and its atmosphere rotate in the GSM frame. The corotation extends in the equatorial plane roughly up to the plasmapause. The angular velocity toward the east is $\Omega_E = 2\pi/24\,\text{h}$. Assuming, again for simplicity, a perfect corotation, the plasma velocity in the GSM frame is

$$\mathbf{V}_{rot} = \Omega_E r\, \mathbf{e}_\phi , \tag{2.27}$$

where $\mathbf{e}_\phi$ is the azimuthal unit vector pointing toward the east. Consequently the motion induced electric field is $\mathbf{E}_{rot} = -\mathbf{V} \times \mathbf{B}$, the potential of which is

$$\varphi_{rot} = \frac{-\Omega_E k_0}{r} = \frac{-\Omega_E B_0 R_E^3}{r} . \tag{2.28}$$

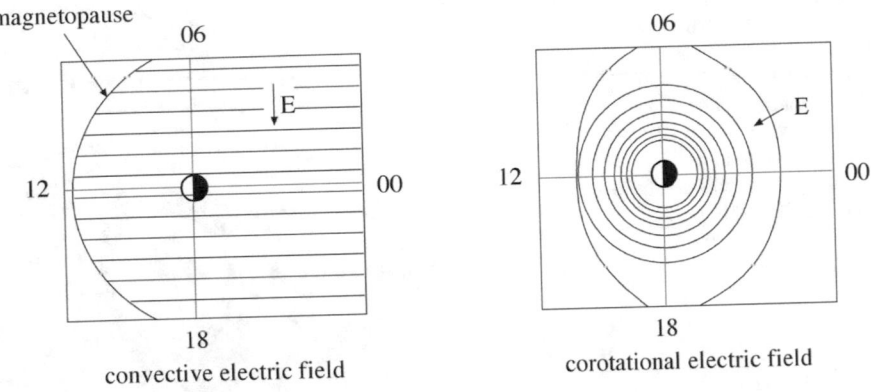

Fig. 2.2 Equipotential lines of convection and corotation electric fields in the equatorial plane. The local time directions are given at the faces of the panels. (From Koskinen 2011, reprinted by permission from SpringerNature)

Here $k_0 = 8 \times 10^{15}\,\mathrm{T\,m^3}$ is the Earth's dipole moment and B_0 the dipole field on the surface of the Earth at the equatorial plane ($B_0 \approx 30\,\mu\mathrm{T}$). The convection and corotation electric fields are illustrated in Fig. 2.2. The total field is the superposition $\mathbf{E}_{conv} + \mathbf{E}_{rot}$.

In addition to the large scale electric field in their frame of reference, the plasma particles are also under the influence of magnetic field gradient and curvature forces. Let us consider particles that move in the equatorial plane of the dipole ($v_\parallel = 0$ or equivalently $\alpha = 90°$). For these particles the curvature effect is zero and and the total drift velocity including the $\mathbf{E} \times \mathbf{B}$ and gradient drifts is

$$\mathbf{v}_D = \frac{1}{B^2}\left[\mathbf{E}_{conv} + \mathbf{E}_{rot} - \nabla\left(\frac{\mu B}{q}\right)\right] \times \mathbf{B} = \frac{1}{B^2}\mathbf{B} \times \nabla\varphi_{eff}\,, \qquad (2.29)$$

where μ is the magnetic moment of the particles. The effective potential is

$$\varphi_{eff} = -E_0 r \sin\phi - \frac{\Omega_E B_0 R_E^3}{r} + \frac{\mu B_0 R_E^3}{q r^3}\,. \qquad (2.30)$$

In the time-independent potential field the particles move along stream lines of constant φ_{eff}. These stream lines depend on the charge and energy of the particles through their magnetic moments. For low-energy particles, whose perpendicular velocity in the moving frame is small ($\mathbf{v}_\perp \approx 0 \Rightarrow \mu \approx 0$), the streamlines are equipotential lines of the combined convection and corotation fields (Fig. 2.3). In this approximation the motion is a pure $\mathbf{E} \times \mathbf{B}$-drift and all particles move with the same velocity.

Fig. 2.3 Orbits of
low-energy particles ($\mu \approx 0$)
in the equatorial plane
assuming $E_0 = 0.3\,\mathrm{mV\,m^{-1}}$.
The distance between
consecutive points is 10 min.
(From Koskinen 2011,
reprinted by permission from
Springer Nature)

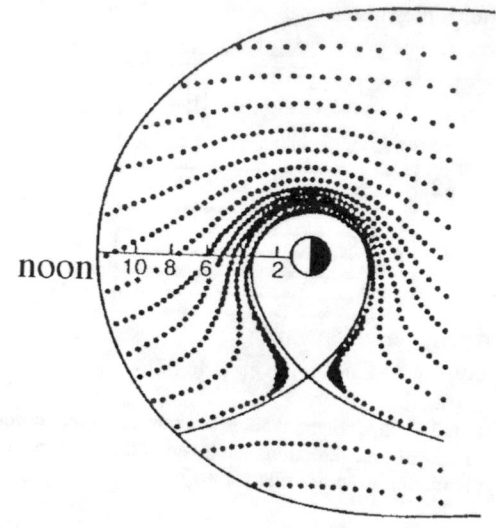

Figure 2.3 illustrates the formation of a *separatrix* that separates the low-energy corotating plasmaspheric plasma from the sunward cold plasma motion in the plasma sheet. In this single-particle model the separatrix is the plasmapause. The separatrix has an electric neutral point ($\mathbf{E} = 0$) at the distance

$$r = \sqrt{\frac{\Omega_E B_0 R_E^3}{E_0}} \tag{2.31}$$

in the direction of 18 MLT. While this model of the plasmasphere is a strong simplification, it explains qualitatively why the plasmasphere is compressed during enhanced magnetospheric activity (Fig. 1.4) and the formation of the plasmaspheric bulge, or plume, in the evening sector (Fig. 1.5). The enhanced energy input enhances the convection velocity and thus the dawn-to-dusk electric field, whereas the rotation of the Earth remains constant and the corotation electric field is always the same. Consequently, the separatrix is pushed toward the Earth when the convection enhances, i.e., E_0 in (2.30) increases.

The real plasma has a finite temperature and the plasma drifting from the plasma sheet is much warmer than the plasmasphere. Thus the gradient drift term in (2.30) must be taken into account. Consequently, the shape, density and actual location of the plasmaspheric plume is determined by the actual plasma environment. The observed plume is widened and often pushed toward the earlier local time (Fig. 1.5). Furthermore, the plasmapause reacts to the changing electric field with some delay, which can lead to detachment of the plume from the plasmasphere and its subsequent disappearance to the dayside magnetopause.

In the finite-temperature plasma sheet, the magnetic field gradient separates the motion of positive and negative charges (remember that we consider here equatorial particles only, for which the curvature drift is zero). To illustrate this effect consider particles whose magnetic moment is so strong that it supersedes the effect of corotation electric field. Now the effective potential, again in the equatorial plane only, is

$$\varphi_{eff} = -E_0 r \sin \phi + \frac{\mu B_0 R_E^3}{q r^3} \ . \tag{2.32}$$

The $r \sin \phi$-dependence implies that far from the Earth the particles follow the convection electric field, but closer in they become dominated by the magnetic drifts. This way the dipole field shields the cold plasmasphere from the hot plasma sheet.

The positive and negative high-energy charges drift as illustrated in Fig. 2.4. They have different separatrices, called *Alfvén layers*. Because the plasma sheet is a finite particle source, a larger fraction of positive charges pass the Earth in the evening sector and a larger fraction of negative charges in the morning sector. This leads to piling of positive space charge in the evening sector and negative charge in the morning sector. The accumulated charges are discharged by magnetic field-aligned currents flowing to the ionosphere from the evening sector and from the ionosphere to the magnetosphere in the morning sector.

Figure 2.4 also gives a qualitative explanation for a mechanism how energetic particles can be lost from the plasmasphere during the main phase of strong magnetospheric storms, and on the other hand, how particles on open drift paths can become trapped again during the recovery phase. As the convective electric field rapidly grows during the main phase, the particles are pushed outward from the radiation belt in the dayside. The particles, whose trajectories are open, intercept

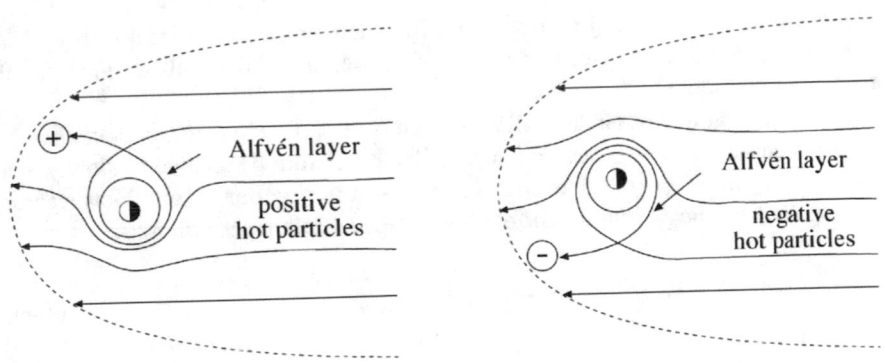

Fig. 2.4 Formation of Alfvén layers of energetic particles. (From Koskinen 2011, reprinted by permission from SpringerNature)

the magnetopause and are lost from the inner magnetosphere. At the same time more particles on the nightside E×B-drift deeper into the ring current and radiation belts. Once the activity ceases, the trapping boundary, i.e., the Alfvén layer, moves outward and thus particles that were originally on open drift paths past the Earth find themselves trapped into the expanding plasmasphere. Note, however, that the actual losses and enhancements of high-energy radiation belt particles depend on several other processes discussed in the subsequent chapters of this book.

2.4 Adiabatic Invariants

The trapped charged particles in the radiation belts perform three nearly periodic motions in the geomagnetic field. These quasi-periodic motions are related to *adiabatic invariants* in the terminology of Hamiltonian mechanics.

In physical systems symmetries correspond to conservation laws. For example, in classical mechanics rotational symmetry is associated with conservation of angular momentum. The angular momentum is invariant in a periodic motion around the axis of symmetry. If the motion is nearly-periodic, such as the gyro motion in the guiding center approximation, adiabatic invariants take the role of conserved quantities. The adiabatic invariants do not need to be the same as conserved quantities in the strictly periodic case and their conservation depends critically of the "slowness" of the variations.

In the framework of *Hamiltonian mechanics* it can be shown that if Q and P are the canonical coordinate and momentum of the system and if the motion is nearly periodic, the integral

$$I = \oint P dQ \tag{2.33}$$

over the (quasi-)period of Q is an adiabatic invariant, i.e., a conserved quantity. This statement requires, of course, a proof (see, e.g., Bellan 2006, or advanced classical mechanics textbooks).

A classic example of an adiabatic invariant is the *Lorentz–Einstein pendulum* in the gravitational field (g). Let the length of the pendulum (l) change so slowly that its frequency $\omega = \sqrt{g/l}$ does not change much during one swing. However, as work is done on the pendulum, the energy of the pendulum per unit mass

$$W = \frac{1}{2}l^2\dot{\theta}^2 + \frac{1}{2}gl\theta^2 , \tag{2.34}$$

where θ is the angle from the vertical direction, is not constant. It is a straightforward exercise to show that in this case the conserved quantity is the ratio W/ω. This is closely analogous to the magnetic moment

$$\mu = \frac{W_\perp}{B} = \frac{q}{m}\frac{W_\perp}{\omega_c}, \tag{2.35}$$

which readily suggests that the magnetic moment is an adiabatic invariant, if the magnetic field changes slowly with respect to the gyro motion.

2.4.1 The First Adiabatic Invariant

A recommendable introduction to the Hamiltonian treatment of radiation belt particles is Ukhorskiy and Sitnov (2013). To show that the magnetic moment is an adiabatic invariant in the Hamiltonian framework recall from the classical electrodynamics that the momentum of a particle in an electromagnetic field is $\mathbf{p} = m\mathbf{v} + q\mathbf{A}$, where $\mathbf{A}$ is the vector potential of the field. We can take the gyro radius vector $\mathbf{r}_L$ as the canonical coordinate in the plane perpendicular to the magnetic field. Then the corresponding canonical momentum is $\mathbf{p}_\perp$. Assuming that the guiding center approximation is valid, i.e., the gyro motion is nearly periodic, the integral

$$
\begin{aligned}
I &= \oint \mathbf{p}_\perp \cdot d\mathbf{r}_L = \oint m\mathbf{v}_\perp \cdot d\mathbf{r}_L + q\int_S (\nabla \times \mathbf{A}) \cdot d\mathbf{S} \\
&= \int_0^{2\pi r_L} mv_\perp dl + q\int_S \mathbf{B} \cdot d\mathbf{S} \\
&= 2\pi m v_\perp r_L - |q|B\pi r_L^2 = \frac{2\pi m}{|q|}\mu
\end{aligned}
\tag{2.36}
$$

is an adiabatic invariant. In plasma physics the magnetic moment is called the *first adiabatic invariant*.

The physical dimension of μ is energy/magnetic field. Its SI-unit is $\mathrm{J\,T^{-1}}$ but in radiation belt physics the non-SI unit $\mathrm{MeV\,G^{-1}}$ is commonly used.

Note that also the magnetic flux enclosed within the gyro orbit

$$\Phi = B\pi r_L^2 = \frac{2\pi m}{q^2}\mu \tag{2.37}$$

is constant. This has important implication to flux conservation in macroscopic plasma physics, which is an expression of the concept of magnetic field lines being frozen-in to plasma motion.

40

To directly show, within Newtonian mechanics, that the magnetic moment, or any of the other invariants discussed below, is an adiabatic invariant is a cumbersome task. Plasma physics textbooks usually treat some special cases only. A fairly complete treatise can be found in Northrop (1963).

For the purpose of this book, it is instructive to look at the invariance of μ in a static magnetic field in the absence of electric field. Since there is no temporal variation of the magnetic field, the energy[1] of the particle $W = W_\parallel + W_\perp$ is constant, i.e.,

$$\frac{dW_\parallel}{dt} + \frac{dW_\perp}{dt} = 0.$$ (2.38)

As $W_\perp = \mu B$,

$$\frac{dW_\perp}{dt} = \mu \frac{dB}{dt} + \frac{d\mu}{dt} B.$$ (2.39)

Now $dB/dt = v_\parallel dB/ds$ is the change of the magnetic field along the GC path. The parallel force is accroding to (2.14)

$$m \frac{dv_\parallel}{dt} = -\mu \frac{dB}{ds}.$$ (2.40)

By multiplying this by $v_\parallel = ds/dt$ we get

$$\frac{dW_\parallel}{dt} = -\mu \frac{dB}{dt}.$$ (2.41)

Thus

$$\frac{dW_\parallel}{dt} + \frac{dW_\perp}{dt} = B \frac{d\mu}{dt} = 0,$$ (2.42)

illustrating that μ is an adiabatic invariant *in this special case*.

If the magnetic field changes slowly in time ($\partial/\partial t \ll \omega_c$), Faraday's law implies the presence of an inductive electric field along the gyro path of the particle, yielding acceleration

$$\frac{dW_\perp}{dt} = q(\mathbf{E} \cdot \mathbf{v}_\perp).$$ (2.43)

[1] Recall that in our notation W refers to kinetic energy unless stated otherwise.

During one gyro period the particle gains energy

$$\triangle W_\perp = q \int_0^{2\pi/\omega_c} \mathbf{E} \cdot \mathbf{v}_\perp dt . \tag{2.44}$$

Due to the slow change we can replace the time integral by a line integral over one closed gyration and use Stokes' law

$$\triangle W_\perp = q \oint_C \mathbf{E} \cdot d\mathbf{l} = q \int_S (\nabla \times \mathbf{E}) \cdot d\mathbf{S} = -q \int_S \frac{\partial \mathbf{B}}{\partial t} \cdot d\mathbf{S} , \tag{2.45}$$

where $d\mathbf{S} = \mathbf{n}\, dS$, $\mathbf{n}$ is the normal vector of the surface S pointing to the direction determined in the right-hand sense by the positive circulation of the loop C. For small variations of the field $\partial \mathbf{B}/\partial t \approx \omega_c \triangle B/2\pi$. Thus

$$\wedge W_\perp = \frac{1}{2}|q|\omega_c r_L^2 \triangle B = \mu \triangle B . \tag{2.46}$$

On the other hand

$$\triangle W_\perp = \mu \triangle B + B \triangle \mu , \tag{2.47}$$

implying $\triangle \mu = 0$. Thus for slow time variation μ is conserved although the inductive electric field accelerates the particle. This is analogous to the Lorentz-Einstein pendulum: The energy is not conserved but μ is.

In Chap. 6 we will discuss how temporal variations with a frequency near the gyro frequency of the particles can break the invariance of the magnetic moment and lead to acceleration, scattering and loss of particles through wave–particle interactions.

Generalization of the magnetic moment to relativistic particles must be done with some care. The relativistic conserved quantity is $p_\perp^2/B$. In Eq. (2.9) we formulated the relativistic magnetic moment dividing this by the constant $2m$ (recall that m is the velocity-independent constant mass of the particle). Now

$$\mu_{rel} = \frac{p_\perp^2}{2mB} = \frac{\gamma^2 m^2 v_\perp^2}{2mB} = \gamma^2 \frac{mv_\perp^2}{2B} = \gamma^2 \mu_{nonrel} . \tag{2.48}$$

When reading the literature it is important to be alert whether the magnetic moment in an equation is the relativistic or non-relativistic quantity. Northrop (1963) pointed out that it is not easy to prove in a general magnetic configuration that this really is the correct generalization of the non-relativistic magnetic moment. Further complications arise if there is a parallel electric field that accelerates the particle and the total momentum is not constant.

Magnetic Mirror and Magnetic Bottle

The invariance of the magnetic moment leads us to important concepts in the motion of radiation belt particles: the magnetic mirror, the magnetic bottle and the loss-cone.

Assume first that the total kinetic energy W and the magnetic moment $\mu = W_\perp/B$ of a charged particle are conserved. Let the guiding center of the particle move along the magnetic field in the direction of a weak positive gradient of B. $W_\perp$ can increase until $W_\parallel \to 0$. The perpendicular velocity is $v_\perp = v \sin \alpha$ and we can write the magnetic moment as

$$\mu = \frac{mv^2 \sin^2 \alpha}{2B} .$$

(2.49)

On the other hand, as now $v^2 \propto W$ is also assumed to be constant, the pitch angles at two different magnetic field intensities are related as

$$\frac{\sin^2 \alpha_1}{\sin^2 \alpha_2} = \frac{B_1}{B_2} .$$

(2.50)

When $W_\parallel \to 0$, $\alpha \to 90°$. The deceleration of the GC motion in an increasing magnetic field (2.14) is said to be due to the *mirror force* $\mathbf{F} = -\mu \nabla_\parallel B$, which finally turns the motion back toward the weakening B. The magnitude of the *mirror field* B_m at the turning point depends on the particle's pitch angle at a reference point B_0. At the mirror field $\alpha_m = 90°$ and we get

$$\sin^2 \alpha_0 = B_0/B_m .$$

(2.51)

As $B_m < \infty$, every mirror field is leaky. Particles having smaller pitch angles than α_0 in the field B_0 pass through the mirror. These particles are said to be in the *loss cone*. Two opposite mirrors form a *magnetic bottle*, which confines particles outside the loss cone of the weaker of the two magnetic mirrors. The quasi-dipolar near-Earth magnetic field is a giant inhomogeneous and temporally variable magnetic bottle trapping the radiation belt and ring current particles.

The mirror force does not need to be the only force affecting the parallel motion of the GC. If there is an electric field with a parallel component $\mathbf{E}_\parallel$ and/or the particle is in a gravitational field, the parallel equation of motion becomes

$$m \frac{d\mathbf{v}_\parallel}{dt} = q\mathbf{E}_\parallel + m\mathbf{g}_\parallel - \mu \nabla_\parallel B .$$

(2.52)

If the non-magnetic forces can be derived from a potential $U(s)$ the equation of motion (2.14) is

$$m \frac{dv_\parallel}{dt} = -\frac{\partial}{\partial s}[U(s) + \mu B(s)] .$$

(2.53)

Now the GC moves in the effective potential $U(s) + \mu B(s)$. Work is done to the charged particle and thus its energy is not conserved but its magnetic moment is. In solar-terrestrial physics, examples of such potentials are the gravitational field on the Sun and parallel electric fields above discrete auroral arcs, of which the latter can have consequences to the radiation belts, affecting the width of the loss cone. Also plasma waves with a parallel electric field component contribute to the dynamics of radiation belt particles.

2.4.2 The Second Adiabatic Invariant

Assume that the energy is conserved, i.e., the speed v of the particle is constant. The *bounce motion* between the magnetic mirrors of a magnetic bottle is nearly periodic if the field does not change much during one *bounce period*, or *bounce time*, τ_b

$$\tau_b = 2 \int_{s_m}^{s'_m} \frac{ds}{v_\parallel(s)} = \frac{2}{v} \int_{s_m}^{s'_m} \frac{ds}{(1 - B(s)/B_m)^{1/2}}, \qquad (2.54)$$

where s is the arc length along the GC orbit and s_m and s'_m are the coordinates of the mirror points. Here we have used Eq. (2.51) and $v_\parallel/v = \cos\alpha = \sqrt{1 - B(s)/B_m}$ to move the constant speed outside the integral, which now depends on the magnetic field configuration only. The bounce period is defined over the entire bounce motion back and forth.

The concept of bounce motion makes sense in the guiding center approximation if $\tau_b \gg \tau_L$. Thus the condition for the bounce motion to be nearly periodic is more restrictive than the condition for nearly periodic gyro motion

$$\tau_b \frac{dB/dt}{B} \ll 1. \qquad (2.55)$$

If this condition is fulfilled, there is an associated adiabatic invariant. In nearly-periodic bounce motion the track of the GC does not enclose any magnetic flux and the canonical momentum reduces to $p_\parallel$. Now the canonical coordinate is the position of the GC along the magnetic field line. Consequently,

$$J = \oint p_\parallel \, ds \qquad (2.56)$$

is an adiabatic invariant, generally known as the *second adiabatic invariant* or *longitudinal invariant*. The invariance of J is broken when the temporal variation of the magnetic field takes place in time scales comparable to or shorter than the bounce motion.

Both μ and J depend on particle momentum. As long as μ is conserved, the momentum can be eliminated from the second invariant by introducing a purely field-geometric quantity K

$$K = \frac{J}{\sqrt{8m\mu}} = I\sqrt{B_m} = \int_{s_m}^{s'_m} [B_m - B(s)]^{1/2} \, ds \, , \tag{2.57}$$

where m is the mass of the particle. The integral I is

$$I = \int_{s_m}^{s'_m} \left[1 - \frac{B(s)}{B_m}\right]^{1/2} ds \tag{2.58}$$

and $J = 2pI$. Since the invariance of J requires the invariance of μ, and m is the constant mass, J and K are physically equivalent. The physical dimension of K is square root of magnetic field times length, which in radiation belt studies is expressed in units of $R_E \, G^{1/2}$.

Neither the bounce period nor the integral I can be expressed in a closed form. We can, however, find useful expressions for particles mirroring close to the minimum of a symmetric magnetic bottle, e.g., the dipole equator, where we can approximate the field by a parabola

$$B(s) \simeq B_0 + \frac{1}{2}a_0 s^2 \, . \tag{2.59}$$

Here $a_0 = \partial^2 B/\partial s^2$ evaluated at the equator, where the magnetic field is B_0. Now the parallel equation of motion $m dv_\parallel/dt = -\mu \partial B/\partial s$ becomes

$$m\frac{d^2 s}{dt^2} = -\mu a_0 s \, . \tag{2.60}$$

This is the familiar equation of motion for a linear pendulum with the period

$$\tau_b = 2\pi \sqrt{\frac{m}{\mu a_0}} = 2\pi \sqrt{\frac{mB}{W_\perp a_0}} \approx \frac{2\pi\sqrt{2}}{v}\sqrt{\frac{B_0}{a_0}} \, , \tag{2.61}$$

where we have approximated $v \approx v_\perp$ and $B \approx B_0$, as we consider particles with the equatorial pitch angle $\alpha_{eq} \approx 90°$.

Now we have also the expression for the bounce period at the limit $\alpha_{eq} \to 90°$ where the integrand in (2.54) diverges at the same time as the path length goes to zero. The bounce time remains finite, even if there is no actual bounce motion. This is analogous to the linear mechanical pendulum at rest (here $v_\parallel = 0$). The period of the pendulum is well-defined whether it swings or not! For the bounce motion about the dipole equator (2.61) this means that the "equatorially mirroring" particles do not violate the guiding center hierarchy $\tau_b \gg \tau_L$. Once the equatorial pitch angle deviates from 90°, if only slightly, the bounce motion takes the time τ_b because the particle spends most of its bounce time close to the mirror point.

For equatorially mirroring particles integrals J, K and I are trivially zero. Using the same expansion as before we can find analytical expression for I for particles mirroring close to the equator. Letting $s = 0$ at the equator the mirror points are found to be

$$s_m \approx \pm \sqrt{\frac{2B_0}{a_0} \left(1 - \frac{B_0}{B_m} \right)} \tag{2.62}$$

and the integral reduces to

$$I \simeq \frac{\pi}{\sqrt{2}} \sqrt{\frac{B_0}{a_0}} \cos^2 \alpha_{eq} . \tag{2.63}$$

This approximation is good if $B_m/B_0 \lesssim 1.1$, corresponding to particles whose equatorial pitch angles are in the range $75° \lesssim \alpha_{eq} \leq 90°$. B_m and a_0 depend on the magnetic field configuration. In the dipole field the integral is found to be (Roederer 1970)

$$I \simeq \frac{\pi}{3\sqrt{2}} L R_E \left(1 - \frac{k_0}{B_m L^3 R_E^3} \right) . \tag{2.64}$$

The bounce periods in the dipole field are calculated in Sect. 2.5.

2.4.3 The Third Adiabatic Invariant

Also the drift across the magnetic field may be nearly-periodic if the drift motion takes place around an axis. This is particularly important in the radiation belts where particles drift around the Earth in the quasi-dipolar magnetic field. The corresponding *third adiabatic invariant* is the magnetic flux Φ through the closed contour defined by the equatorial track of the GC given by

$$\Phi = \oint \mathbf{A} \cdot \mathbf{dl}, \tag{2.65}$$

where $\mathbf{A}$ is the vector potential of the field and $\mathbf{dl}$ is the arc element along the drift path of the GC, which now can be taken as the canonical momentum and coordinate. In order for Φ to be an adiabatic invariant, the *drift period* τ_d has to fulfil $\tau_d \gg \tau_b \gg \tau_L$. Consequently, the invariant is weaker than μ and J and much slower changes in the magnetic field can break the invariance of Φ.

46

2 Charged Particles in Near-Earth Space

Table 2.1 Summary of adiabatic invariants

Invariant	Velocity	Time-scale	Validity
Magnetic moment μ	Gyro motion $\mathbf{v}_\perp$	Gyro period $\tau_L = 2\pi/\omega_c$	$\tau \gg \tau_L$
Longitudinal-invariant J	Parallel velocity of GC $\mathbf{V}_\parallel$	Bounce period τ_b	$\tau \gg \tau_b \gg \tau_L$ and μ constant
Flux invariant Φ	Perp. velocity of GC $\mathbf{V}_\perp$	Drift period τ_d	$\tau \gg \tau_d \gg \tau_b \gg \tau_L$ and μ and J constant

Table 2.1 summarizes the functions $\{\mu, J, \Phi\}$ and the conditions for their adiabatic invariance.

In three-dimensional space there are at most three independent adiabatic invariants but the actual number can be smaller, even zero. In radiation belts μ is often a good invariant and J is invariant for particles that spend at least some time in the magnetic bottle of the nearly-dipolar field. Under steady conditions also Φ is constant, but its invariance can be broken by spatial inhomogeneities and ultra low frequency (ULF) oscillations of the magnetospheric magnetic field, as will be discussed in Chap. 6.

In the language of Hamiltonian mechanics the functions $\{\mu, J, \Phi\}$ defined in Eqs. (2.36), (2.56), and (2.65), whether invariant or not, form a set of *canonical action variables* or *action integrals* of the canonical electromagnetic momentum $\mathbf{p} + q\mathbf{A}$ over the periods of the canonical coordinates $\mathbf{s}_i$

$$J_i = \frac{1}{2\pi} \oint_i (\mathbf{p} + q\mathbf{A}) \cdot d\mathbf{s}_i \tag{2.66}$$

each of which having an associated *phase angle* φ_i: the gyro phase, bounce phase and drift phase. $\{J_i, \varphi_i\}$ is a convenient set of six independent variables, e.g., when discussing velocity distribution functions, also known as phase space densities (Sect. 3.5), of radiation belt particles.

Note that the set of action integrals and their adiabatic invariance depend on the symmetries of the magnetic field configuration. For example, in a two-dimensional picture of the current sheet in the Earth's magnetotail the motion of charged particles may be symmetric about the center of the sheet with an associated adiabatic invariant. Such a motion is known as *Speiser motion*. The violation of the invariance in this case leads to the chaotization of the particle motion with consequences to the stability of the current sheet (e.g., Büchner and Zelenyi 1989).

2.4.4 Betatron and Fermi Acceleration

When particles drift adiabatically in an inhomogeneous magnetic field, their energies and/or pitch angles are affected. Consider the rate of change of the kinetic energy W of a charged particle in a general time-dependent magnetic field $\mathbf{B}$. The time derivative in a moving frame of reference is $d/dt = \partial/\partial t + \mathbf{V} \cdot \nabla$, where $\mathbf{V}$ is the velocity of the frame of reference, i.e., the velocity of the GC In the GCS the energy equation is

$$\frac{dW_{GCS}}{dt} = \mu \frac{dB}{dt} = \mu \left(\frac{\partial B}{\partial t} + \mathbf{V}_\perp \cdot \nabla_\perp B + V_\parallel \frac{\partial B}{\partial s} \right). \tag{2.67}$$

This equation can be transformed to the energy equation in the frame of reference of the observer (OBS)

$$\frac{dW_{OBS}}{dt} = \frac{dW_{GCS}}{dt} + \frac{d}{dt} \left(\frac{1}{2}mV_\parallel^2 \right) + \frac{d}{dt} \left(\frac{1}{2}mV_\perp^2 \right)$$

$$= \mu \frac{\partial B}{\partial t} + q\mathbf{V} \cdot \mathbf{E}, \tag{2.68}$$

where non-electromagnetic forces have been neglected.

The first term in the right hand side of (2.68) describes the gyro betatron effect on charged particles, in which the changing magnetic field at the position of the GC leads to *gyro betatron acceleration* of the charged particle through the inductive electric field. The increase in the perpendicular velocity of the particle increases also the gradient drift velocity of the GC that depends on the perpendicular energy (2.15).

The second term describes the action of the electric field on the motion of the GC covering both magnetic field-aligned acceleration (if $E_\parallel \neq 0$) and another betatron effect, called *drift betatron acceleration*. The second term is zero if the velocity of the GC and electric field are perpendicular to each other, i.e., the drift betatron acceleration requires that the electric field has a component in the direction of particle's drift velocity. Therefore, the drifts due to the electric field do not contribute to drift betatron acceleration.

When the GC advects adiabatically in a quasi-static electric field toward an increasing field ($B_2 > B_1$), the invariance of μ implies

$$\frac{W_{\perp 2}}{W_{\perp 1}} = \frac{B_2}{B_1}. \tag{2.69}$$

Thus $W_{\perp 2} > W_{\perp 1}$. This leads to an anisotropic "pancake" velocity space distribution (Sect. 3.4) of particles transported by the convection electric field from the magnetotail to the inner magnetosphere. Pancake distributions drive two of the most important wave modes in radiation belt dynamics, the electromagnetic ion cyclotron waves and the whistler-mode chorus waves (Sect. 5.2). The betatron effect driven by the inductive electric field is discussed in Sect. 2.6 in the context of the drift shells in a quasi-dipolar magnetic field.

A special case of betatron acceleration is the drift of a particle in J-conserving bounce motion toward a region where the mirror field increases. This corresponds to moving the mirror points closer to each other and, consequently, decreasing $\oint ds$. To compensate this $v_\parallel$ of the particle and thus its parallel energy $W_\parallel$ in the observer's frame must increase. This mechanism is known as *Fermi acceleration*. A mechanical analogue of Fermi acceleration is hitting a tennis ball with a racket. In the spectator's frame the ball is accelerated but in the racket's frame it just mirrors. The concept of Fermi acceleration is often used in the context of moving astrophysical shock fronts, the solar-terrestrial examples of which are interacting CME shocks or an ICME shock approaching the bow shock of the Earth.

2.5 Charged Particles in the Dipole Field

Figure 2.5 illustrates the motion of charged particles in the guiding center approximation in a dipole field introduced in Sect. 1.2.1.

The guiding center approximation can be applied in a static dipole field if the particle's gyro radius is much smaller than the curvature radius R_C of the field. In terms of the particle's *rigidity* $p_\perp/|q|$ the condition for the guiding center approximation is

$$r_L \left| \frac{\nabla_\perp B}{B} \right| = \frac{p_\perp}{|q|R_C B} \propto \frac{p_\perp}{|q|r_0 B} \ll 1 , \tag{2.70}$$

i.e., the GC approximation is valid if

$$\frac{p_\perp}{|q|} \ll r_0 B , \tag{2.71}$$

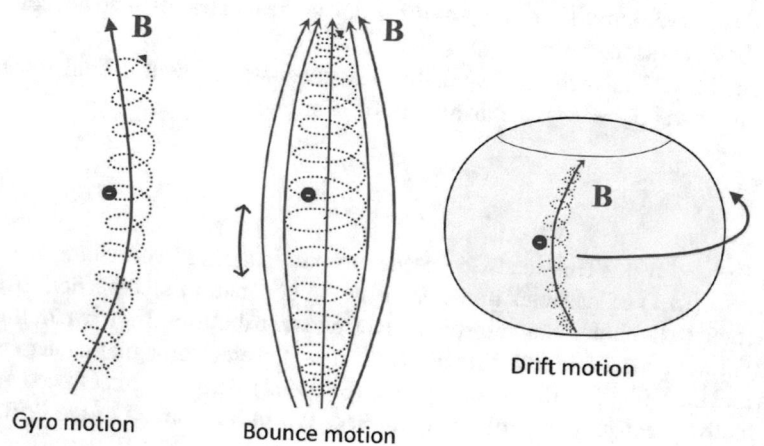

Gyro motion   Bounce motion   Drift motion

Fig. 2.5 Motion of an electron in a dipole field

where r_0 is the Earth-centered distance to point where the field line crosses the dipole equator. The conversion of particle energy to rigidity is given by the relativistic relationship between momentum and kinetic energy (A.16)

$$p_\perp^2 = \frac{1}{c^2}(W^2 + 2mc^2 W)\sin^2 \alpha \,. \tag{2.72}$$

Rigidity describes how the magnetic field affects a charged particle. A particle with a large rigidity is less affected by the ambient magnetic field than a particle with smaller rigidity. The rigidities of most cosmic rays are so large that they cannot be trapped by the Earth's dipole field and just move through the radiation belts being deflected by the magnetic field. The deflection depends on the rigidity of the particle and on the angle it arrives to the magnetosphere.

Some of the cosmic ray particles penetrate to the atmosphere, the most energetic ones even to the ground. In cosmic ray physics it is common to use the units familiar from elementary particle physics, setting $c = 1$. Thus particle mass, momentum and energy are all given in energy units, typically with GeV, which is of the order of proton mass $m_p = 0.931$ GeV. Consequently, the unit of rigidity is GV.

Let λ_m be the mirror latitude of a particle trapped in the dipole field and B_0 the magnetic field at the equatorial plane. The equatorial pitch angle of the particle is

$$\sin^2 \alpha_{eq} = \frac{B_0}{B(\lambda_m)} = \frac{\cos^6 \lambda_m}{(1 + 3\sin^2 \lambda_m)^{1/2}} \,. \tag{2.73}$$

Denote the latitude, where the field line intersects the surface of the Earth, by λ_e. If $\lambda_e < \lambda_m$, the particle hits the Earth before mirroring and is lost from the bottle. Actually the loss takes place already in the upper atmosphere through collisions with atmospheric atoms and molecules. This can, of course, happen within a wide range of altitudes. In this book we mainly discuss particles with energies $\gtrsim 100$ keV, most of which deposit their energy at altitudes around 100 km or below. In any case 100 km is small as compared to 1 R_E and we can approximate the equatorial half-width of the loss cone by

$$\sin^2 \alpha_{eq,l} = L^{-3}(4 - 3/L)^{-1/2} = (4L^6 - 3L^5)^{-1/2} \,, \tag{2.74}$$

where L is the geocentric distance to the point where the magnetic field line crosses the dipole equator expressed in units of R_E, i.e., McIlwain's L-parameter (Sect. 1.2.1). The particle is in the loss cone if $\alpha_{eq} < \alpha_{eq,l}$.

As shown in Fig. 2.6, the half-width of the equatorial loss cone in the outer radiation belt is a few degrees and widens quickly along the field lines when the magnetic latitude exceeds 30°. In addition to the loss-cone width, the curves in the figure indicate the magnetic latitudes where the dipole field lines cross the surface of the Earth, e.g., $L = 4$ at $\lambda = 60°$.

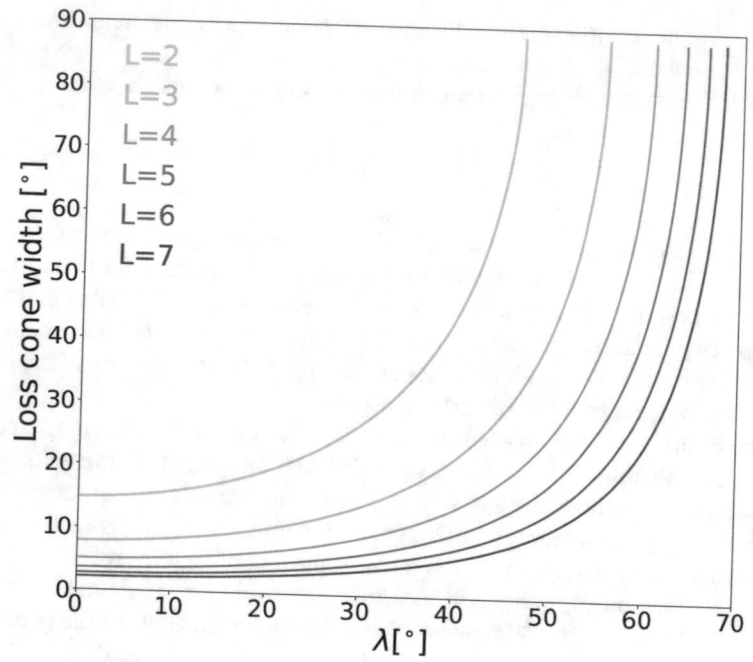

Fig. 2.6 The half-width of the atmospheric loss cone as a function of magnetic latitude for *L*-parameters 2–7

The bounce period in a dipolar bottle is

$$\tau_b = 4 \int_0^{\lambda_m} \frac{\mathrm{d}s}{v_\parallel} = 4 \int_0^{\lambda_m} \frac{\mathrm{d}s}{\mathrm{d}\lambda} \frac{\mathrm{d}\lambda}{v_\parallel}$$

$$= \frac{4r_0}{v} \int_0^{\lambda_m} \frac{\cos\lambda (1 + 3\sin^2\lambda)^{1/2}}{1 - \sin^2\alpha_{eq}(1 + 3\sin^2\lambda)^{1/2}/\cos^6\lambda} \, \mathrm{d}\lambda$$

$$= \frac{4r_0}{v} T(\alpha_{eq}). \tag{2.75}$$

$T(\alpha_{eq})$ is known as the *bounce function*, which needs to be integrated numerically. At the limit $\lambda_m = 90°$ ($\alpha_{eq} = 0$) the integral can be given in a closed form yielding $T(0) = (S/r_0)/2 = 1.38$, where S is the length of the field line (1.7). For nearly equatorially mirroring particles ($\alpha_{eq} \approx 90°$), the dipole field can be approximated as a parabola (2.59), yielding $T(\pi/2) = (\pi/6)\sqrt{(2)} \approx 0.74$. Between these extremes ($0 < \alpha_{eq} < 90°$) a good approximation is (Schulz and Lanzerotti 1974)

$$T(\alpha_{eq}) \approx 1.3802 - 0.3198 \left(\sin\alpha_{eq} + \sqrt{\sin\alpha_{eq}}\right). \tag{2.76}$$

For $\alpha_{eq} \gtrsim 40°$ a little less accurate approximation

$$T(\alpha_{eq}) \approx 1.30 - 0.56 \sin \alpha_{eq} \tag{2.77}$$

is often good enough in practice (as used by, e.g., Roederer 1970; Lyons and Williams 1984). As $T(\alpha_{eq})$ is of the order of 1, $\tau_b \approx 4r_0/v$ is a pretty good approximation in back-of-the-envelope calculations.

Finally, we investigate the drift time around the Earth. In the terrestrial dipole field (where $\nabla \times \mathbf{B} = 0$), ions drift to the west and electron to the east with the combined curvature and gradient drift velocity (2.24)

$$
\begin{aligned}
v_{GC} &= \frac{W}{qBR_C}(1 + \cos^2 \alpha) \\
&= \frac{3mv^2 r_0^2}{2qk_0} \frac{\cos^5 \lambda (1 + \sin^2 \lambda)}{(1 + 3\sin^2 \lambda)^2} \left[2 - \sin^2 \alpha \frac{(1 + 3\sin^2 \lambda)^{1/2}}{\cos^6 \lambda} \right],
\end{aligned}
\tag{2.78}
$$

where the Eqs. (1.4) and (1.8) for the dipole field $B(\lambda)$ and curvature radius $R_C(\lambda)$ have been inserted.

In the drift motion around the Earth, v_{GC} is usually less interesting than the bounce-averaged azimuthal speed $\langle d\phi/dt \rangle = \langle v_{GC}/r \cos \lambda \rangle$ that gives the drift rate of the guiding center around the dipole axis. A straightforward calculation yields

$$
\begin{aligned}
\left\langle \frac{d\phi}{dt} \right\rangle &= \frac{4}{v\tau_b} \int_0^{\lambda_m} \frac{v_{GC}(\lambda)(1 + 3\sin^2 \lambda)^{1/2}}{\cos^2 \lambda \cos \alpha(\lambda)} \, d\lambda \\
&\equiv \frac{3mv^2 r_0}{2qk_0} g(\alpha_{eq}) = \frac{3mv^2 R_E L}{2qk_0} g(\alpha_{eq}),
\end{aligned}
\tag{2.79}
$$

where

$$g(\alpha_{eq}) = \frac{1}{T(\alpha_{eq})} \int_0^{\lambda_m} \frac{\cos^3 \lambda (1 + \sin^2 \lambda)[1 + \cos^2 \alpha(\lambda)]}{(1 + 3\sin^2 \lambda)^{3/2} \cos \alpha(\lambda)} \, d\lambda. \tag{2.80}$$

Similarly to $T(\alpha_{eq})$ also $g(\alpha_{eq})$ is of the order of 1. For equatorial pitch angles larger than 30° $g(\alpha_{eq})$ can be approximated as

$$g(\alpha_{eq}) \approx 0.7 + 0.3 \sin(\alpha_{eq}) \tag{2.81}$$

yielding for non-relativistic equatorial particles ($\alpha_{eq} = 90°$)

$$\left(\frac{d\phi}{dt} \right)_0 = \frac{3mv^2 R_E L}{2qk_0} = \frac{3\mu}{qr_0^2}. \tag{2.82}$$

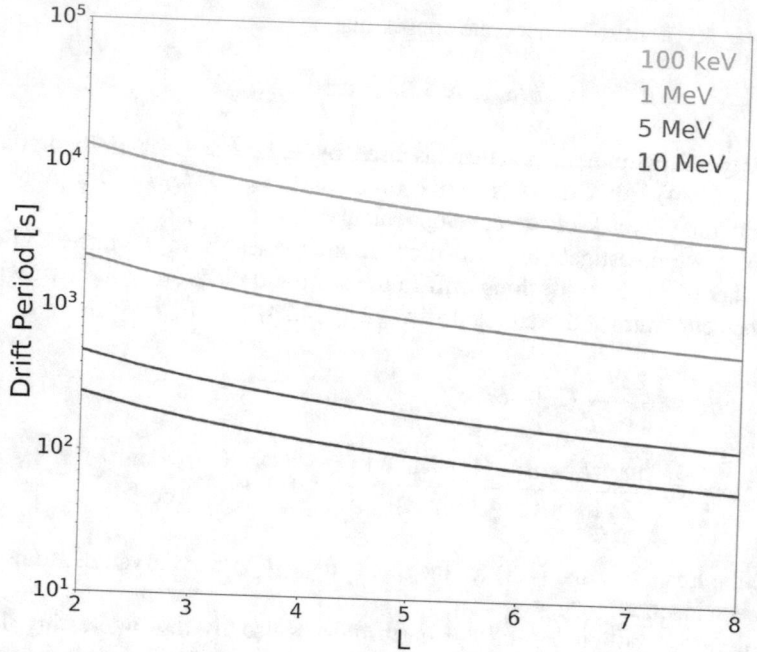

Fig. 2.7 Drift periods of radiation belt electrons with energies 100 keV, 1, 5, and 10 MeV as a function of L calculated from Eq. (2.83)

For relativisic equatorial particles this is modified to

$$\left(\frac{d\phi}{dt}\right)_0 = \frac{3mc^2 R_E L}{2qk_0}\gamma\beta^2 = \frac{3\mu_{rel}}{\gamma qr_0^2},\qquad(2.83)$$

where μ_{rel} is the relativistic first invariant $p_\perp^2/(2mB)$ (2.48) and $\beta = v/c$. The coefficient $3mc^2 R_E/(2qk_0)$ is

- 0.035 degrees per second for electrons and
- 64.2 degrees per second for protons.

Note that the drift period decreases with increasing L. While this may, at the first sight, seem counterintuitive, recall that according to the expression for the gradient–curvature drift velocity in current-free magnetic field (2.24) the drift speed is proportional to $1/(BR_C)$. The dipole field scales as L^{-3} and the curvature radius as L. Thus $v_{GC} \propto L^2$, $(d\phi/dt)_0 = v_{GC}/L \propto L$ and $\tau_d \propto L^{-1}$.

In Fig. 2.7 drift periods of radiation belt electrons are plotted for selected energies in the range 100 keV–10 MeV at L-shells 2–8 and Table 2.2 gives examples of electron gyro, bounce and drift periods at $L = 2$, $L = 4$ and $L = 6$. The bounce times are calculated from (2.75) for $\alpha_{eq} = 80°$ correspoinding to the bounce

Table 2.2 Examples of approximate electron gyro, bounce and equatorial drift periods for selected energies. Note that the magnetic drift period of 10-keV electrons deep in the plasmasphere ($L = 2$) is longer than the corotation time (24 h). Thus their physical drift motion around the Earth is determined by the corotational electric field. Note also that the ultra-relativistic particles move practically with the speed of light, thus their bounce times at a given L-shell are almost identical

		10 keV	100 keV	1 MeV	5 MeV	10 MeV
$L = 2$	τ_L	9.71 μs	11,4 μs	28.1 μs	103 μs	196 μs
	τ_b	0.64 s	0.23 s	0.14 s	0.13 s	0.13 s
	τ_d	44.2 h	3.65 h	36.9 min	8.09 min	4.19 min
$L = 4$	τ_L	0.08 ms	0.09 ms	0.26 ms	0.82 ms	1.57 ms
	τ_b	1.27 s	0.46 s	0.27 s	0.26 s	0.26 s
	τ_d	22.1 h	1.83 h	18.5 min	4.05 min	2.09 min
$L = 6$	τ_L	0,26 ms	0.31 ms	0.76 ms	2.88 ms	5.29 ms
	τ_b	1.91 s	0.69 s	0.41 s	0.38 s	0.38 s
	τ_d	12.3 h	1.22 h	12.3 min	2.70 min	1.40 min

function $T \approx 0.75$. The drift periods are calculated from (2.83) for particles with $\alpha_{eq} = 90°$.

2.6 Drift Shells

The guiding center of a charged particle traces a *drift shell* when the particle drifts around the Earth (right-hand picture in Fig. 2.5). In a symmetric static magnetic field in the absence of external forces when μ and W are conserved, the drift shell can be uniquely identified by

$$I = \text{constant}$$
$$B_m = \text{constant} \tag{2.84}$$

where I is the integral (2.58) and B_m the mirror field. In a dipolar configuration the magnetic field and the drift shells are symmetric around the dipole axis. The drift shells are in this case defined by constant L and commonly referred to as *L–shells*.

Beyond about 3–4 R_E temporal and spatial asymmetries of the magnetospheric magnetic field affect the formation and evolution of the drift shells. The geomagnetic field is compressed on the dayside by the Chapmann–Ferraro current (1.14), which together with the stretching of the nightside magnetic field causes azimuthal asymmetry. The asymmetry increases with increasing dynamic pressure of the solar wind, in particular, when a solar wind pressure pulse or interplanetary shock compresses the dayside configuration, or the nightside field stretches during the substorm growth phase.

For a distorted dipole the L-parameter can be generalized by defining

$$L^* = \frac{2\pi k_0}{\Phi R_E}.$$

(2.85)

L^* is sometimes called *Roederer's L-parameter* to distinguish it from the original McIlwain's L-parameter of the dipole field. L^* is inversely proportional to the magnetic flux (Φ) enclosed by particle's drift contour. Thus L^* is an alternative way to express the third action integral. If changes in the field are slower than the drift period of the particle, L^* remains invariant.

L^* is equal to McIlwain's L-parameter only in a purely dipolar field. Otherwise, L^* corresponds to the radial distance to the equatorial points of the symmetric L-shell on which the particle would be found if all non-dipolar contributions to the magnetic field were turned off adiabatically. This method can also be applied if internal field perturbations closer to the Earth are taken into account.

2.6.1 Bounce and Drift Loss Cones

Charged particles that complete one or more drift cycles around the Earth are said to be *stably-trapped*. Trapping is lost if the particle reaches an altitude where it is lost through collisions with atmospheric particles, or if the particle's drift shell encounters the magnetopause and the particle is lost there. Particles that are able to perform a number of bounces but are lost before a complete drift cycle are said to be *pseudo-trapped*.

If the particle does not fulfil a complete bounce cycle in the magnetic bottle of the inner magnetosphere before it is lost into the atmosphere, it is said to be in the *atmospheric* (or *bounce*) *loss cone*. In the dipole field the width of the atmospheric loss cone depends on L and the magnetic latitude (Fig. 2.6). As bounce periods of radiation belt electrons are only fractions of a second (Table 2.2), electrons close to the edge of the bounce loss cone can be lost rapidly from the belts due to wave–particle interactions whereas scattering from larger equatorial pitch angles takes more time (Chap. 6).

The asymmetric deviations from the dipole introduce complications to this picture. Although the field inside of about 3–4 R_E is very close to that of a dipole, the dipole is displaced from the center of the Earth, which introduces an asymmetry in the Earth-centered frame of reference. Now the distance from the dipole to the upper atmosphere varies as a function of geographic latitude and longitude.

Let us consider an electron in the Earth's displaced dipole field, freshly injected to a field line where it is not in a bounce loss cone. Assume that the electron mirrors from the field B_m at an altitude somewhat above 100 km, and drifts around the Earth. If the dipole field were symmetric, the electron would stay trapped unless something (e.g., wave–particle interaction) lowers its pitch angle letting it to escape from the magnetic bottle. But in the case of a displaced dipole the electron can drift to a field line that is connected to the atmosphere above a region that is farther from the dipole. The field at 100 km is now smaller than B_m and the particle may reach a lower altitude and be lost from the magnetic bottle. The particle has performed a number of bounces but was not able to complete a full drift around the Earth. It is said to have been in the *drift loss cone*. The drift loss cone is as wide as the widest bounce loss cone on a given drift shell. The effect is strongest above the South Atlantic Anomaly, where the minimum surface magnetic field is 22 μT, which is much smaller than 35 μT of Earth-centered 11°-tilted dipole in the same region where dipole latitude is about −20°.

2.6.2 Drift Shell Splitting and Magnetopause Shadowing

In an azimuthally symmetric static magnetic field particles with different pitch angles, which at a given longitude are on a common drift shell, will remain on the same shell. This is called *shell degeneracy*. It is a consequence of condition (2.84), according to which drift shells are defined by constant mirror magnetic field magnitude B_m and constant integral I. In a pure dipole field I can be replaced by the L-parameter.

In an azimuthally asymmetric field charged particles with different pitch angles, whose guiding centers are on a joint magnetic field line at some longitude, do not stay on a common field line when they move to another longitude. The reason is that they mirror at different field strengths B_m and, consequently, their I-integrals are different. This is known as *drift shell splitting*. The magnetosphere is always compressed on the dayside and stretched on the nightside, and the asymmetry increases during increased solar wind forcing and geomagnetic storms. Figure 2.8 illustrates the shell splitting in the noon–midnight cross section under weakly asymmetric configuration. The particles with different equatorial pitch angles on a common drift shell in the nightside are found on different shells in the dayside (top). Vice versa, particles on a common drift shell in the dayside are on different shells in the nightside (bottom).

In Fig. 2.8 the drift shell splitting is most pronounced at equatorial distances beyond about 6–7 R_E. The upper panel indicates that the drift shells of particles with large equatorial pitch angles (small $\cos \alpha_0$) extend in the dayside further out than of particles with smaller pitch angles. Depending on the amount of compression of the dayside magnetosphere these particles may hit the magnetopause and be lost to the magnetosheath before they pass the sub-solar direction. The phenomenon where particles are lost at the magnetopause is known as *magnetopause shadowing*. Particles lost this way are said to *pseudo-trapped*, as they remain trapped during a part of their drift around the Earth.

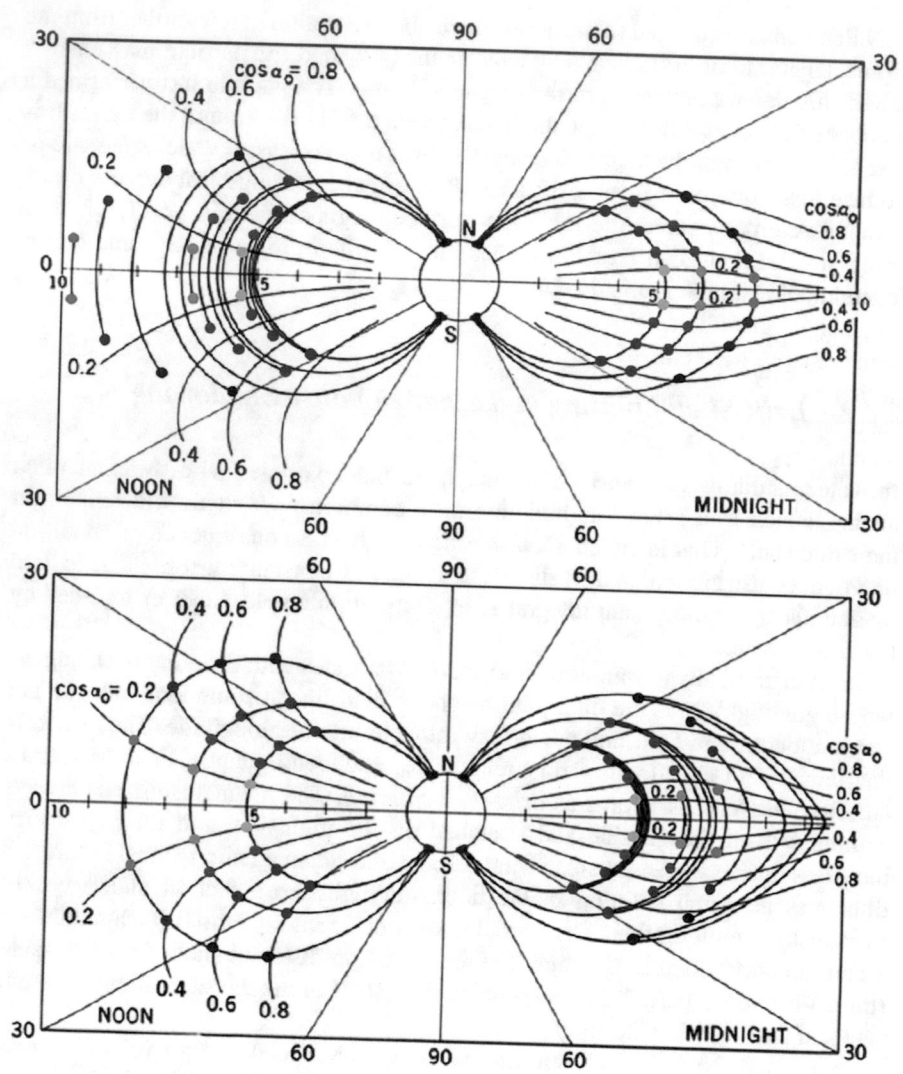

Fig. 2.8 Illustration of drift shell splitting from nightside to dayside (top) and from dayside to nightside (bottom) in a weakly compressed/stretched magnetic field configuration. The dots indicate the mirror points for different equatorial pitch-angle cosines. (From Roederer 1970, the colors of dots have been added to guide the eye) (Reprinted by permission from Springer Nature)

The magnetopause shadowing, where the drift shell splitting plays a role, thus leads to a loss of particles nearly 90° pitch angles, while the scattering of particles to the atmosphere results in the loss of particles with small pitch angles. In two-dimensional velocity space ($v_\parallel$, $v_\perp$) this leads to a butterfly shape of the particle distribution function (Sect. 3.4.2).

Magnetopause shadowing is an important particle loss mechanism from the outer electron belt (Sect. 6.5.1). The shadowing is most important during strong magnetospheric activity when the dayside magnetopause is most compressed/eroded and the nightside field is stretched already at the distance of the outer radiation belt. Note that although the bulk plasma is frozen-in the magnetospheric field, the gyro radii of the energetic radiation belt particles are large enough to break the freezing close to the magnetopause boundary layer.

Correspondingly, common drift shells in the dayside are split when particles drift toward the nightside (Fig. 2.8, bottom). In this case the drift shells of particles with larger equatorial pitch angles are closer to the Earth in the nightside and the particles remain stably trapped. However, now particles with *small pitch angles*, reaching farther out to the magnetotail, may lose their magnetic field guidance, when crossing the current sheet, before they drift through the midnight meridian and again become pseudo-trapped.

Note that particles observed at a given point on the dayside come from different locations in the nightside. As there may well be more high-energy particles with large pitch angles in the nightside closer to the Earth than farther away, the observed pitch-angle distribution in the dayside may have a shape of a pancake (Sect. 3.4.1)

The compression of the dayside magnetopause due to solar wind pressure produces further local quasi-trapping regions in the high-latitude dayside magnetosphere. While the pure dipole field along a given field line has a minimum on the equatorial plane, the compression of the dayside magnetopause enhances the equatorial magnetic field causing a local maximum at the dayside equator. Now the equatorial minimum bifurcates to off-equatorial local minima (Fig. 2.9).

These local magnetic bottles in the northern and southern hemispheres are continually reformed due to the changing angle between of the dipole axis and the Sun-Earth line and due to the changes in the solar wind pressure. This leads to complicated charged particle orbits known as *Shabansky orbits* (Antonova and Shabansky 1968). For example, a fraction of the particles drifts across the noon sector bouncing in the northern hemisphere, another fraction bouncing in the southern hemisphere. (For examples of Shabansky orbits, see, e.g, McCollough et al. 2012).

2.7 Adiabatic Drift Motion in Time-Dependent Nearly-Dipolar Field

As an example of drift betatron acceleration (Sect. 2.4.4) we consider the effects of slow temporal changes of the quasi-dipolar field, examples of which are due to changes in solar wind pressure or approaching dipolarization fronts from the magnetotail. To keep the presentation simple we limit the discussion to bounce-averaged motion in the equatorial plane. Furthermore, we assume that there is no background electric field.

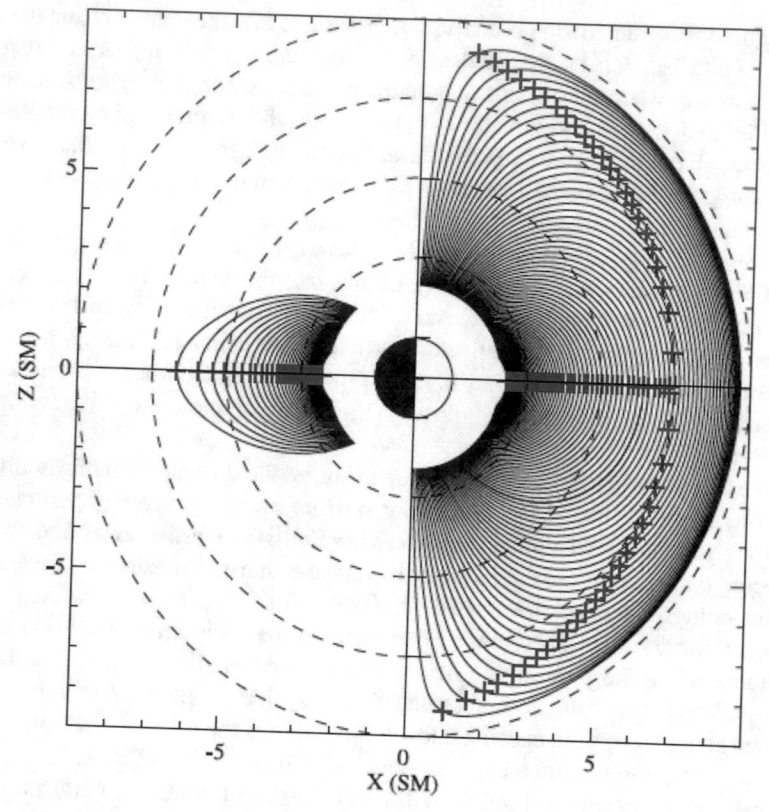

Fig. 2.9 Bifurcation of the local magnetic minima in the compressed dayside dipole field. The crosses indicate the local minima on different field lines in the noon–midnight meridional plane. (From McCollough et al. 2012, reprinted by permission from American Geophysical Union)

We start with the *fully adiabatic motion*[2] conserving all three adiabatic invariants, i.e., we assume that the temporal changes in the magnetic field are slower than the drift motion around the dipole

$$\frac{B}{\mathrm{d}B/\mathrm{d}t} \gg \tau_d \,, \qquad (2.86)$$

where τ_d is the bounce-averaged drift period (2.83). Temporal increase or decrease of the magnetic field is associated with an azimuthal inductive electric field $\mathbf{E}_i$. In the equatorial plane this leads to an outward or inward $\mathbf{E}_i \times \mathbf{B}$ drift across the magnetic field lines, corresponding to positive or negative $\mathrm{d}L^*/\mathrm{d}t$.

---

[2] The term "adiabatic motion" is often used to refer to conservation of the first adiabatic invariant without the requirement of conserving the other action integrals.

Particles can thus gain (lose) energy when they are transported by the $E_i \times B$ drift across the region of increasing (decreasing) magnetic field magnitude. The radial location and energy of the particles change during such a slow contraction or expansion of the magnetic field but the fully adiabatic process is reversible and the initial state of the particles is recovered when the field returns to its original strength. Particles that were on a common drift shell at the beginning of the magnetic field perturbation are transferred to an outer drift shell and return to the original shell when the magnetic configuration recovers its configuration before the perturbation. Due to the slow temporal changes all particles experience the same amount of inductive electric field in their orbit around the Earth, and therefore the same amount of inward/outward displacement. This kind of reversible process is observed during weak and moderate magnetic storms when the ring current enhances slowly in the main phase and then recovers in the recovery phase. This causes weakening and subsequent strengthening of the equatorial magnetic field. As the changes of the equatorial magnetic field strength are displayed in the Dst index, the reversible process is called the Dst effect.

The change of the particle energy during the expansion and contraction of the drift shells can be calculated by bounce-averaging the energy equation (2.68)

$$\left\langle \frac{dW}{dt} \right\rangle_b = \mu \left\langle \frac{\partial B}{\partial t} \right\rangle_b + q \left\langle V \cdot E_i \right\rangle_b \qquad (2.87)$$

$$= \mu \left\langle \frac{\partial B}{\partial t} \right\rangle_b + q \left\langle V_0 \right\rangle_b \cdot E_{i0},$$

where V_0 is the drift velocity of the GC and E_{i0} is the inductive electric field, both at the equator. In a slowly changing magnetic field the inductive electric field increases the kinetic energy of the particle both in its gyro motion (gyro betatron acceleration) and in its drift motion (drift betatron acceleration). The particle has in this case time to gradient drift considerably around the Earth before $E_i \times B$ drift takes it significantly inward, i.e., the drift velocity of the GC consists now primarily on the azimuthal gradient drift velocity and thus, the drift betatron term in, i.e., the second term on the RHS of (2.87), is finite. The $E_i \times B$ drift, being perpendicular to E_i, does not contribute to the betatron acceleration.

In the inner magnetosphere the temporal changes often are faster than the azimuthal drift motion but slower than the bounce motion ($\tau_b \ll B/(dB/dt) \ll \tau_d$). Bounce-averaging of the energy equation is still possible but now the $E_i \times B$ drift is faster than the azimuthal drift. Consequently, the charged particles follow the contracting or expanding field lines. Now in the second term on the RHS of (2.87) the velocity of the GC consist primarily of the $E_i \times B$ drift that is naturally perpendicular to the inductive electric field E_i. The drift betatron acceleration is zero because the $E_i \times B$ drift displaces the particle inward/outward so quickly that it does not have time to gradient drift much around the Earth. The increase/decrease in particle's energy is therefore due to gyro-betatron acceleration/deceleration.

Let us then consider a sudden impulse (SI, Sect. 1.4.2) in which a solar wind shock suddenly compresses the magnetosphere and the field thereafter relaxes slowly without no major storm development in the ring current. Assume that the compression is fast enough to violate the third invariant. Since the compression is now also asymmetric, particles at different longitudes, i.e., at different phases in their drift motion around the Earth respond differently to the disturbance. The compression is strongest on the dayside and particles there experience largest E_i and are mostly transported toward the Earth by the $E_i \times B$ drift. Since compression is now fast compared to the drift motion and the frozen-in condition applies, this can be considered in terms of particles following field lines as they are compressed toward the Earth. On the other hand, particles in the nightside at the time of the shock arrival more or less retain their original drift orbits. As a consequence, particles originally on the same drift shell but at different longitudes are transported to different distances toward the Earth and gain, in irreversible manner, different amount of energy.

Assuming that after the initial rapid compression the inductive electric field effect vanishes adiabatically, the particles find themselves on different drift shells. The change of the drift shell due to the inductive electric field depends on the longitude, i.e., the magnetic local time, where the particles are at the time of the asymmetric compression, and dL^*/dt is positive for some particles and negative for others. After the return to the original state the original distribution has diffused to a wider range of L^*. This is an example of *radial diffusion* that will be discussed further in the context of ULF waves in Chap. 6. For violation of L^* the disturbance has thus to be both fast when compared to the drift motion around the Earth and azimuthally asymmetric. Otherwise, all particles initially at the same drift shell would experience the same amount of displacement due to EixB-drift and there would be no diffusion in L^*.

Let us finally consider the case where the shock compression is followed by a magnetic storm. If the main phase (the rapid decrease of the Dst index) is not too fast, particles are subject to the fully adiabatic Dst effect described above. Faster and asymmetric changes can result, e.g., from substorm related $\partial \mathbf{B}/\partial t$ and corresponding inductive electric field, and ULF waves. They can lead to further violations of L^* and further spreading of drift shells compared to the initial situation. Furthermore, the radiation belt particles are accelerated and scattered in pitch-angle by various wave–particle interactions (Chap. 6). Thus the reversible Dst effect can be, while partially identifiable in observations, obscured by other processes.

Another inductive electric field effect is related to the injection of particles from the magnetotail into the inner magnetosphere during substorms. If the dipolarization is faster than the azimuthal drift motion, the injected particles are transported by the $E_i \times B$ drift toward the Earth gaining energy. At some point their energy becomes large enough for the gradient and curvature drifts to take over and the particles become trapped in the ring current and radiation belts.

Open Access This chapter is licensed under the terms of the Creative Commons Attribution 4.0 International License (http://creativecommons.org/licenses/by/4.0/), which permits use, sharing, adaptation, distribution and reproduction in any medium or format, as long as you give appropriate credit to the original author(s) and the source, provide a link to the Creative Commons license and indicate if changes were made.

The images or other third party material in this chapter are included in the chapter's Creative Commons license, unless indicated otherwise in a credit line to the material. If material is not included in the chapter's Creative Commons license and your intended use is not permitted by statutory regulation or exceeds the permitted use, you will need to obtain permission directly from the copyright holder.

Chapter 3
From Charged Particles to Plasma Physics

In this chapter we move from single particle motion to the statistical description of a large number of charged particles, the plasma. This discussion provides the basis for the rich flora of plasma waves that are essential for understanding the sources and losses of radiation belt particles through wave–particle interactions.

3.1 Basic Plasma Concepts

While we assume the reader to be familiar with basic plasma physics, we start with a brief review of concepts that we are using in the subsequent chapters. According to our favorite characterization space plasma is *quasi-neutral gas with so many free charges that collective electromagnetic phenomena are important to its physical behavior.*

3.1.1 Debye Shielding

The first attribute in the above characterization is the *quasi-neutrality.*[1] Space plasmas consist of positive and negative charges with roughly the same charge density

$$\rho_{q,tot} = \sum_\alpha \rho_{q\alpha} = \sum_\alpha n_\alpha q_\alpha \approx 0 , \tag{3.1}$$

---

[1] Also non-neutral plasmas are considered in plasma physics but they are of no interest in our treatise.

© The Author(s) 2022
H. E. J. Koskinen, E. K. J. Kilpua, *Physics of Earth's Radiation Belts*,
Astronomy and Astrophysics Library, https://doi.org/10.1007/978-3-030-82167-8_3

where α labels the different charge species and n_α is the number density. In the radiation belt context we consider high-energy electrons and ions in different spatial domains. However, the total amount of radiation belt particles is so small compared to the background plasma that they do not violate the quasi-neutrality of the entire system.

A significant fraction of *free charges* makes plasma electrically conductive. In space plasmas temperature is relatively high and particle density low. Thus interparticle collisions are rare and the classical collisional resistivity is very small. If an external electric field is applied to the plasma, electrons are quickly rearranged to neutralize the external field. As a consequence, no significant large-scale electric fields exist in the *rest frame of the plasma*. Recall, however, that the electric field is a coordinate-dependent quantity. The large-scale plasma motion across the magnetic field corresponds to an electric field $\mathbf{E} = -\mathbf{V} \times \mathbf{B}$ in the *frame of reference of the observer*, e.g., in an Earth-centered non-rotating frame such as the GSM coordinate system (Sect. 2.3). Another example of macroscopic electric fields are those arising from magnetic field-aligned electric potential differences above the auroral ionosphere related to the field-aligned electric currents coupling the ionosphere and magnetosphere to each other.

Although plasma is neutral in large scales, deviations from charge neutrality appear in smaller scales. Suppose that a positive test charge q_T is embedded into an otherwise quasi-neutral plasma. The Coulomb potential of the test charge is $\varphi_T = q_T/(4\pi\epsilon_0 r)$, where r is the distance from q_T. Electrons are attracted to q_T producing a localized polarization charge density ρ_{pol} that forms a neutralizing cloud around q_T. This is called *Debye shielding*.

The total charge density of the system is $\rho_{tot}(\mathbf{r}) = q_T\delta(\mathbf{r} - \mathbf{r}_T) + \rho_{pol}$, where $\mathbf{r}_T$ is the location of the test charge and δ is the *Dirac delta*. The shielded potential of q_T is found by solving the Poisson equation

$$\nabla^2\varphi = -\frac{\rho_{tot}(\mathbf{r})}{\epsilon_0} \tag{3.2}$$

with the boundary condition that $\varphi \to \varphi_T$ when $\mathbf{r} \to \mathbf{r}_T$.

Assuming that the plasma population α is in thermal equilibrium, the density is given by the *Boltzmann distribution*

$$n_\alpha = n_{0\alpha} \exp\left(-\frac{q_\alpha\varphi}{k_B T_\alpha}\right), \tag{3.3}$$

where k_B is the *Boltzmann constant*, $n_{0\alpha}$ the equilibrium number density in the absence of q_T, and T_α the equilibrium temperature. We will return to the concept of plasma temperature in Sect. 3.2.2.

For a gas to be in the plasma state a sufficient amount of electrons and ions must not be bound to each other. In other words, the random thermal energy must be much greater than the average electrostatic energy, $k_B T_\alpha \gg q_\alpha \varphi$. With this in mind we can expand Eq. (3.3) as

$$n_\alpha \simeq n_{0\alpha} \left(1 - \frac{q_\alpha \varphi}{k_B T_\alpha} + \frac{1}{2} \frac{q_\alpha^2 \varphi^2}{k_B^2 T_\alpha^2} + \cdots \right). \tag{3.4}$$

The leading term of the polarization charge density now becomes

$$\rho_{pol} = \sum_\alpha n_\alpha q_\alpha \approx \sum_\alpha n_{\alpha 0} q_{\alpha 0} - \sum_\alpha \frac{n_{0\alpha} q_\alpha^2}{k_B T_\alpha} \varphi = - \sum_\alpha \frac{n_{0\alpha} q_\alpha^2}{k_B T_\alpha} \varphi, \tag{3.5}$$

where $\sum_\alpha n_{\alpha 0} q_{\alpha 0} = 0$ due to quasi-neutrality. Inserting (3.5) into the Poisson equation (3.2) and solving the equation in spherical coordinates, the potential is found to be

$$\varphi = \frac{q_T}{4\pi \epsilon_0 r} \exp \left(-\frac{r}{\lambda_D} \right), \tag{3.6}$$

where λ_D is the *Debye length*

$$\lambda_D^{-2} = \sum_\alpha \lambda_{D,\alpha}^{-2} = \epsilon_0^{-1} \sum_\alpha \frac{n_{0\alpha} q_\alpha^2}{k_B T_\alpha}. \tag{3.7}$$

In many practical cases the thermal velocity of ions is much smaller than the thermal velocity of electrons and it is customary to consider the electron Debye length only. It can be estimated from

$$\lambda_D(\text{m}) \approx 7.4 \sqrt{\frac{T(\text{eV})}{n(\text{cm}^{-3})}}. \tag{3.8}$$

Intuitively, λ_D is the limit beyond which the thermal speed of a plasma particle is high enough to escape from the Coulomb potential of q_T.

The Debye length gives us a convenient way of describing the plasma state. For the collective phenomena to dominate the plasma behavior there must be a large number of particles in the *Debye sphere* of radius λ_D: $(4\pi/3)n_0\lambda_D^3 \gg 1$. Because plasma must also be quasi-neutral, its characteristic size $L \sim \mathcal{V}^{1/3}$, where $\mathcal{V}$ is the volume of the plasma, must be larger than λ_D. Thus for a plasma

$$\frac{1}{\sqrt[3]{n_0}} \ll \lambda_D \ll L. \tag{3.9}$$

Debye shielding is a property of the background plasma. The Debye sphere is strictly spherical only for test particles at rest in the plasma frame. If the speed of the test particle approaches the thermal speed of the surrounding plasma, the sphere becomes distorted. High-energy radiation belt particles move so fast through the background that no Debye sphere is formed around them.

Large number of particles inside the Debye sphere further implies that plasma in thermal equilibrium resembles an *ideal gas*. We discuss the interpretation of this in collisionless plasmas in Sect. 3.2.2.

3.1.2 Plasma Oscillation

Much of our further discussion concerns the great variety of plasma waves. The most fundamental wave phenomenon is the *plasma oscillation*, which is found by considering freely moving cold ($T_e \approx 0$) electrons and fixed background ions in a non-magnetized plasma. A typical first exercise problem in elementary plasma physics course is to show that a small perturbation in the electron density leads to a local electric field, which gives rise to a restoring force to pull the electrons back toward the equilibrium. This results in a standing oscillation of the density at the *plasma frequency*

$$\omega_{pe}^2 = \frac{n_0 e^2}{\epsilon_0 m_e} .$$

(3.10)

Note that the term "plasma frequency" is frequently used to refer to both the angular frequency ω_{pe} and the oscillation frequency $f_{pe} = \omega_{pe}/2\pi$.

Plasma frequency is proportional to the square root of the density divided by the mass of the oscillating particles. A useful rule of thumb to calculate the electron oscillation frequency is

$$f_{pe}(\text{Hz}) \approx 9.0 \sqrt{n(\text{m}^{-3})} .$$

(3.11)

The plasma oscillation determines a natural length scale in the plasma known as the *electron inertial length* c/ω_{pe}, where c is the speed of light. Physically it gives the length scale for attenuation of an electromagnetic wave with the frequency ω_{pe} penetrating to plasma. It is analogous to the *skin depth* in classical electromagnetism.

Similarly, the *ion plasma frequency* is defined by

$$\omega_{pi}^2 = \frac{n_0 e^2}{\epsilon_0 m_i} .$$

(3.12)

The corresponding *ion inertial length* is c/ω_{pi}. Both ω_{pe} and ω_{pi} are important parameters in plasma wave propagation (Chap. 4).

3.2 Basic Plasma Theories

Depending on the temporal and spatial scales of the processes investigated different theoretical approaches can be used. Every step from the motion of individual particles toward statistical and macroscopic theories involves approximations that must be understood properly.

3.2.1 Vlasov and Boltzmann Equations

We begin the discussion of statistical plasma physics by introducing the *Vlasov equation* for the *distribution function* $f_\alpha(\mathbf{r}, \mathbf{p}, t)$ of the particle species α. As the background plasma in the magnetosphere is nonrelativistic, we set the Lorentz factor $\gamma = 1$. The nonrelativistic distribution function $f_\alpha(\mathbf{r}, \mathbf{v}, t)$ expresses the number density of particles in a volume element $dx\, dy\, dz\, dv_x\, dv_y\, dv_z$ of a six-dimensional *phase space* $(\mathbf{r}, \mathbf{v})$ at the time t. In the following we use the normalization

$$\int_{\mathscr{V}_{ps}} f_\alpha(\mathbf{r}, \mathbf{v}, t)\, d^3r\, d^3v = N\,, \tag{3.13}$$

where the integration is over the phase space volume $\mathscr{V}_{ps}$ and N the number of all particles in this volume.[2] In the $(\mathbf{r}, \mathbf{v})$-space the SI-units of f are $\mathrm{m}^{-6}\,\mathrm{s}^3$. If the distribution function is given in the $(\mathbf{r}, \mathbf{p})$-space, as often is the case, its SI units are $\mathrm{m}^{-6}\,\mathrm{kg}^{-3}\,\mathrm{s}^3$.

Assuming that the only force acting on plasma particles is the Lorentz force, the time evolution of the distribution function $\partial f_\alpha/\partial t$ is given by the *Vlasov equation*

$$\frac{\partial f_\alpha}{\partial t} + \mathbf{v} \cdot \frac{\partial f_\alpha}{\partial \mathbf{r}} + \frac{q_\alpha}{m_\alpha}(\mathbf{E} + \mathbf{v} \times \mathbf{B}) \cdot \frac{\partial f_\alpha}{\partial \mathbf{v}} = 0\,. \tag{3.14}$$

The Vlasov equation actually states that in the six-dimensional phase space the total time derivative of the distribution function is zero

$$\frac{df_\alpha}{dt} = \left(\frac{\partial f_\alpha}{\partial t} + \mathbf{r} \cdot \nabla f_\alpha + \mathbf{a} \cdot \nabla_\mathbf{v} f_\alpha \right) = 0\,, \tag{3.15}$$

---

[2] In plasma theory the distribution function is often normalized to 1, when it describes the probability of finding a particle in a given location of the phase space at a given time. This is convenient when plasma is assumed to be homogeneous and the average density can be moved outside of the velocity integrals. We will apply the normalization to unity in the derivation of kinetic dispersion equations in Chaps. 4 and 5.

where **a** is the acceleration due to the Lorentz force and ∇_v is the gradient in the velocity space. Consequently f_α is often called the *phase space density* (PSD).[3] The Vlasov equation is a conservation law for f_α in the phase space. It is formally similar to the continuity equation

$$\frac{\partial n}{\partial t} + \nabla \cdot (n\mathbf{V}) = 0 \qquad (3.16)$$

for particle density in three-dimensional configuration space.

In a collisionless plasma, which is a good first approximation for magnetospheric plasma above the ionosphere, the Vlasov theory is a very accurate starting point. Here "collisionless" means that head-on collisions between the plasma particles are so infrequent that their contribution to the plasma dynamics vanishes in comparison with the effect of long-range Coulomb collisions, which, in turn, give rise to the Lorentz force term in the Vlasov equation.

However, there are some important collisional processes in the inner magnetosphere that break the conservation of f_α. In particular, the *charge exchange collisions* between ring current ions and exospheric neutrals contribute to the loss of energetic current carriers and the decay of the ring current. In a charge exchange collision the ring current ion captures an electron from a background atom. The newly born neutral particle maintains the velocity of the energetic ion and becomes an *energetic neutral atom* (ENA) that escapes unaffected by the magnetic field far from ring current region. The newly ionized particle, in turn, has a much lower energy and does not carry a significant amount of current. Using remote observations of ENAs, images of the ring current and the plasmasphere can be constructed. Charge exchange collisions have also a role in the energy loss of inner belt protons but it is a very slow process at energies $\gtrsim 100$ keV.

Short-range collisions can be accommodated in the phase space description by replacing the Vlasov equation by the *Boltzmann equation*

$$\frac{\partial f_\alpha}{\partial t} + \mathbf{v} \cdot \frac{\partial f_\alpha}{\partial \mathbf{r}} + \frac{q_\alpha}{m_\alpha}(\mathbf{E} + \mathbf{v} \times \mathbf{B}) \cdot \frac{\partial f_\alpha}{\partial \mathbf{v}} = \left(\frac{d f_\alpha}{dt}\right)_c, \qquad (3.17)$$

where the collision term $(d f_\alpha/dt)_c$ typically is a complicated function of velocity depending on the type of the particle interactions. In collisionless plasmas it is sometimes practical to separate the electromagnetic fluctuations from the background fields and describe them formally as a collision term, as we will see in the context of quasi-linear theory (Chap. 6).

In addition to conserving the number of particles N, the Vlasov equation has several other important properties. For example, it conserves entropy defined by

$$S = -\sum_\alpha \int f_\alpha \ln f_\alpha \, d^3r \, d^3v, \qquad (3.18)$$

---

[3] The same acronym is frequently used also for power spectral density. Both are important concepts in radiation belt physics, but the risk of confusion is small.

which is readily seen by calculating

$$\frac{\mathrm{d}S}{\mathrm{d}t} = -\sum_\alpha \int \left(\frac{\mathrm{d}f_\alpha}{\mathrm{d}t} \ln f_\alpha + \frac{\mathrm{d}f_\alpha}{\mathrm{d}t} \right) \mathrm{d}^3 r \, \mathrm{d}^3 v = 0 \,. \tag{3.19}$$

This is important when we discuss the important Landau damping mechanism in Sect. 4.2.

Particularly important in space plasma physics is that the Vlasov equation has many equilibrium solutions. In statistical physics of collisional gases Boltzmann's *H-theorem* states that there is a unique equilibrium in the collisional time scale, the Maxwellian distribution. The relevant time scales in radiation belt physics are much shorter than the average collision time and we can set $\partial f/\partial t|_c \rightarrow 0$. Thus non-Maxwellian distributions can survive much longer than the physical processes we are investigating.

3.2.2 Macroscopic Variables and Equations

We define the *macroscopic* plasma quantities as *velocity moments* of the distribution function

$$\int f \, \mathrm{d}^3 v \; ; \; \int \mathbf{v} f \, \mathrm{d}^3 v \; ; \; \int \mathbf{v}\mathbf{v} f \, \mathrm{d}^3 v \,.$$

The average density in a spatial volume $\mathscr{V}$ is $\langle n \rangle = N/\mathscr{V}$. The *particle density* n is, in turn, a function of space and time. It can be expressed as the zero order velocity moment of the distribution function

$$n(\mathbf{r}, t) = \int f(\mathbf{r}, \mathbf{v}, t) \, \mathrm{d}^3 v \,. \tag{3.20}$$

In a plasma different particle populations (labeled by α) may have different distributions and thus have different velocity moments ($n_\alpha(\mathbf{r}, t)$, etc.). If the particles of a given species have the charge q_α, the *charge density* of the species is

$$\rho_\alpha = q_\alpha n_\alpha \,. \tag{3.21}$$

The first-order moment yields the *particle flux*

$$\Gamma_\alpha(\mathbf{r}, t) = \int \mathbf{v} f_\alpha(\mathbf{r}, \mathbf{v}, t) \, \mathrm{d}^3 v \,. \tag{3.22}$$

Dividing this by particle density we get the *macroscopic velocity*

$$\mathbf{V}_\alpha(\mathbf{r}, t) = \frac{\int \mathbf{v} f_\alpha(\mathbf{r}, \mathbf{v}, t)\, \mathrm{d}^3 v}{\int f_\alpha(\mathbf{r}, \mathbf{v}, t)\, \mathrm{d}^3 v} \ . \tag{3.23}$$

With these we can write the *electric current density* as

$$\mathbf{J}_\alpha(\mathbf{r}, t) = q_\alpha \boldsymbol{\Gamma}_\alpha = q_\alpha n_\alpha \mathbf{V}_\alpha \ . \tag{3.24}$$

The second order moment defines the *pressure tensor*

$$\mathsf{P}_\alpha(\mathbf{r}, t) = m_\alpha \int (\mathbf{v} - \mathbf{V}_{\boldsymbol{\alpha}})(\mathbf{v} - \mathbf{V}_{\boldsymbol{\alpha}}) f_\alpha(\mathbf{r}, \mathbf{v}, t)\, \mathrm{d}^3 v \ , \tag{3.25}$$

which in a spherically symmetric case reduces to the *scalar pressure*

$$P_\alpha(\mathbf{r}, t) = \frac{m_\alpha}{3} \int (\mathbf{v} - \mathbf{V}_\alpha)^2 f_\alpha(\mathbf{r}, \mathbf{v}, t)\, \mathrm{d}^3 v = n_\alpha k_B T_\alpha \ . \tag{3.26}$$

Here we have introduced the Boltzmann constant k_B and the *temperature* T_α. In the frame moving with the velocity $\mathbf{V}_\alpha$ the temperature is given by

$$\frac{3}{2} k_B T_\alpha(\mathbf{r}, t) = \frac{m_\alpha}{2} \frac{\int v^2 f_\alpha(\mathbf{r}, \mathbf{v}, t)\, \mathrm{d}^3 v}{\int f_\alpha(\mathbf{r}, \mathbf{v}, t)\, \mathrm{d}^3 v} \ . \tag{3.27}$$

For a *Maxwellian distribution* T_α is the temperature of classical thermodynamics. In thermal equilibrium a plasma with a large number of particles in the Debye sphere can be considered as an ideal gas with the *equation of state* given by (3.26). In collisionless plasmas equilibrium distributions may, however, be far from the Maxwellian making temperature a non-trivial concept in plasma physics.

The ratio of particle pressure to *magnetic pressure* (*magnetic energy density*, $B^2/2\mu_0$) is the *plasma beta*

$$\beta = \frac{2\mu_0 \sum_\alpha n_\alpha k_B T_\alpha}{B^2} \ . \tag{3.28}$$

If $\beta > 1$, plasma governs the evolution of the magnetic field. If $\beta \ll 1$, the magnetic field determines the plasma dynamics. In the magnetosphere the smallest beta values ($\beta \sim 10^{-6}$) are found on the auroral region magnetic field lines at altitudes of a few Earth radii. In the tail plasma sheet β is of the order of one, but in the tail lobes some 4 orders of magnitude smaller.

The equations between the macroscopic quantities can be derived by taking velocity moments of the Vlasov or Boltzmann equation. The procedure is described in most advanced plasma physics textbooks (e.g., Koskinen 2011). The technical details are of secondary interest for our subsequent discussion.

The zeroth moment yields the *equation of continuity*

$$\frac{\partial n_\alpha}{\partial t} + \nabla \cdot (n_\alpha \mathbf{V}_\alpha) = 0 . \tag{3.29}$$

The equation of continuity is an example of a *conservation law*

$$\frac{\partial F}{\partial t} + \nabla \cdot \mathbf{G} = 0 , \tag{3.30}$$

where F is the density of a physical quantity and $\mathbf{G}$ the associated flux. Continuity equations for charge or mass densities are obtained by multiplying (3.29) by q_α or m_α, respectively.

The continuity equation contains $\mathbf{V}_\alpha$, which is the first-order velocity moment of the distribution function. Calculating the first-order moment of the Vlasov/Boltzmann equation leads to the continuity equation for macroscopic momentum density $\rho_{m\alpha} \mathbf{V}_\alpha$, i.e., the macroscopic *equation of motion*, where $\rho_{m\alpha}$ is the *mass density*. This equation contains the second-order moment, the pressure tensor. By continuing the moment integration of the Boltzmann equation to the second order we get an *energy equation* relating the temporal evolution of plasma and magnetic field energy densities to divergence of the third moment of f, the *heat flux*.

This chain, where the conserved quantity depends on a quantity of a higher order, continues *ad infinitum*. To obtain a tractable and useful macroscopic theory, the chain of equations has to be truncated at some level. In the inner magnetosphere the divergence of heat flux can be neglected. Thus we can replace the energy equation by introducing an *equation of state* that relates the scalar pressure P_α to the number density n_α and temperature T_α as

$$P_\alpha = n_\alpha k_B T_\alpha . \tag{3.31}$$

In thermal equilibrium this is the ideal gas law. As (3.31) contains *three functions*, their mutual dependencies need to be given to reflect the actual thermodynamic process. This can be done by specifying an appropriate *polytropic index* γ_p

$$P = P_0 \left(\frac{n}{n_0}\right)^{\gamma_p} \; ; \; T = T_0 \left(\frac{n}{n_0}\right)^{\gamma_p - 1} . \tag{3.32}$$

For an adiabatic process in d-dimensional space $\gamma_p = (d+2)/d$. In this form the equation of state also applies to isothermal ($\gamma_p = 1$) and isobaric ($\gamma_p = 0$) processes. In this sense collisionless magnetospheric plasma physics is simpler than classical gas or fluid dynamics, where the moment calculations often must be continued to higher orders.[4]

---

[4] In the description of astrophysical plasma environments, including the Sun and its atmosphere, thermal transport and radiative effects often are highly important.

3.2.3 *Equations of Magnetohydrodynamics*

Above we have introduced separate macroscopic equations for each plasma species. In the magnetosphere several plasma species co-exist; in addition to electrons and protons, there may be heavier ions as well as neutral particles that interact with charged particles. Sometimes it is also necessary to consider different populations of identical particles as different species. For example, in a given spatial volume there may be two electron populations of widely different temperatures or macroscopic velocities. Depending on the temporal and spatial scales of the phenomena under investigation the appropriate theoretical framework may be the Vlasov theory, a multi-species macroscopic theory, the single-fluid *magnetohydrodynamics* (MHD), or some combination of these. Examples of the combinations are hybrid approaches where electrons are treated as a fluid in the configuration space and ions either as (quasi-)particles or as a Vlasov fluid in phase space.

In MHD the plasma is considered as a single fluid in the center-of-mass (CM) frame. The single-fluid equations are obtained summing up the macroscopic equations of different particle species. A single-fluid description is a well-motivated approach in collision-dominated gases, where the collisions constrain the motion of individual particles and thermalize the distribution toward a Maxwellian. Single-fluid MHD works remarkably well also in collisionless tenuous space plasmas, but great care needs to be exercised with the validity of the approximations.

After a lengthy procedure of summing up the macroscopic equations for each particle species and with several—not always quite obvious—approximations (see, e.g., Koskinen 2011) we can write the MHD equations, supplemented with Faraday's and Ampère's laws of electromagnetism in the form

$$\frac{\partial \rho_m}{\partial t} + \nabla \cdot (\rho_m \mathbf{V}) = 0 \tag{3.33}$$

$$\rho_m \left(\frac{\partial}{\partial t} + \mathbf{V} \cdot \nabla \right) \mathbf{V} + \nabla P - \mathbf{J} \times \mathbf{B} = 0 \tag{3.34}$$

$$\mathbf{E} + \mathbf{V} \times \mathbf{B} = \mathbf{J}/\sigma \tag{3.35}$$

$$P = P_0 \left(\frac{n}{n_0} \right)^{\gamma_p} \tag{3.36}$$

$$\frac{\partial \mathbf{B}}{\partial t} = -\nabla \times \mathbf{E} \tag{3.37}$$

$$\nabla \times \mathbf{B} = \mu_0 \mathbf{J} . \tag{3.38}$$

Summing over charges of all particle species yields the continuity equation for charge density

$$\frac{\partial \rho_q}{\partial t} + \nabla \cdot \mathbf{J} = 0 . \tag{3.39}$$

This equation is actually redundant because in the MHD approximation the displacement current in the Ampère–Maxwell law is neglected and we need only Ampère's law (3.38), yielding $\nabla \cdot \mathbf{J} = 0$. Because the charge density is a coordinate-dependent quantity, similarly as the electric field, this does not mean that ρ_q would have to be zero in *a given frame of reference*. When needed, it can be obtained as the divergence of the electric field calculated in the appropriate frame of reference.

The momentum equation (3.34) corresponds to the Navier–Stokes equation of hydrodynamics. In the context of MHD the viscosity is neglected whereas the Lorentz force is essential. Note that in the MHD approximation the electric force $\rho_q \mathbf{E}$ is negligible compared to the magnetic force $\mathbf{J} \times \mathbf{B}$.

In Ohm's law (3.35) we have retained finite *conductivity* (σ) although we will mostly operate within the *ideal MHD*, where the resistivity is assumed zero, corresponding to $\sigma \to \infty$, and thus

$$\mathbf{E} + \mathbf{V} \times \mathbf{B} = 0 . \tag{3.40}$$

This is the foundation of the *frozen-in magnetic field* concept, meaning that plasma elements connected by a magnetic field line maintain the connection when plasma moves with velocity $\mathbf{V}$ (Sect. 1.4.1).

In collisionless space plasmas the first refinement of the ideal Ohm's law often is the inclusion of the *Hall electric field*

$$\mathbf{E} + \mathbf{V} \times \mathbf{B} - \frac{1}{ne}\mathbf{J} \times \mathbf{B} = 0 . \tag{3.41}$$

The Hall term is particularly important in the presence of thin current sheets and the current sheet disruption in the magnetic reconnection process. It decouples the electron motion from the ion motion, after which the magnetic field becomes frozen-in the electron flow $\mathbf{E} = -\mathbf{V}_e \times \mathbf{B}$. Another example of this decoupling will be encountered in Sect. 4.4.1, where we discuss the splitting of the low-frequency MHD shear Alfvén wave mode to the electromagnetic ion cyclotron and whistler-mode waves at higher frequencies.

3.3 From Particle Flux to Phase Space Density

The function $f(\mathbf{r}, \mathbf{p}, t)$ can be considered as the plasma theorist's distribution function whose first order velocity moment is the particle flux. However, the distribution function cannot be measured directly. Instead, the observable is the particle flux to a detector. The empirical approach is to determine the flux from observations and thereafter relate the flux to the distribution function.

We start by defining the *differential unidirectional flux j* as the number of particles dN coming from a given incident direction (unit vector $\mathbf{i}$) that hit a surface of unit area dA, oriented perpendicular to the particles' direction of incidence, per unit time dt, unit solid angle $d\Omega$ and unit kinetic energy dW. Hence we write

$$dN = j\, dA\, dt\, d\Omega\, dW \ . \tag{3.42}$$

In ideal world the differential unidirectional flux

$$j = j(\mathbf{r}, \mathbf{i}, W, t) \tag{3.43}$$

contains full information on the particles' spatial $(\mathbf{r})$, angular $(\mathbf{i})$ and energy (W) distribution at a given time. The flux j is a quantity measured by an ideal directional instrument. It is commonly given in units $cm^{-2}\, s^{-1}\, ster^{-1}\, keV^{-1}$, also in the literature otherwise using SI-units. Depending on the energy range of observed particles the energy scale may be sorted in keV, MeV, or GeV. Thus it is important to pay attention to the powers of 10 in data presentations.

Real particle detectors are not planar surfaces. They may consist of a complicated assembly of time-of-flight measuring arrangements, electric and magnetic deflectors, stacks of detector plates, etc. Furthermore, real detectors do not sample infinitesimal solid angles or energy intervals. Thus the conversion from the *detector counting rate* to flux requires consideration of sensitivity, resolution and configuration of the instrument, and of course careful calibration.

A real detector has a low-energy cut-off. If there is nothing that would limit the higher energies to reach the detector, the flux is convenient to represent as an *integral directional flux* as[5]

$$j_{>E} = \int_{E}^{\infty} j\, dW \ . \tag{3.44}$$

Other important concepts are the *omnidirectional differential flux* J defined by

$$J = \int_{4\pi} j\, d\Omega \tag{3.45}$$

and the corresponding *omnidirectional integral flux*

$$J_{>E} = \int_{E}^{\infty} J\, dW \ . \tag{3.46}$$

In radiation belts the particles are moving in the Earth's magnetic field. Assume that the particle distribution function is smooth in a locally homogeneous magnetic field $\mathbf{B}$. The magnetic field direction gives a natural axis for the frame of reference. The direction of incidence $\mathbf{i}$ is given by the particle's pitch angle α (2.6) and the

---

[5] Here the flux of particles with so high energies, that they pass through the detector without leaving a trace, is assumed to be negligible and the upper limit of integration can be set to infinity.

azimuthal angle ϕ around **B**. If particles are uniformly distributed in the gyro phase, i.e., the distribution is *gyrotropic*, the angle of incidence and j depend directionally only on α. Considering particles, whose pitch angles lie within the interval $(\alpha, \alpha + d\alpha)$, arriving from all azimuthal directions, the solid angel element is $d\Omega = 2\pi \sin\alpha d\alpha$. The number of particles crossing a given point per unit time per unit perpendicular area and energy, can now be expressed as

$$\frac{dN}{dA\,dW\,dt} = 2\pi j \,\sin\alpha\,d\alpha = -2\pi j\,d(\cos\alpha)\,. \qquad (3.47)$$

The flux is called *isotropic* if the number of incoming particles depends only on the size of the solid angle of acceptance and is independent of the direction of incidence, i.e., j is constant with respect to α

$$\frac{dN}{d(\cos\alpha)} = \text{const}\,. \qquad (3.48)$$

Consequently, in an isotropic distribution equal numbers of particles arrive to the detector from equal intervals of pitch angle cosines[6]

$$j = j(\mathbf{r}, d(\cos\alpha), W, t)\,, \qquad (3.49)$$

the omnidirectional flux being

$$J = 4\pi \int_0^1 j\,d(\cos\alpha) = 4\pi j\,. \qquad (3.50)$$

In the absence of sources and losses *Liouville's theorem* of statistical physics states that the *phase space density* $f_p(\mathbf{r}, \mathbf{p}, t)$ is constant along any dynamical trajectory in phase space

$$f_p = \frac{dN}{dx\,dy\,dz\,dp_x\,dp_y\,dp_z} = \text{const}\,. \qquad (3.51)$$

Let the z axis be along the velocity vector. Then $dx\,dy = dA$, $dz = v\,dt$, and $dp_x\,dp_y\,dp_z = p^2\,dp\,\sin\alpha\,d\alpha\,d\phi = p^2\,dp\,d\Omega$. Furthermore, $v\,dp = dW$, and using (3.47) the relation of the differential unidirectional flux and the phase space density is

$$f_p = \frac{dN}{p^2\,dA\,dt\,d\Omega\,dW} = \frac{j}{p^2}\,. \qquad (3.52)$$

---

[6] In the literature the notation $\mu = \cos\alpha$ is frequently used. As we have reserved μ for the magnetic moment, we prefer to write $d(\cos\alpha)$ here.

For non-relativistic particles $f_p \approx j/2mW$. We retain the velocity space distribution function of the previous sections by writing $f = m^3 f_p$, where m is the mass of the particle. Thus in the velocity space we can write

$$j = \frac{v^2}{m} f(v).$$
(3.53)

3.4 Important Distribution Functions

While much of basic plasma theory is presented either at the limit of cold plasma or in the MHD approximation for isotropic Maxwellian distribution functions (Fig. 3.1) of the form

$$f(\mathbf{v}) = n \left(\frac{m}{2\pi k_B T} \right)^{3/2} \exp\left(-\frac{mv^2}{2k_B T} \right),$$
(3.54)

practical observations can only seldom be presented as such.

In any location there are particles with different past histories, carrying information of their origin, of the acceleration processes they have experienced, etc. In magnetized plasmas the magnetic field introduces anisotropy, as the particle motion is different along the magnetic field and perpendicular to it. In radiation belts the leakage of particles from the magnetic bottle leads to loss-cone features of distribution functions.

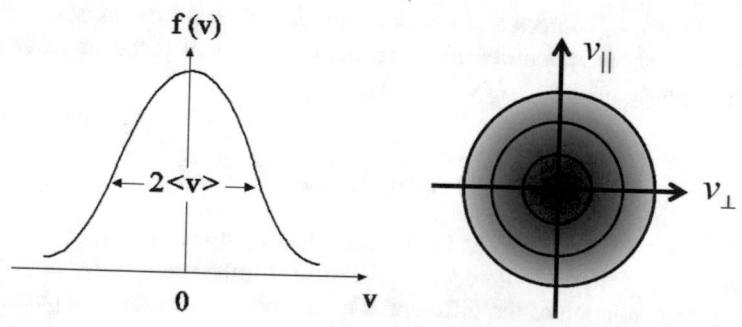

Fig. 3.1 Isotropic Maxwellian velocity distribution function. The right-hand picture shows contours of constant f in the two-dimensional velocity space

3.4.1 Drifting and Anisotropic Maxwellian Distributions

In the case where the whole plasma population is moving with velocity $\mathbf{V}_0$ in the frame of the observer, the three-dimensional Maxwellian distribution function is of the form

$$f(\mathbf{v}) = n \left(\frac{m}{2\pi k_B T} \right)^{3/2} \exp\left(-\frac{m(\mathbf{v} - \mathbf{V}_0)^2}{2k_B T} \right). \tag{3.55}$$

A typical example of such a motion is the E×B drift, and the distribution is called the *drifting Maxwellian* distribution.

Anisotropy introduced by the magnetic field can be illustrated by considering an ideal non-leaking magnetic bottle, assuming that in the center of the bottle the distribution is Maxwellian. If the bottle is contracted in the direction of the magnetic field, the mirror points move slowly closer to each other. To conserve the second adiabatic invariant (2.56), the decreasing field line length between the mirror points means that parallel velocity must increase (i.e., the pitch angle must decrease), corresponding to the Fermi mechanism (Sect. 2.4.4). The distribution is elongated parallel to the magnetic field to a *cigar-shaped* distribution. In the opposite case, where the bottle is stretched, the mirror points move away from each other and the distribution is stretched perpendicular to the magnetic field forming a *pancake* distribution.

In both cases the distribution remains Maxwellian both parallel and perpendicular to the magnetic field but with different temperatures $T_\parallel$ and $T_\perp$. As the parallel space is one-dimensional and the perpendicular space two-dimensional, the total *bi-Maxwellian* distribution function is

$$f(v_\perp, v_\parallel) = \frac{n}{T_\perp T_\parallel^{1/2}} \left(\frac{m}{2\pi k_B} \right)^{3/2} \exp\left(-\frac{mv_\perp^2}{2k_B T_\perp} - \frac{mv_\parallel^2}{2k_B T_\parallel} \right). \tag{3.56}$$

Here the distribution has been assumed to be *gyrotropic*, i.e., it looks the same in all perpendicular directions. In inhomogeneous plasmas this is not necessarily the case.

If the anisotropic Maxwellian plasma moves across the magnetic field, for example, due to the E×B drift, the distribution is given by

$$f(v_\perp, v_\parallel) = \frac{n}{T_\perp T_\parallel^{1/2}} \left(\frac{m}{2\pi k_B} \right)^{3/2} \exp\left(-\frac{m(\mathbf{v}_\perp - \mathbf{v}_{0\perp})^2}{2k_B T_\perp} - \frac{mv_\parallel^2}{2k_B T_\parallel} \right). \tag{3.57}$$

The population may also have been accelerated along the magnetic field forming a *plasma beam*

$$f(v_\perp, v_\parallel) = \frac{n}{T_\perp T_\parallel^{1/2}} \left(\frac{m}{2\pi k_B} \right)^{3/2} \exp\left(-\frac{mv_\perp^2}{2k_B T_\perp} - \frac{m(v_\parallel - v_{0\parallel})^2}{2k_B T_\parallel} \right). \tag{3.58}$$

Fig. 3.2 Thermally
anisotropic pancake
distribution drifting
perpendicular to the magnetic
field

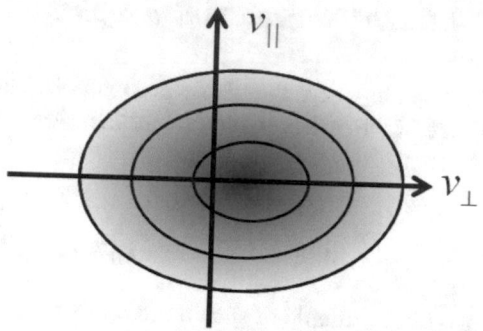

The distribution (3.57) is known as the *drifting pancake distribution* (Fig. 3.2). In radiation belts drifting pancake distributions of electrons and protons are of particular interest. The plasma injected from the tail drifts toward increasing magnetic field conserving the first two invariants. As a result the particles' perpendicular energies ($W_\perp$) increase according to Eq. (2.69) at the expense of $W_\parallel$. Although the distance of the mirror points at the same time becomes shorter, which increases $W_\parallel$, the net result is pancake-shaped temperature anisotropy ($T_\perp > T_\parallel$) because $W_\perp$ scales as B^3 whereas the field line length scales as B. Anisotropic proton distributions drive electromagnetic ion cyclotron waves and anisotropic electron distributions drive whistler-mode chorus waves (Chap. 5), both of which have central roles in acceleration and losses of radiation belt particles (Chap. 6).

The differences in E×B drifts of thermal ions and energy-dependent drifts of suprathermal ions injected from the magnetotail can lead to formation of ring-shaped distribution functions in the ($v_\parallel$, $v_\perp$)-space, known as *ion ring distributions*. Furthermore, the charge-exchange collisions between the drifting ions and exospheric neutrals can contribute to the development of the ring distribution by depleting the small-velocity core of the distribution. These distributions lead to instabilities that are able to drive, e.g., magnetosonic waves in the inner magnetosphere close to the equator (Sect. 5.3.2).

3.4.2 Loss Cone and Butterfly Distributions

In the real world all magnetic bottles are leaky due to the finite magnetic field at the end of the bottle. In the absence of a mechanism that would replenish the lost particles the distribution becomes the *loss cone distribution* where the loss cone in the velocity space is around the direction of the background magnetic field (Fig. 3.3). The half-width of the loss cone at the equator of a dipole field varies within the radiation belts from 2° to 16° (Eq. (2.74)), resolving of which experimentally requires very good angular resolution and precise alignment of the detector in the direction of the magnetic field. Further away from the equatorial

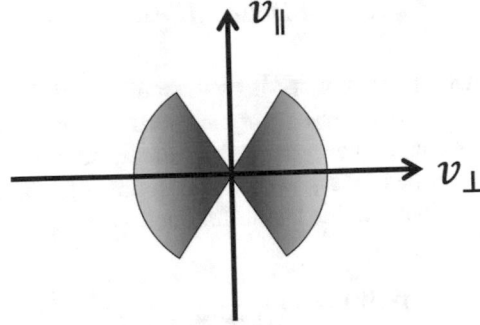

Fig. 3.3 Loss cone distribution

plane the loss cone widens and becomes easier to detect (Fig. 2.6). The edges of the loss cone are of particular importance because various plasma waves can scatter particles into the loss cone. This is a key particle loss mechanism in the radiation belts (Chap. 6).

The magnetopause shadowing caused by drift shell splitting (Sect. 2.6.2) can, in turn, lead to loss of particles near 90° pitch angle. Combination of parallel and perpendicular loss cones results in a velocity space distribution that resembles the wings of a butterfly and is termed accordingly the *butterfly distribution*. Note, however, that while the magnetopause shadowing takes place at the outskirts of the radiation belts, butterfly distributions have also been observed inside $L = 6$ (Fig. 3.4), where the smaller flux around the perpendicular direction cannot be explained by magnetopause loss, except during extreme dayside compression. If the formation of butterfly distribution is not due to magnetopause shadowing, it may result from wave–particle interactions that preferentially accelerate particles at medium pitch angles. The high-resolution data from the *Van Allen Probes* has made it possible to study this in detail (e.g., Xiao et al. 2015, and references therein).

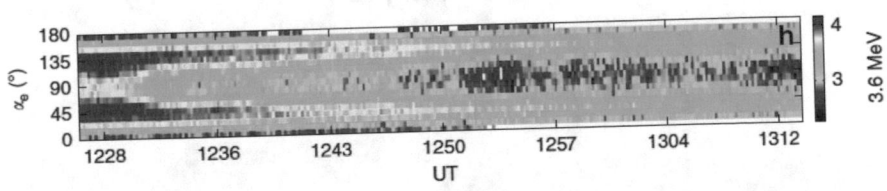

Fig. 3.4 Time series of butterfly distribution of relativistic (3.6-MeV) electrons observed by the REPT instrument of *Van Allen Probe* A on 29 June 2013. During the shown period the spacecraft was moving outward close to $L = 4.8$. The flux maxima were in the pitch angle ranges 30°–60° and 120°–150° (From Xiao et al. 2015, Creative Commons Attribution 4.0 International License)

3.4.3   *Kappa Distribution*

An important non-Maxwellian distribution in space plasmas is the *kappa distribution*. The observed particle spectra often are nearly Maxwellian at low energies but have high-energy tails where the flux decreases more slowly. The tail is customarily described by a power law in contrast to the exponential decay of the Maxwellian distribution. The kappa distribution has the form

$$f_\kappa(W) = n \left(\frac{m}{2\pi\kappa W_0} \right)^{3/2} \frac{\Gamma(\kappa+1)}{\Gamma(\kappa-1/2)} \left(1 + \frac{W}{\kappa W_0} \right)^{-(\kappa+1)} , \tag{3.59}$$

where W_0 is the energy at the peak of the particle flux and Γ is the gamma function of mathematics. When $\kappa \gg 1$, the kappa distribution approaches a Maxwellian. When κ is smaller but yet >1, the distribution has a high-energy tail. The smaller κ, i.e., the less negative the power law index is, the *harder* the particle spectrum is said to be.

Figure 3.5 illustrates the presentation of the distribution as a function of particle energy instead of velocity. For a Maxwellian velocity distribution

$$f(v) = n \left(\frac{m}{2\pi k_B T} \right)^{3/2} \exp\left(-\frac{W}{k_B T} \right) \tag{3.60}$$

the transformation to the energy distribution $g(W)$ is given by

$$g(W) = 4\pi \left(\frac{2W}{m^3} \right)^{1/2} f(v) . \tag{3.61}$$

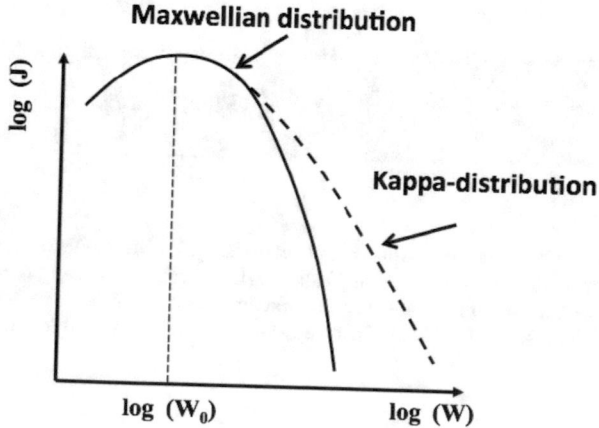

Fig. 3.5 Maxwellian and kappa distributions as functions of energy. J is the omnidirectional differential particle flux described in Sect. 3.5

3.5 Action Integrals and Phase Space Density

In radiation belt studies it is common to express the phase space density as a function of the action integrals discussed in Chap. 2

$$J_i = \frac{1}{2\pi} \oint (\mathbf{p}_i + q\mathbf{A}) \cdot d\mathbf{s}_i$$

with associated phase angles ϕ_i. For the set of integrals $\{\mu, J, \Phi\}$ related to the magnetic moment, the bounce motion and the magnetic flux enclosed by the particle's drift path, the phase angles are the gyro phase, the bounce phase, and the drift phase.

If all action integrals μ, J and Φ are adiabatic invariants, the phase space density can be averaged over the phase angles and the six-dimensional phase space reduces to three-dimensional space with coordinates $\{\mu, J, \Phi\}$. Let us denote the phase angle averaged phase space density, for the time being, by

$$\overline{f} = \overline{f}(\mu, J, \Phi; t) . \tag{3.62}$$

The function $\overline{f}$ does not, in general, satisfy the Liouville theorem, i.e., it is not constant along particle trajectories because it represents a phase average over particles that have followed different dynamical trajectories to the point of observation.

While the triplet $\{\mu, J, \Phi\}$ may seem the most natural set of coordinates in the nearly dipolar magnetic field of the inner magnetosphere, it is not always the most practical. As discussed in Sect. 2.4.2, both μ and J depend on particle momentum. It is customary to replace J by the purely field-geometrical quantity K (Eq. (2.57))

$$K = \frac{J}{\sqrt{8m\mu}} = \int_{s_m}^{s_m'} [B_m - B(s)]^{1/2} \, ds .$$

Furthermore, in radiation belt studies Φ is often replaced by L or L^* (Eq. (2.85)). Note, however, that L^* depends on the dipole moment and thus evolves over long time periods due to the slow, *secular*, variation of the geomagnetic field. This can be seen in radiation belt data from the lifetime of long-lived satellites, e.g., SAMPEX, which returned data from almost two full solar cycles 1992–2012. For this reason the coordinates $\{\mu, K, \Phi\}$ may be recommendable for radiation belt models (see, e.g., Schulz 1996). In studies of individual events the triplet $\{\mu, K, L^*\}$ is fully appropriate.

The phase space density (PSD) $f(\mu, K, L^*)$ is a powerful and widely used tool in studies of particle acceleration and transport processes, in evaluation of magnetic field models and also in cross-calibration of instruments. However, its accurate determination from particle observations is far from trivial (see, e.g., Green and Kivelson 2004; Morley et al. 2013). Incompletely observed fluxes, spatio-temporal limitations of observations and inaccuracies of magnetic field models all contribute to error bars and call for caution in the interpretation of the calculated PSDs. Of the invariant coordinates only μ can readily be determined from *in situ* data, whereas the calculation of K and L^* requires the use of a magnetic field model.

From the observed flux as a function of kinetic energy, pitch angle, position and time $j(W, \alpha, \mathbf{r}, t)$ the phase space density $f(\mu, K, L^*, t)$ can be determined through the following procedure:

1. The observed flux shall first be converted to the PSD as a function of $\{W, \alpha, \mathbf{r}, t\}$ as discussed in the derivation of Eq. (3.52)

$$f(W, \alpha, \mathbf{r}, t) = \frac{j(W, \alpha, \mathbf{r}, t)}{p^2}, \qquad (3.63)$$

where p is the relativistic momentum (A.16)

$$p^2 = (W^2 + 2mc^2 W)/c^2.$$

2. The next step is to determine $K(\alpha, \mathbf{r}, t)$. Using a magnetic field model the pitch angles can be given as $\alpha(K, \mathbf{r}, t)$. Here the accuracy of the applied model becomes critical. After this step the PSD can be transformed to $f(W, K, \mathbf{r}, t)$.
3. The relativistic magnetic moment can be written as a function of the pitch angle $\alpha(K)$ as

$$\mu = \frac{p^2 \sin^2 \alpha(K)}{2mB}, \qquad (3.64)$$

from which the kinetic energy can be derived as a function of μ and K using (A.16). The PSD is now expressed as $f(\mu, K, \mathbf{r}, t)$.
4. The last step is to replace $\mathbf{r}$ by L^*. The magnetic moment and K already contain the information of gyro and bounce phase averaged position of the particle. Thus the only missing piece of information is the drift shell. Again a magnetic field model is needed to calculate the drift path of the particle around the Earth and the enclosed magnetic flux Φ, from which L^* is obtained as $L^* = 2\pi k_0/(\Phi R_E)$.

In this procedure observational inaccuracies and deviations of the model magnetic field from the actual field are propagated from one step to the next, which makes it difficult to estimate the error bars in the phase space density. Matching phase space densities calculated from measurements of inter-calibrated instruments with sufficient energy and pitch angle resolution and coverage, the errors due to the used magnetic field can be estimated. Using observations of the two *Van Allen Probes* during several *L*-shell conjunctions on 8–9 October 2012 Reeves et al. (2013) concluded (in the supplementary material of their article) that most of the PSD values were within a factor of 1.4 and all values matched better than a factor of 2. Furthermore, Morley et al. (2013) found that, of the several models they tested, the TS04 model (Tsyganenko and Sitnov 2005) captured the inner magnetospheric configuration best in the phase space matching procedure.

Open Access This chapter is licensed under the terms of the Creative Commons Attribution 4.0 International License (http://creativecommons.org/licenses/by/4.0/), which permits use, sharing, adaptation, distribution and reproduction in any medium or format, as long as you give appropriate credit to the original author(s) and the source, provide a link to the Creative Commons license and indicate if changes were made.

The images or other third party material in this chapter are included in the chapter's Creative Commons license, unless indicated otherwise in a credit line to the material. If material is not included in the chapter's Creative Commons license and your intended use is not permitted by statutory regulation or exceeds the permitted use, you will need to obtain permission directly from the copyright holder.

Chapter 4
Plasma Waves in the Inner Magnetosphere

Understanding the role of plasma waves, extending from magnetohydrodynamic (MHD) waves at ultra-low-frequency (ULF) oscillations in the millihertz range to very-low-frequency (VLF) whistler-mode emissions at frequencies of a few kHz, is necessary in studies of sources and losses of radiation belt particles. In order to make this theoretically heavy part of the book accessible to a reader, who is not familiar with wave–particle interactions, we have divided the treatise into three chapters. In the present chapter we introduce the most important wave modes that are critical to the dynamics of radiation belts. The drivers of these waves are discussed in Chap. 5 and the roles of the wave modes as sources and losses of radiation belt particles are dealt with in Chap. 6.

Basic plasma wave concepts such as dispersion equation, wave vector, index of refraction, phase and group velocities, etc., are summarized in Appendices A.2 and A.3.

4.1 Wave Environment of Radiation Belts

We begin with ULF waves. They can be observed directly in space and as *geomagnetic pulsations* on the ground. The ground-based observations are particularly useful when local space observations are not available or the low frequency of the waves makes them difficult to identify by using instruments onboard fast moving satellites. Ground-based magnetometers can also capture ULF waves more globally through the wide longitudinal and latitudinal coverage of magnetometer stations. On the other hand, not all ULF wave modes reach the ground and those that do so may become distorted in the ionosphere.

In studies of geomagnetic pulsations the ULF waves are traditionally grouped as irregular (Pi) and continuous (Pc) pulsations and further according to the observed

© The Author(s) 2022
H. E. J. Koskinen, E. K. J. Kilpua, *Physics of Earth's Radiation Belts*,
Astronomy and Astrophysics Library, https://doi.org/10.1007/978-3-030-82167-8_4

Table 4.1 Periods and frequencies of Pc1–Pc5 and Pi1 and Pi2 pulsations (for further details, see Jacobs et al. 1964)

	Pc1	Pc2	Pc3	Pc4	Pc5	Pi1	Pi2
Period (s)	0.2–5	5–10	10–45	45–150	150–600	1–40	40–50
Freq. (Hz)	5–0.2	0.2–0.1	0.1–0.02	0.02–0.007	0.007–0.0017	1–0.025	0.025–0.007

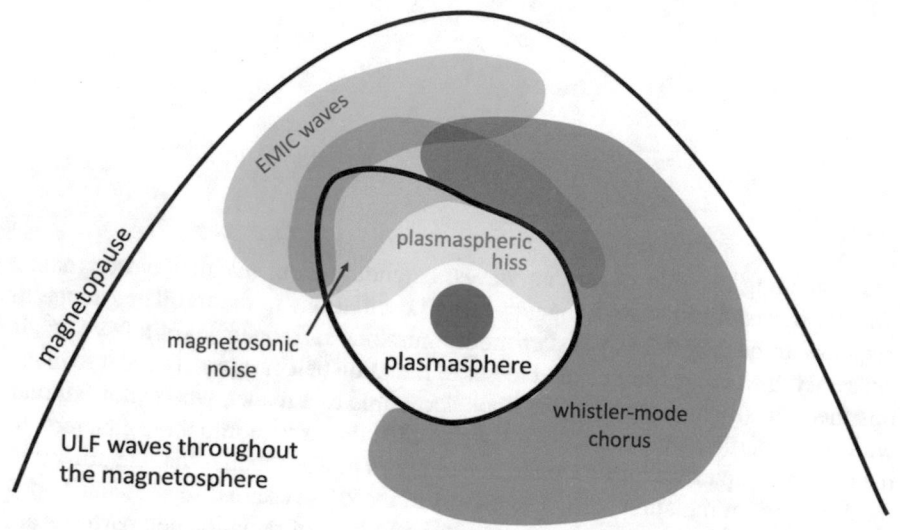

Fig. 4.1 Schematic map of the equatorial occurrence of the wave modes that are most important to the radiation belt electrons. Note that the occurrence of different modes varies depending on the magnetospheric activity and availability of free energy to drive the waves, and, e.g., chorus waves and EMIC waves can be observed at all local times, although less frequently than in the domains indicated here. More detailed empirical maps are presented in Chap. 5

periods. Table 4.1 summarizes the pulsation periods most frequently encountered in magnetospheric physics.

In the radiation belt context the most important ULF waves have frequencies in the ranges of Pc1, Pc4, and Pc5 pulsations. Pc4 and Pc5 waves are global-scale magnetohydrodynamic waves (Sect. 4.4). Their role is particularly important in radial diffusion and transport of radiation belt electrons (Chap. 6). The Pc1 range includes *electromagnetic ion cyclotron* (EMIC) waves, also known as *Alfvén ion cyclotron waves*, whose frequencies are below the local ion gyro frequency but higher than those of Pc4 and Pc5 waves. The dispersion equation of EMIC waves can be found by solving the cold plasma dispersion equation (Sect. 4.3), although determining their growth and decay rates requires calculation based on Vlasov theory. EMIC waves play an important role in the ring current and in the loss of ultra-relativistic radiation belt electrons.

Figure 4.1 illustrates the most common equatorial domains of waves whose interactions with charged particles can lead to acceleration, transport and loss of radiation belt electrons. EMIC waves are predominantly observed in the afternoon sector close to the plasmapause and beyond. The next wave mode in the order of increasing frequency is the *equatorial magnetosonic noise*, observed from a few Hz to a few hundreds of Hz. Magnetosonic noise is found both inside and outside the dayside plasmapause. The *plasmaspheric hiss* can be found all over the plasmasphere with highest occurrence rates on the dayside as indicated in Fig. 4.1. The frequency of hiss emissions extends to several kHz. However, their interaction with radiation belt electrons is most efficient at frequencies below 100 Hz. The highest-frequency waves in Fig. 4.1 are the VLF *whistler-mode chorus* emissions from about 0.5 kHz to 10 kHz. They are observed outside the plasmasphere, most commonly from the dawn sector to the dayside.

4.2 Waves in Vlasov Description

The basic characteristics of the most important wave modes in radiation belt physics can be found from reduced plasma descriptions, such as cold plasma theory (EMIC, whistler-mode chorus, plasmaspheric hiss) or magnetohydrodynamics (ULF waves). However, these theories are not sufficient to describe how the waves are driven nor how the waves accelerate, scatter and transport plasma particles. To understand the source and loss mechanisms of energetic particles in radiation belts a more detailed treatment is needed. For this reason we start our discussion of plasma waves from the elements of Vlasov theory and move thereafter to the cold plasma and MHD descriptions.

4.2.1 *Landau's Solution of the Vlasov Equation*

The Vlasov equation for particle species α (3.14)

$$\frac{\partial f_\alpha}{\partial t} + \mathbf{v} \cdot \frac{\partial f_\alpha}{\partial \mathbf{r}} + \frac{q_\alpha}{m_\alpha}(\mathbf{E} + \mathbf{v} \times \mathbf{B}) \cdot \frac{\partial f_\alpha}{\partial \mathbf{v}} = 0$$

is not easy to solve. It has to be done under the constraint that the electromagnetic field fulfils Maxwell's equations, whose source terms $(\rho, \mathbf{J})$ are determined by the distribution function, which, in turn, evolves according to the Vlasov equation. When looking for analytical solutions the background plasma and magnetic field must in practice be assumed homogeneous. In space physics this is a problem at various boundary layers, where the wavelengths become comparable to the thickness of the boundary. Furthermore, the force term in the Vlasov equation is nonlinear and the Vlasov equation can be solved analytically only for small perturbations when linearization is possible. This is sometimes a serious limitation in radiation belts where the wave amplitudes are known to grow to the nonlinear regime as will be discussed in the subsequent chapters.

We start by writing the distribution function for plasma species α and the electromagnetic field as sums of equilibrium solutions (subscript 0) and small perturbations (subscript 1)

$$f_\alpha = f_{\alpha 0} + f_{\alpha 1}$$
$$\mathbf{E} = \mathbf{E}_0 + \mathbf{E}_1$$
$$\mathbf{B} = \mathbf{B}_0 + \mathbf{B}_1$$

and linearize the Vlasov equation by considering only the first-order terms in perturbations. The problem still remains difficult. For example, the general linearized solution for homogeneous plasma in a homogeneous background magnetic field was not presented until late 1950s by Bernstein (1958). Inclusion of spatial inhomogeneities rapidly leads to problems that require numerical methods.

Lev Landau (1946) found the solution to the Vlasov equation in the absence of background fields. At the first sight, this may seem irrelevant in the context of radiation belts where particle dynamics is controlled by the magnetospheric magnetic field. However, the wave–particle interactions described by Landau's solution are important also in magnetized plasma and lay the foundation for the transfer of energy from plasma waves to charged particles, and vice versa.

Let us consider homogeneous plasma without ambient electromagnetic fields ($\mathbf{E}_0 = \mathbf{B}_0 = 0$) in the *electrostatic approximation*, in which the electric field perturbation is given as the gradient of a scalar potential $\mathbf{E}_1 = -\nabla \varphi_1$ and the magnetic field perturbation $\mathbf{B}_1 = 0$. The linearized Vlasov equation is now

$$\frac{\partial f_{\alpha 1}}{\partial t} + \mathbf{v} \cdot \frac{\partial f_{\alpha 1}}{\partial \mathbf{r}} - \frac{q_\alpha}{m_\alpha} \frac{\partial \varphi_1}{\partial \mathbf{r}} \cdot \frac{\partial f_{\alpha 0}}{\partial \mathbf{v}} = 0 \,, \tag{4.1}$$

where

$$\nabla^2 \varphi_1 = -\frac{1}{\epsilon_0} \sum_\alpha n_\alpha q_\alpha \int f_{\alpha 1} \, \mathrm{d}^3 v \,. \tag{4.2}$$

Here it is convenient to normalize the distribution function to 1. As we assume the plasma being homogeneous, the constant *background* density n_α has in (4.2) been moved outside the integral.

Vlasov tried to solve these equations at the end of the 1930s using Fourier transformations in space and time. He ended up with the integral

$$\int_{-\infty}^{\infty} \frac{\partial f_{\alpha 0}/\partial v}{\omega - kv} \, \mathrm{d}v \,,$$

which has a singularity along the path of integration. Vlasov did not find the way how to deal with the singularity.

Landau realized that, because the perturbation must begin at some point in time, the problem should be treated as an initial value problem and, instead of a Fourier transform, a Laplace transform is to be applied in the time domain. In this approach the initial perturbations turned out to be transients that fade away with time and the *asymptotic solution* gives the intrinsic properties of the plasma, i.e., the dispersion equation between the frequency and the wave number. The Laplace transformation makes the frequency a complex quantity $\omega = \omega_r + i\omega_i$, which, when inserted in the plane wave expression $\exp(i(\mathbf{k} \cdot \mathbf{r} - \omega t))$, leads to a term proportional to $\exp(\omega_i t)$ that either grows ($\omega_i > 0$) or decays ($\omega_i < 0$) exponentially as a function of time.

After Fourier transforming in space and Laplace transforming in time the perturbations $f_{\alpha 1}$ and φ_1, the asymptotic solution leads to the *dispersion equation*

$$K(\omega, \mathbf{k}) = 0 , \tag{4.3}$$

where

$$K(\omega, \mathbf{k}) = 1 + \frac{1}{\epsilon_0} \sum_\alpha \frac{n_\alpha q_\alpha^2}{m_\alpha} \frac{1}{k^2} \int \frac{\mathbf{k} \cdot \partial f_{\alpha 0}/\partial \mathbf{v}}{\omega - \mathbf{k} \cdot \mathbf{v}} d^3 v . \tag{4.4}$$

K is called the *dielectric function* because it describes the dielectric behavior of the plasma, i.e., it formally relates the electric field to the electric displacement $\mathbf{D} = K\epsilon_0 \mathbf{E}$. Now the frequency is $\omega = ip$, where p is the coordinate in Laplace transformed time domain $\exp(-pt)$.

Because $K(\omega, \mathbf{k})$ contains the information of the relation between frequency and wave vector, we do not usually need to make the inverse transformations back to the $(t, \mathbf{r})$-space. However, it is important to know, how the inverse Laplace transformation is to be done in order to correctly treat the pole in (4.4). This is a non-trivial exercise in complex integration. The procedure can be found in advanced plasma physics textbooks (e.g., Koskinen 2011). Here we skip the technical details.

Non-magnetized homogeneous plasma is essentially one-dimensional. We can simplify the notation by selecting one of the coordinate axes in the direction of $\mathbf{k}$ and write the one-dimensional distribution function as

$$F_{\alpha 0}(u) \equiv \int f_{\alpha 0}(\mathbf{v}) \, \delta \left(u - \frac{\mathbf{k} \cdot \mathbf{v}}{|k|} \right) d^3 v , \tag{4.5}$$

where $\delta(x)$ is Dirac's delta.

Careful analysis of the inverse Laplace transform indicates that the integral in (4.4) must be calculated along a contour that is closed in the upper half of the complex plane and passes below the pole. The integration path is called the *Landau contour*, denoted by $\int_L$, and the dispersion equation is

$$K(\omega, k) \equiv 1 - \sum_\alpha \frac{\omega_{p\alpha}^2}{k^2} \int_L \frac{\partial F_{\alpha 0}(u)/\partial u}{u - \omega/|k|} du = 0 . \tag{4.6}$$

The pole in the integral leads to a complex solution of (4.6)

$$\omega(k) = \omega_r(k) + i\omega_i(k) .$$

(4.7)

If $\omega_i < 0$, the electrostatic potential φ_1 is damped and the distribution function is *stable*. If $\omega_i > 0$, φ_1 grows corresponding to an *instability*.

Recall that this analysis is done by assuming small perturbations and the result is valid at the asymptotic limit. Consequently, the solution is valid when $|\omega_i| \ll |\omega_r|$. Such solutions are called *normal modes*. Larger $|\omega_i|$ leads either to an overdamped wave or to a perturbation growing to the nonlinear regime.

4.2.2 Landau Damping of the Langmuir Wave

The Landau integration can be performed analytically for some specific distribution functions only. Already the Maxwellian distribution leads to technical complications.

Assume again $\mathbf{E}_0 = \mathbf{B}_0 = 0$ and consider the one-dimensional Maxwellian

$$F_{\alpha 0}(u) = \sqrt{\frac{m_\alpha}{2\pi k_B T_\alpha}} \exp(-u^2/v_{th,\alpha}^2) ,$$

(4.8)

where the *thermal speed* $v_{th,\alpha}$ is defined as

$$v_{th,\alpha} = \sqrt{\frac{2k_B T_\alpha}{m_\alpha}} .$$

(4.9)

A difficulty, although manageable, with the Landau contour is the calculation of the closure of the integration path of

$$\int \frac{\partial F_{\alpha 0}/\partial u}{u - \omega/|k|} \, du \propto \int \frac{u F_{\alpha 0}}{u - \omega/|k|} \, du$$

in the complex plane when $u \to \infty$. The result is commonly expressed in terms of the *plasma dispersion function*

$$Z(\zeta) = \frac{1}{\sqrt{\pi}} \int_{-\infty}^{\infty} \frac{\exp(-x^2)}{x - \zeta} \, dx \; ; \; \text{Im}(\zeta) > 0$$

(4.10)

and its derivatives. $Z(\zeta)$ is related to the error function of mathematics and must, in practise, be computed numerically.

Considering the electron oscillations only, similar to the case of cold plasma oscillation (Sect. 3.1.2) but assuming now a finite temperature, the dispersion equation turns out to be

$$1 - \frac{\omega_{pe}^2}{k^2 v_{th,e}^2} Z'\left(\frac{\omega}{k v_{th,e}}\right) = 0 , \qquad (4.11)$$

where Z' denotes the derivative of the plasma dispersion function with respect to its argument.

For normal modes ($|\omega_i| \ll \omega_r$) the dispersion equation can be expanded around $\omega = \omega_r$ as

$$1 - \sum_\alpha \frac{\omega_{p\alpha}^2}{k^2}\left(1 + i\omega_i \frac{\partial}{\partial \omega_i}\right)\left[\text{P} \int \frac{\partial F_{\alpha 0}/\partial u}{u - \omega_r/|k|} \, du + \pi i \left(\frac{\partial F_{\alpha 0}}{\partial u}\right)_{u = \omega_r/|k|}\right] = 0 . \qquad (4.12)$$

Here P indicates the Cauchy principal value. The second term in the brackets comes from the residue at the pole. Because the pole in this case is on the real axis, the residue is multiplied by πi instead of $2\pi i$. Using this expression we can find solutions for the dispersion equation at long and short wavelengths. These correspond to series expansions of the dispersion function Z for large and small arguments, respectively. At intermediate wavelengths numerical computation of Z cannot be avoided.

The most fundamental normal mode is the propagating variant of the fundamental plasma oscillation (Sect. 3.1.2), known as the *Langmuir wave*. It can be found as the long wavelength ($\omega/k \gg v_{th}$) solution of (4.12). At this limit

$$-\text{P} \int \frac{\partial F_{\alpha 0}/\partial u}{u - \omega_r/|k|} \, du = \int \frac{\partial F_{\alpha 0}}{\partial u}\left(\frac{1}{\omega/|k|} + \frac{u}{(\omega/|k|)^2} + \frac{u^2}{(\omega/|k|)^3} + \dots\right) du . \qquad (4.13)$$

By using this expansion, considering electron dynamics only, and inserting a Maxwellian electron distribution function, we find the dispersion equation for the Langmuir wave. The real part of the frequency is

$$\omega_r \approx \omega_{pe}(1 + 3k^2\lambda_{De}^2)^{1/2} \approx \omega_{pe}\left(1 + \frac{3}{2}k^2\lambda_{De}^2\right) \qquad (4.14)$$

and the imaginary part

$$\omega_i \approx -\sqrt{\frac{\pi}{8}}\frac{\omega_{pe}}{|k^3\lambda_{De}^3|}\exp\left(-\frac{1}{2k^2\lambda_{De}^2} - \frac{3}{2}\right) . \qquad (4.15)$$

The finite temperature of the Maxwellian distribution makes the standing cold plasma oscillation to propagate. Furthermore, the negative imaginary part of the frequency indicates that the wave is damped. This phenomenon is known as *Landau damping*.

Landau damping is not limited to electrostatic waves. As will be discussed in Chap. 6, it is also important in the context of electromagnetic waves causing resonant scattering of electrons with pitch angles close to 90°.

4.2.3 Physical Interpretation of Landau Damping

Landau's solution was met with scepticism until it was experimentally verified in laboratory experiments in the 1960s (Malmberg and Wharton 1964). The problem was that the Vlasov equation conserves entropy, whereas the Landau solution does not seem to do so. The electric field of the Langmuir wave interacts with electrons accelerating those whose velocity is slightly less than the phase velocity of the wave, and decelerating those that move a little faster. In a Maxwellian distribution $\partial f/\partial v < 0$, meaning that there are more slower than faster electrons around the phase velocity (Fig. 4.2). Figuratively speaking the wave forces the particles near the phase velocity to "glide down" along the slope of the distribution function until the population is warm enough to damp the oscillation below the observable level. Thus, there is a net energy transfer from the wave to the particles.

Although Landau damping looks like a dissipative process, the entropy is conserved in Vlasov theory and no information must be lost from the combined system consisting of both the distribution function and the electrostatic potential. The apparent contradiction can be resolved by carefully considering what happens to the distribution function in the damping process (for a detailed discussion, see,

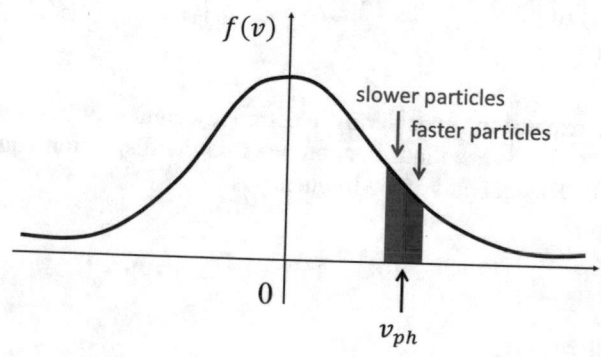

Fig. 4.2 In a Maxwellian plasma $\partial f/\partial v < 0$ and there are more particles that are accelerated by the Langmuir wave in the vicinity of the phase velocity v_{ph} than those that lose energy to the wave. Thus the wave is damped and the electron population is heated

e.g., Krall and Trivelpiece 1973). At the time-asymptotic limit an extra term appears to the distribution function in the Fourier space

$$f_{\alpha k} = \hat{f}_{\alpha b} \exp(-i\mathbf{k} \cdot \mathbf{v}t) + \sum_{\omega_k} \hat{f}_{\alpha k} \exp(-i\omega_k t) , \qquad (4.16)$$

where ω_k are the solutions of the dispersion equation and $\hat{f}_{\alpha b}$ and $\hat{f}_{\alpha k}$ are time-independent amplitudes. The terms in the sum over ω_k are damped at the same rate as the perturbed potential $\varphi_k(t)$. In the first term on the RHS of (4.16) the subscript b stands for *ballistic*. The ballistic term is a consequence of the Liouville theorem, according to which the Vlasov equation conserves entropy. As the system is deterministic, every particle "remembers" its initial perturbation wherever it moves in the phase space.

When t increases, the ballistic term becomes increasingly oscillatory in the velocity space and its contribution to $\varphi_k(t)$ behaves at the limit $t \to \infty$ as

$$k^2 \varphi_k = \frac{1}{c_0} \sum_{\alpha} q_\alpha n_\alpha \int \hat{f}_{\alpha b} \exp(-i\mathbf{k} \cdot \mathbf{v}t) d^3 v \to 0 . \qquad (4.17)$$

That is, at the time-asymptotic limit the ballistic terms of each particle species contain the information of the initial perturbation but they do not contribute to the *observable* electric field.

The existence of ballistic terms leads to an observable *nonlinear* phenomenon called the *Landau echo*. Assume that an initial perturbation took place at time t_1 and its spectrum was narrow near wave number k_1. Wait until the perturbation has been damped below the observable limit and only the ballistic term superposed on the equilibrium distribution remains. Then launch another narrow-band wave near k_2 at time t_2 and wait until it also is damped. At time $t = t_3$ defined by

$$k_1(t_3 - t_1) - k_2(t_3 - t_2) = 0 \qquad (4.18)$$

the oscillations in the ballistic terms interfere positively. This beating of the ballistic terms of the first two perturbations produces a new observable fluctuation, which is the Landau echo. Also the echo is transient because the condition (4.18) is satisfied only for a short while and Landau damping acts on this fluctuation as well. The effect has been verified in laboratories and shows that the Landau damping does not violate the conservation of entropy in the timescale shorter than the collisional time.

As collisional timescales in tenuous space plasmas often are very long compared to the relevant timescales of interesting plasma phenomena, the existence of Landau echoes indicates that even in the case of small-amplitude perturbations there can be *nonlinear mixing* of wave modes at the microscopic level. This is one viewpoint to *plasma turbulence*.

4.2.4 Solution of the Vlasov Equation in Magnetized Plasma

Magnetospheric plasma is embedded in a background magnetic field and we need to look for a more general description including the background fields $\mathbf{E}_0(\mathbf{r}, t)$ and $\mathbf{B}_0(\mathbf{r}, t)$. The linearized Vlasov equation is written as

$$\left[\frac{\partial}{\partial t} + \mathbf{v} \cdot \frac{\partial}{\partial \mathbf{r}} + \frac{q_\alpha}{m_\alpha} (\mathbf{E}_0 + \mathbf{v} \times \mathbf{B}_0) \cdot \frac{\partial}{\partial \mathbf{v}} \right] f_{\alpha 1} = - \frac{q_\alpha}{m_\alpha} (\mathbf{E}_1 + \mathbf{v} \times \mathbf{B}_1) \cdot \frac{\partial f_{\alpha 0}}{\partial \mathbf{v}} .$$

(4.19)

This is possible to solve by employing the *method of characteristics*, which can be described as "integration over unperturbed orbits". Define new variables $(\mathbf{r}', \mathbf{v}', t')$ as

$$\frac{d\mathbf{r}'}{dt'} = \mathbf{v}' ; \quad \frac{d\mathbf{v}'}{dt'} = \frac{q_\alpha}{m_\alpha} \left[\mathbf{E}_0(\mathbf{r}', t') + \mathbf{v}' \times \mathbf{B}_0(\mathbf{r}', t') \right] ,$$

(4.20)

where the acceleration is determined by the background fields and the boundary conditions are

$$\mathbf{r}'(t' = t) = \mathbf{r}$$
$$\mathbf{v}'(t' = t) = \mathbf{v} .$$

(4.21)

Consider $f_{\alpha 1}(\mathbf{r}', \mathbf{v}', t')$ and use (4.19) to calculate its total time derivative

$$\frac{d f_{\alpha 1}(\mathbf{r}', \mathbf{v}', t')}{dt'} = \frac{\partial f_{\alpha 1}(\mathbf{r}', \mathbf{v}', t')}{\partial t'} + \frac{d\mathbf{r}'}{dt'} \cdot \frac{\partial f_{\alpha 1}(\mathbf{r}', \mathbf{v}', t')}{\partial \mathbf{r}'} + \frac{d\mathbf{v}'}{dt'} \cdot \frac{\partial f_{\alpha 1}(\mathbf{r}', \mathbf{v}', t')}{\partial \mathbf{v}'}$$

$$= - \frac{q_\alpha}{m_\alpha} \left[\mathbf{E}_1(\mathbf{r}', t') + \mathbf{v}' \times \mathbf{B}_1(\mathbf{r}', t') \right] \cdot \frac{\partial f_{\alpha 0}(\mathbf{r}', \mathbf{v}')}{\partial \mathbf{v}'} .$$

(4.22)

The boundary conditions (4.21) imply that $f_{\alpha 1}(\mathbf{r}', \mathbf{v}', t') = f_{\alpha 1}(\mathbf{r}, \mathbf{v}, t)$ at time $t' = t$. Thus the solution of (4.22) at $t' = t$ is a solution of the Vlasov equation. The point of this procedure is that (4.22) can be calculated by a direct integration because its LHS is an *exact differential*. The formal solution is

$$f_{\alpha 1}(\mathbf{r}, \mathbf{v}, t) = - \frac{q_\alpha}{m_\alpha} \int_{-\infty}^{t} \left[\mathbf{E}_1(\mathbf{r}', t') + \mathbf{v}' \times \mathbf{B}_1(\mathbf{r}', t') \right] \cdot \frac{\partial f_{\alpha 0}(\mathbf{r}', \mathbf{v}')}{\partial \mathbf{v}'} dt'$$

$$+ f_{\alpha 1}(\mathbf{r}'(-\infty), \mathbf{v}'(-\infty), t'(-\infty)) .$$

(4.23)

The procedure can be interpreted in the following way: The perturbation of the distribution function $f_{\alpha 1}$ has been found by integrating the Vlasov equation from $-\infty$ to t along the path in the $(\mathbf{r}, \mathbf{v})$-space that at each individual time coincides with the orbit of a charged particle in the background fields $\mathbf{E}_0$ and $\mathbf{B}_0$. This, of course, requires that the deviation from the background orbit at each step in the integration is small. Consequently, the method is limited to linear perturbations.

From $f_{\alpha 1}$ we can calculate $n_{\alpha 1}(\mathbf{r}, t)$ and $\mathbf{V}_{\alpha 1}(\mathbf{r}, t)$ and insert these in Maxwell's equations

$$\nabla \times \mathbf{E}_1 = -\frac{\partial \mathbf{B}_1}{\partial t} \tag{4.24}$$

$$\nabla \cdot \mathbf{E}_1 = \frac{1}{\epsilon_0} \sum_\alpha q_\alpha n_{\alpha 1} \tag{4.25}$$

$$\nabla \times \mathbf{B}_1 = \frac{1}{c^2} \frac{\partial \mathbf{E}_1}{\partial t} + \mu_0 \sum_\alpha q_\alpha (n_\alpha \mathbf{V}_\alpha)_1 . \tag{4.26}$$

This set of equations can now (in principle) be solved as an initial value problem in the same way as the Landau solution. Accepting that the Landau contour is the correct way to deal with the resonant integrals, assuming that the waves are plane waves $\mathbf{E}_1(\mathbf{r}, t) = \mathbf{E}_{\mathbf{k}\omega} \exp(i\mathbf{k} \cdot \mathbf{r} - i\omega t)$, and $f_{\alpha 1}(\mathbf{r}', \mathbf{v}', t \to -\infty) \to 0$, the growing solutions $(\mathrm{Im}(\omega) > 0)$ are found to be

$$f_{\alpha \mathbf{k}} = -\frac{q_\alpha}{m_\alpha} \int_{-\infty}^{0} (\mathbf{E}_{\mathbf{k}\omega} + \mathbf{v}' \times \mathbf{B}_{\mathbf{k}\omega}) \cdot \frac{\partial f_{\alpha 0}(\mathbf{v}')}{\partial \mathbf{v}'} \exp[i(\mathbf{k} \cdot \mathbf{R} - \omega\tau)] \, d\tau , \tag{4.27}$$

where $\tau = t' - t$, $\mathbf{R} = \mathbf{r}' - \mathbf{r}$. The damped solutions $(\mathrm{Im}(\omega) < 0)$ are found by analytic continuation of $f_{\alpha \mathbf{k}}$ to the lower half-plane. By inserting this into Maxwell's equations in the $(\omega, \mathbf{k})$ space and eliminating $\mathbf{B}_{\mathbf{k}\omega}$ we get the wave equation

$$\mathbf{K} \cdot \mathbf{E} = 0 . \tag{4.28}$$

Now the dielectric function is the *dielectric tensor* or *dispersion tensor* $\mathbf{K}$. It is even in a homogeneous background magnetic field a complicated function. Let us start by considering the field-free isotropic case $(\mathbf{E}_0 = \mathbf{B}_0 = 0$ and $f_0 = f_0(v^2))$. Define $F_{\alpha 0}(u) = \int f_{\alpha 0} \delta(u - \mathbf{k} \cdot \mathbf{v}/|\mathbf{k}|) \, d^3v$ and denote the component of the wave electric field in the direction of wave propagation by $E_{\mathbf{k}} = (\mathbf{k} \cdot \mathbf{E})/|\mathbf{k}|$ and the transverse component by $\mathbf{E}_\perp = (\mathbf{k} \times \mathbf{E})/|\mathbf{k}|$. The wave equation now becomes

$$\begin{bmatrix} K_\perp & 0 & 0 \\ 0 & K_\perp & 0 \\ 0 & 0 & K_{\mathbf{k}} \end{bmatrix} \begin{bmatrix} E_{\perp 1} \\ E_{\perp 2} \\ E_{\mathbf{k}} \end{bmatrix} = 0 , \tag{4.29}$$

where

$$K_\perp = 1 - \frac{k^2 c^2}{\omega^2} - \sum_\alpha \frac{\omega_{p\alpha}^2}{\omega} \int \frac{F_{\alpha 0}}{\omega - |k|u} \, du \tag{4.30}$$

$$K_{\mathbf{k}} = 1 + \sum_\alpha \frac{\omega_{p\alpha}^2}{\omega} \int_L \frac{F_{\alpha 0}/\partial u}{\omega/|k| - u} \, du \ . \tag{4.31}$$

These give

electrostatic modes : $K_{\mathbf{k}} = 0 \ (\mathbf{E}_\perp = 0)$

electromagnetic modes : $K_\perp = 0 \ (\mathbf{E}_{\mathbf{k}} = 0)$.

The electrostatic solution is Landau's solution familiar from above. The dispersion equation for the electromagnetic modes is

$$\omega^2 = k^2 c^2 + \sum_\alpha \omega_{p\alpha}^2 \int_{-\infty}^{\infty} \frac{\omega F_{\alpha 0}}{\omega - |k|u} \, du \ . \tag{4.32}$$

This has propagating solutions if $\omega \gg k v_{th,e}$ and we find the electromagnetic wave in non-magnetized cold plasma

$$\omega^2 \approx k^2 c^2 + \omega_{pe}^2 \ . \tag{4.33}$$

The propagation is limited to frequencies higher than ω_{pe}.

Include next a homogeneous background magnetic field $\mathbf{B}_0 = B_0 \mathbf{e}_z$, but keep the background electric field $\mathbf{E}_0$ zero. Assume that the background particle distribution function is gyrotropic but may be anisotropic $f_{\alpha 0} = f_{\alpha 0}(v_\perp^2, v_\|)$. Already in this very symmetric configuration the derivation of the dielectric tensor is a tedious procedure. A lengthy calculation, first presented by Bernstein (1958), leads to the dielectric tensor in the form

$$\mathsf{K}(\omega, \mathbf{k}) = \left(1 - \sum_\alpha \frac{\omega_{p\alpha}^2}{\omega^2} \right) \mathsf{I} - \sum_\alpha \sum_{n=-\infty}^{\infty} \frac{2\pi \omega_{p\alpha}^2}{n_{\alpha 0} \omega^2} \times \tag{4.34}$$

$$\int_0^\infty \int_{-\infty}^\infty v_\perp dv_\perp dv_\| \left(k_\| \frac{\partial f_{\alpha 0}}{\partial v_\|} + \frac{n \omega_{c\alpha}}{v_\perp} \frac{\partial f_{\alpha 0}}{\partial v_\perp} \right) \frac{\mathsf{S}_{n\alpha}(v_\|, v_\perp)}{k_\| v_\| + n \omega_{c\alpha} - \omega} \ .$$

I is the unit tensor and the tensor $\mathsf{S}_{n\alpha}$ is

$$
\mathsf{S}_{n\alpha}(v_{\parallel}, v_{\perp}) =
\begin{bmatrix}
\dfrac{n^2\omega_{c\alpha}^2}{k_{\perp}^2} J_n^2 & \dfrac{inv_{\perp}\omega_{c\alpha}}{k_{\perp}} J_n J_n' & \dfrac{nv_{\parallel}\omega_{c\alpha}}{k_{\perp}} J_n^2 \\[2ex]
-\dfrac{inv_{\perp}\omega_{c\alpha}}{k_{\perp}} J_n J_n' & v_{\perp}^2 J_n'^2 & -iv_{\parallel}v_{\perp} J_n J_n' \\[2ex]
\dfrac{nv_{\parallel}\omega_{c\alpha}}{k_{\perp}} J_n^2 & iv_{\parallel}v_{\perp} J_n J_n' & v_{\parallel}^2 J_n^2
\end{bmatrix} .
\tag{4.35}
$$

Here J_n are the ordinary Bessel functions of the first kind with the argument $k_{\perp}v_{\perp}/\omega_{c\alpha}$, and $J_n' = dJ_n/d(k_{\perp}v_{\perp}/\omega_{c\alpha})$.

Finite $\mathbf{B}_0$ makes the plasma behavior anisotropic. The temperature may now be different in parallel and perpendicular directions as, e.g., in the case of a *bi-Maxwellian* distribution

$$
f_{\alpha 0} = \frac{m_\alpha}{2\pi k_B T_{\alpha\perp}} \sqrt{\frac{m_\alpha}{2\pi k_B T_{\alpha\parallel}}} \exp\left[-\frac{m_\alpha}{2k_B}\left(\frac{v_{\perp}^2}{T_{\alpha\perp}} + \frac{v_{\parallel}^2}{T_{\alpha\parallel}} \right) \right] .
\tag{4.36}
$$

When this is inserted into the elements of K, the resonant integrals in the direction of $v_{\parallel}$ can be expressed in terms of the plasma dispersion function Z (4.10).

The wave modes are the non-trivial solutions of

$$
\mathsf{K} \cdot \mathbf{E} = 0 .
\tag{4.37}
$$

The mode structure is now more complex than in non-magnetized plasma:

- The distinction between electrostatic and electromagnetic modes is no more exact; there still are electrostatic modes fulfilling $\mathbf{E} \parallel \mathbf{k}$ as an approximation but also the electromagnetic modes may have an electric field component along $\mathbf{k}$.
- The Bessel functions introduce harmonic mode structure organized according to $\omega = n\omega_{c\alpha}$ for each particle species α.
- The Landau resonance $\omega = \mathbf{k} \cdot \mathbf{v}$ of the isotropic plasma is replaced by

$$
\omega - n\omega_{c\alpha} = k_{\parallel}v_{\parallel} .
\tag{4.38}
$$

Thus only the velocity component along $\mathbf{B}_0$ is associated with Landau damping ($n = 0$) and only for waves with $k_{\parallel} \neq 0$.

Parallel Propagation

Let us first look at the solutions for wave modes propagating parallel to the background magnetic field ($k_{\perp} = 0$). At the lowest frequencies ($\omega \ll \omega_{ci}$) we find the parallel propagating *Alfvén wave*

$$
\omega_r = \frac{k_{\parallel}v_A}{\sqrt{1 + v_A^2/c^2}} ,
\tag{4.39}
$$

where $v_A = B_0/\sqrt{\rho_m\mu_0}$ is the *Alfvén speed*. This is an MHD mode to be discussed further in Sect. 4.4. Note that (4.39) contains a "cold plasma correction" (v_A^2/c^2) in the denominator, which is not found in MHD. This due to the inclusion of the displacement current into Ampère's law in Vlasov and cold plasma descriptions, in contrast to standard MHD. When the Vlasov equation is solved together with the full set of Maxwell's equations, the solutions include the cold plasma and MHD approximations as limiting cases.

In Vlasov theory Alfvén waves are damped, which is not found in ideal MHD. The damping rate is very small at low frequencies. When $\omega \to \omega_{ci}$, the mode approaches the ion gyro resonance (see Fig. 4.4 in the discussion of cold plasma waves, Sect. 4.3), and the damping rate increases. At this limit the mode is the left-hand (L) circularly polarized *electromagnetic ion cyclotron* (EMIC) *wave*, which is damped not only by the resonant ions but also by relativistic electrons with sufficiently large Lorentz factor γ and Doppler shift $k_\parallel v_\parallel$ of the frequency. This is an important loss mechanism of ultra-relativistic radiation belt electrons (Sect. 6.5.4).

We return to the right-hand (R) and left-hand (L) circularly polarized electromagnetic modes in cold plasma theory (Sect. 4.3), where they can be described in a more transparent manner. The most important right-hand polarized wave mode in radiation belts is the *whistler mode*. Again Vlasov theory is needed to describe the damping and growth of the whistler-mode waves leading to acceleration and pitch-angle scattering of radiation belt electrons as discussed in Chap. 6. Near the electron gyro frequency the whistler mode goes over to the *electromagnetic electron cyclotron wave*.

In the linear approximation the parallel propagating electromagnetic waves do not have harmonic structure. However, if the amplitude grows to nonlinear regime, the representation of the wave, e.g., as a Fourier series contains higher harmonics.

Perpendicular Propagation

For perpendicular propagation ($k_\parallel = 0$) the wave equation reduces to

$$\begin{bmatrix} K_{xx} & K_{xy} & 0 \\ K_{yx} & K_{yy} & 0 \\ 0 & 0 & K_{zz} \end{bmatrix} \cdot \begin{bmatrix} E_x \\ E_y \\ E_z \end{bmatrix} = 0 , \qquad (4.40)$$

where the z-axis is along the background magnetic field.

Assuming an isotropic background distribution function one component of the dispersion equation is

$$K_{zz} = 1 - \frac{k^2 c^2}{\omega^2} - \frac{2\pi}{\omega} \sum_\alpha \sum_n \omega_{p\alpha}^2 \int\limits_{-\infty}^{\infty} dv_\parallel \int\limits_{0}^{\infty} \frac{J_n^2 f_{\alpha 0} v_\perp}{\omega - n\omega_{c\alpha}} \, dv_\perp = 0 . \qquad (4.41)$$

One solution of this equation is the so-called *ordinary mode* (*O*-mode), which is also found in cold plasma approximation. $K_{zz} = 0$ furthermore gives a series of modes with narrow bands slightly *above* the harmonics of the cyclotron frequency

$$\omega = n\omega_{c\alpha} \left\{ 1 + \mathcal{O} \left[\frac{\omega_{p\alpha}^2}{k^2 c^2} (k r_{L\alpha})^{2n} \right] \right\}, \tag{4.42}$$

where $r_{L\alpha}$ is the gyro radius of species α and $\mathcal{O}$ indicates terms of the order of its argument, in this case small as compared to 1. These modes are *electrostatic cyclotron waves*. Both electrons and all ion species have their own families of electrostatic cyclotron modes.

The remaining perpendicular propagating modes are found from the determinant

$$\begin{vmatrix} K_{xx} & K_{xy} \\ -K_{xy} & K_{yy} \end{vmatrix} = 0. \tag{4.43}$$

This equation covers the electromagnetic modes for which $|\mathbf{E} \cdot \mathbf{k}| \ll |\mathbf{E} \times \mathbf{k}|$. They are called *extraordinary modes* (*X*-modes), which are also found in cold plasma theory.

At frequencies below the cold plasma *lower hybrid resonance frequency* ω_{LHR} (defined by Eq. 4.70 below) the *X*-mode is often called the *magnetosonic mode*. It is an extension of the perpendicular propagating magnetosonic mode of MHD (Sect. 4.4) from frequencies below the ion gyro frequency to higher frequencies. In the finite-temperature Vlasov theory the *X*-mode has *quasi-resonances*, where the group velocity of the wave (A.28) $\partial\omega/\partial\mathbf{k} \to 0$, at the multiples of the ion gyro frequency $n\omega_{ci}$ for $n \geq 1$ up to ω_{LHR}. This gives to the *X*-mode wave an observable banded structure, an example of which is shown in Fig. 5.14. These bands were first identified in Bernstein's dielectric tensor (4.34) and they are, consequently, known as *Bernstein modes*.

Another set of Bernstein-mode solutions of (4.43) is found at short wavelengths. These modes are quasi-electrostatic ($|\mathbf{E} \cdot \mathbf{k}| \gg |\mathbf{E} \times \mathbf{k}|$) and they are found both for electrons and all ion species. The exactly perpendicular modes are not Landau damped. If the modes have finite $k_\parallel$, they experience *cyclotron damping* when $n \neq 0$.

Propagation to Arbitrary Directions

A convenient way to illustrate the wave solutions at arbitrary directions of the wave vector $\omega = \omega(k_\parallel, k_\perp)$ is to represent them as *dispersion surfaces* in three-dimensional $(\omega, k_\parallel, k_\perp)$-space. An example of dispersion surfaces is given in Fig. 4.3. The surface has been calculated by solving Eq. (4.37) for plasma parameters corresponding to the inner magnetosphere slightly outside the plasmapause using the numerical dispersion equation solver WHAMP (Waves in homogeneous

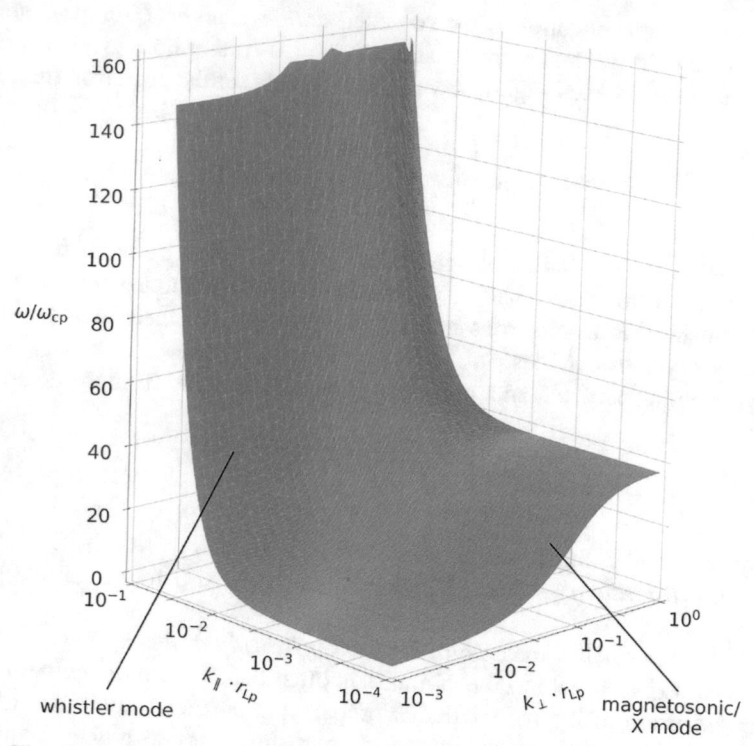

Fig. 4.3 The dispersion surface that contains the parallel propagating right-hand polarized whistler mode and perpendicular propagating X-mode. The frequency on the vertical axis is normalized to the local proton gyro frequency and the parallel and perpendicular wave numbers are normalized to the proton gyro radius (Figure courtesy: Yann Pfau-Kempf)

anisotropic magnetized plasmas) originally written by Kjell Rönnmark.[1] Figure 4.3 illustrates how the parallel propagating right-hand polarized whistler mode joins the perpendicular propagating X-mode when the direction of the wave vector is rotated from parallel toward perpendicular direction, For further examples of dispersion surfaces, see, e.g., André (1985) or Koskinen (2011).

The growth/damping rate ω_i varies from one point to another on a dispersion surface and the solution may in some domains of the surfaces be strongly damped. Depending on the local plasma parameters and characteristics of the particle populations, there may be free energy to drive the instabilities leading to growing solutions of the dispersion equation. These are discussed using practical examples in Chap. 5.

---

[1] WHAMP is available from GitHub: https://github.com/irfu/whamp.

4.3 Cold Plasma Waves

While the Vlasov theory is necessary for the treatment of the growth and damping of plasma waves, the real part of the dispersion equation for several of the most important linear wave modes in radiation belt physics can be derived from the much simpler cold plasma theory.

4.3.1 Dispersion Equation for Cold Plasma Waves in Magnetized Plasma

From Maxwell's equations and Ohm's law $\mathbf{J} = \sigma \cdot \mathbf{E}$, where σ is generally a tensor, it is straightforward to derive a *wave equation* in the form

$$\mathbf{k} \times (\mathbf{k} \times \mathbf{E}) + \frac{\omega^2}{c^2} \mathsf{K} \cdot \mathbf{E} = 0 \,, \tag{4.44}$$

where

$$\mathsf{K} = \mathsf{I} + \frac{\mathrm{i}}{\omega \epsilon_0} \sigma \tag{4.45}$$

is the *dielectric tensor* and I the unit tensor. K is a dimensionless quantity and we can relate it to the electric permittivity of the dielectric medium familiar from classical electrodynamics as

$$\mathbf{D} = \epsilon \cdot \mathbf{E} = \epsilon_0 \mathsf{K} \cdot \mathbf{E} \,. \tag{4.46}$$

In case of no background fields ($\mathbf{E}_0 = \mathbf{B}_0 = 0$) K reduces to a scalar

$$K = 1 - \frac{\omega_{pe}^2}{\omega^2} \equiv n^2 \,, \tag{4.47}$$

i.e., K is the square root of the refractive index n defined in Appendix A (A.24). The wave equation has the already familiar solutions

$$\mathbf{k} \parallel \mathbf{E} \Rightarrow \omega^2 = \omega_{pe}^2 \qquad \text{longitudinal standing plasma oscillation}$$
$$\mathbf{k} \perp \mathbf{E} \Rightarrow \omega^2 = k^2 c^2 + \omega_{pe}^2 \quad \text{electromagnetic wave in plasma}$$

Consider small perturbations $\mathbf{B}_1$ to a homogeneous background magnetic field $\mathbf{B}_0$ ($B_1 \ll B_0$). In the cold plasma approximation all particles of species α are assumed to move at their macroscopic fluid velocity $\mathbf{V}_\alpha(\mathbf{r}, t)$. Thus the total plasma current is

$$\mathbf{J} = \sum_\alpha n_\alpha q_\alpha \mathbf{V}_\alpha . \tag{4.48}$$

Assuming that $\mathbf{V}_\alpha(\mathbf{r}, t)$ oscillates sinusoidally $\propto \exp(-i\omega t)$ the first-order macroscopic equation of motion is

$$-i\omega \mathbf{V}_\alpha = q_\alpha (\mathbf{E} + \mathbf{V}_\alpha \times \mathbf{B}_0) . \tag{4.49}$$

It is convenient to consider the plane perpendicular to $\mathbf{B}_0$ as a complex plane and use the basis of unit vectors $\{\sqrt{1/2}(\mathbf{e}_x + i\mathbf{e}_y), \sqrt{1/2}(\mathbf{e}_x - i\mathbf{e}_y), \mathbf{e}_z\}$ where $\mathbf{B}_0 \parallel \mathbf{e}_z$. Denote the components in this basis by integers $d = \{-1, 1, 0\}$ and express the plasma and gyro frequencies as

$$X_\alpha = \frac{\omega_{p\alpha}^2}{\omega^2} , \quad Y_\alpha = \frac{s_\alpha \omega_{c\alpha}}{\omega} . \tag{4.50}$$

Note that $\omega_{c\alpha}$ is an unsigned quantity and the sign of the charge is indicated by s_α. In this basis the components of the current are

$$J_{d,\alpha} = i\epsilon_0 \omega \frac{X_\alpha}{1 - dY_\alpha} E_d \tag{4.51}$$

and the dielectric tensor (4.45) is diagonal

$$\mathbf{K} = \begin{bmatrix} 1 - \sum_\alpha \dfrac{X_\alpha}{1 - Y_\alpha} & 0 & 0 \\ 0 & 1 - \sum_\alpha \dfrac{X_\alpha}{1 + Y_\alpha} & 0 \\ 0 & 0 & 1 - \sum_\alpha X_\alpha \end{bmatrix} . \tag{4.52}$$

The components of the tensor are denoted by letters R, L and P:

$$R = 1 - \sum_\alpha \frac{\omega_{p\alpha}^2}{\omega^2} \left(\frac{\omega}{\omega + s_\alpha \omega_{c\alpha}} \right) \tag{4.53}$$

$$L = 1 - \sum_\alpha \frac{\omega_{p\alpha}^2}{\omega^2} \left(\frac{\omega}{\omega - s_\alpha \omega_{c\alpha}} \right) \tag{4.54}$$

$$P = 1 - \sum_\alpha \frac{\omega_{p\alpha}^2}{\omega^2} . \tag{4.55}$$

The component R has a singularity when $\omega = \omega_{ce}$ and $s_\alpha = -1$. At this frequency the wave is in *resonance* with the gyro motion of the electrons. Thus

R corresponds to the *right-hand circularly polarized* wave. Similarly *L* has a resonance with positive ions and corresponds to the *left-hand circularly polarized* wave. Recall that different from optics, where the handedness is given by the sense of the rotation of the wave electric field approaching the observer, the left- and right-handedness in magnetized plasmas correspond to the sense of the gyro motion of charged particles around the background magnetic field in the frame of reference of the guiding center. If the observer looks into the direction to which magnetic field points, the left-hand polarized wave rotates in the same sense as the gyro motion of positively charged particle. If the observer looks against the magnetic field, the rotation of the wave appears right-handed.

The component *P* corresponds to a standing plasma oscillation in the cold plasma approximation. As discussed in the context of Vlasov theory above, a finite temperature makes the plasma oscillation a propagating Langmuir wave.

Transforming **K** back to the $\{x, y, z\}$-basis we get

$$\mathbf{K} = \begin{bmatrix} S & -iD & 0 \\ iD & S & 0 \\ 0 & 0 & P \end{bmatrix}, \tag{4.56}$$

where $S = (R + L)/2$ and $D = (R - L)/2$.

The wave equation can be written in terms of the *wave normal vector* $\mathbf{n} = c\mathbf{k}/\omega$ as

$$\mathbf{n} \times (\mathbf{n} \times \mathbf{E}) + \mathbf{K} \cdot \mathbf{E} = 0. \tag{4.57}$$

Recall that $\mathbf{B}_0$ is in the z-direction. Select the x-axis so that $\mathbf{n}$ is in the xz-plane. The angle θ between $\mathbf{n}$ and $\mathbf{B}_0$ is the *wave normal angle* (WNA). In these coordinates the wave equation is

$$\begin{bmatrix} S - n^2 \cos^2\theta & -iD & n^2 \cos\theta \sin\theta \\ iD & S - n^2 & 0 \\ n^2 \cos\theta \sin\theta & 0 & P - n^2 \sin^2\theta \end{bmatrix} \begin{bmatrix} E_x \\ E_y \\ E_z \end{bmatrix} = 0. \tag{4.58}$$

The solutions of the wave equation are the non-trivial roots of the dispersion equation

$$An^4 - Bn^2 + C = 0, \tag{4.59}$$

where

$$\begin{aligned} A &= S \sin^2\theta + P \cos^2\theta \\ B &= RL \sin^2\theta + PS(1 + \cos^2\theta) \\ C &= PRL. \end{aligned} \tag{4.60}$$

It is convenient to solve the dispersion equation (4.59) for $\tan^2 \theta$ as

$$\tan^2 \theta = \frac{-P(n^2 - R)(n^2 - L)}{(Sn^2 - RL)(n^2 - P)}. \tag{4.61}$$

With this equation it is straightforward to discuss the propagation of the waves to different directions with respect to the background magnetic field. The modes propagating in the direction of the magnetic field ($\theta = 0$) and perpendicular to it ($\theta = \pi/2$) are called the *principal modes*

$$\theta = 0: \quad P = 0, \; n^2 = R, \; n^2 = L$$
$$\theta = \pi/2: \quad n^2 = RL/S, \; n^2 = P.$$

These modes have *cut-offs*

$$n^2 \to 0 \; (v_p \to \infty, \; k \to 0, \; \lambda \to \infty)$$
$$P = 0, \; R = 0, \; \text{or} \; L = 0$$

and *resonances*

$$n^2 \to \infty \; (v_p \to 0, \; k \to \infty, \; \lambda \to 0)$$
$$\tan^2 \theta = -P/S \; (\text{under the condition } P \neq 0).$$

When the wave approaches a region where it has a cut-off ($n^2 \to 0$), it cannot propagate further and is reflected. At a resonance the wave energy is absorbed by the plasma.

4.3.2   Parallel Propagation ($\theta = 0$)

Figure 4.4 presents the solutions of the cold plasma dispersion equation for parallel propagation.

The resonance frequency of the right-hand polarized mode

$$n_R^2 = R = 1 - \sum_i \frac{\omega_{pi}^2}{\omega(\omega + \omega_{ci})} - \frac{\omega_{pe}^2}{\omega(\omega - \omega_{ce})} \tag{4.62}$$

is $\omega = \omega_{ce}$. The left-hand polarized mode

$$n_L^2 = L = 1 - \sum_i \frac{\omega_{pi}^2}{\omega(\omega - \omega_{ci})} - \frac{\omega_{pe}^2}{\omega(\omega + \omega_{ce})} \tag{4.63}$$

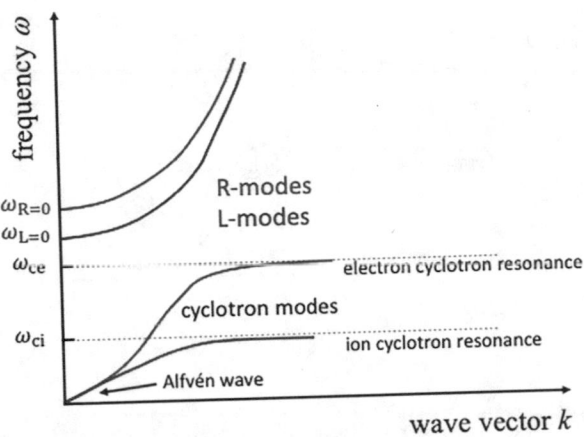

Fig. 4.4 Parallel propagation in high plasma density approximation, which is a good approximation in the radiation belts. The red lines indicate the R-modes and the blue lines the L-modes. Cut-offs are found where the dispersion curve meets the frequency axis ($k = 0$) and resonances at the limit $k \to \infty$

has resonances $\omega = \omega_{ci}$ for each ion species of different masses.

The low-frequency branches of the left- and right-hand modes propagating below their respective cyclotron frequencies are of particular importance in radiation belt physics. At the low frequency limit ($\omega \to 0$) $n^2 \to c^2/v_A^2$ and L- and R-modes merge to parallel propagating MHD waves at the Alfvén speed $v_A = \omega/k$ (Sect. 4.4).

With increasing k the phase velocities of the L- and R-modes become different. As a linearly polarized wave can be expressed as a sum of left- and right-hand polarized components, this leads to the *Faraday rotation* of the polarization of linearly polarized waves.

Electromagnetic Ion Cyclotron Wave

The parallel propagating left-hand polarized waves below ω_{ci} of each ion species are the *electromagnetic ion cyclotron* (EMIC) *waves*. In the inner magnetosphere the most important ion species are protons and singly charged helium and oxygen ions, the last two being of ionospheric origin. Figure 4.5 is an example of simultaneous observation of hydrogen and helium ion cyclotron waves in the dayside magnetosphere.

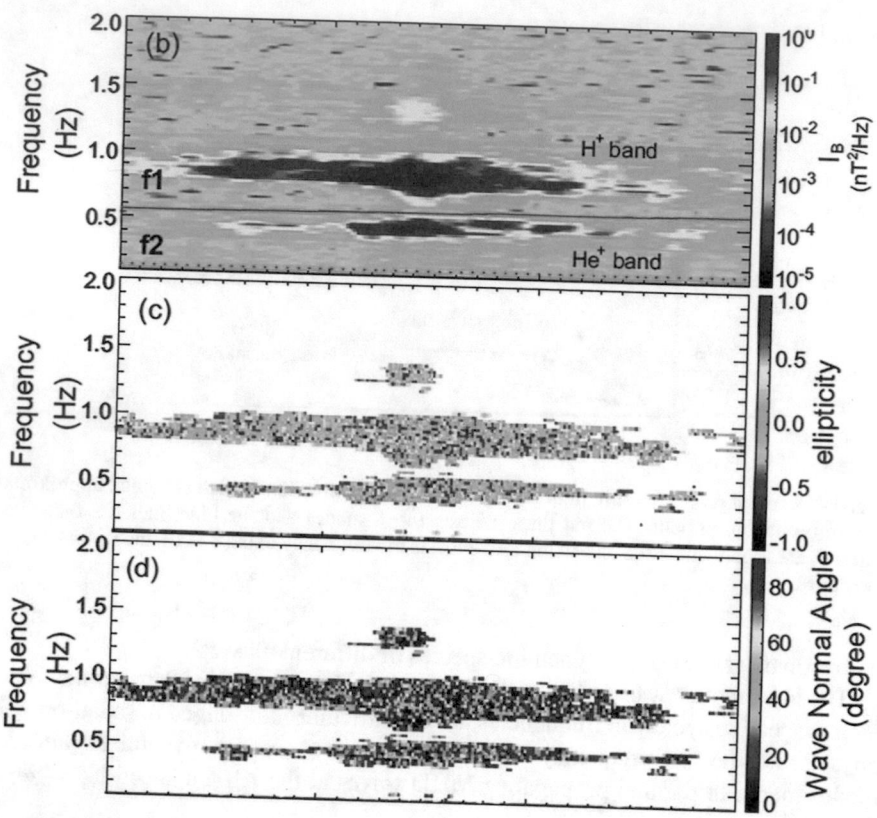

Fig. 4.5 Multi-band EMIC wave observation by *Van Allen Probe* A over a period of 30 min on 14 April 2014, in the noon sector (MLT ≈ 11 and L ≈ 5.7). The uppermost panel shows the magnetic power spectrum in H$^+$ and He$^+$ bands. In the panel the He$^+$ gyro frequency is indicated by the red line. The middle and lower panels indicate that the waves are circularly polarized (ellipticity close to 0) and propagating along the magnetic field (small WNA) (From Fu et al. 2018, reprinted by permission from COSPAR)

Whistler Mode

The *R*-mode propagating at frequencies between ω_{ci} and ω_{ce} is known as the *whistler mode*. If $\omega_{ci} \ll \omega \ll \omega_{ce}$ the dispersion equation can be approximated by

$$k = \frac{\omega_{pe}}{c} \sqrt{\frac{\omega}{\omega_{ce}}} \qquad (4.64)$$

giving the phase and group velocities

$$v_p = \frac{\omega}{k} = \frac{c\sqrt{\omega_{ce}}}{\omega_{pe}}\sqrt{\omega} \tag{4.65}$$

$$v_g = \frac{\partial\omega}{\partial k} = \frac{2c\sqrt{\omega_{ce}}}{\omega_{pe}}\sqrt{\omega}. \tag{4.66}$$

The dispersive whistler mode was identified for the first time during the First World War as descending whistling tones induced into telecommunication cables in the frequency band around 10 kHz. The origin of the signals was not understood until Storey (1953) suggested that the waves originated from wide-band electromagnetic emissions of lightning strokes. A fraction of the wave energy is ducted along the magnetic field as a *whistler wave* to the other hemisphere. The time of arrival depends on the frequency as

$$t(\omega) = \int \frac{ds}{v_g} = \int \frac{\omega_{pe}(s)}{2c\sqrt{\omega\,\omega_{ce}(s)}}\,ds \propto \frac{1}{\sqrt{\omega}} \tag{4.67}$$

implying that the higher frequencies arrive before the lower tones, resulting in the whistling sound when replayed as an audio signal. This explanation was not accepted immediately because it requires a higher plasma density in the inner magnetosphere than was known at the time. Storey actually found the plasmasphere, which has thereafter been thoroughly studied using radio wave propagation experiments and *in situ* satellite observations.

There are all the time thunderstorms somewhere in the atmosphere and thus the *lightning-generated whistlers* are continuously observed in the recordings of ground-based VLF receivers. To avoid confusion, sometimes also misunderstanding, it is advisable to dedicate the term "whistler" to the descending-tone lightning-generated signals and call all right-hand polarized waves in this frequency range generally "whistler-mode waves". For example, the man-made VLF signals from naval communication transmitters, which are known to affect the radiation belts, do not whistle because they are narrow-band signals from the beginning. Also the *whistler-mode chorus* (Sect. 5.2) and *plasmaspheric hiss* (Sect. 5.3) waves, which are most important in radiation belt physics, are different from lightning-generated whistlers. For example, the chorus is composed of *rising* tones, reflecting the local nonlinear physics in the inner magnetosphere rather than long-distance propagation.

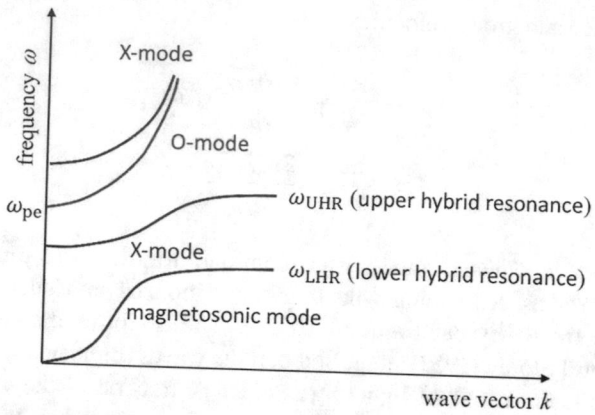

Fig. 4.6 Sketch of perpendicular propagating waves in cold plasma approximation. The red curve is the O-mode, which is the same as the electromagnetic wave in cold non-magnetized plasma. The X-mode has three different branches of which the lowermost is the most important in radiation belt context

4.3.3 Perpendicular Propagation ($\theta = \pi/2$)

The perpendicular propagating ordinary and extraordinary electromagnetic waves were already introduced in Vlasov theory (Sect. 4.2.4).[2] Figure 4.6 shows their dispersion curves.

The *ordinary (O)* mode is the mode whose index of refraction is

$$n_O^2 = P = 1 - \frac{\omega_{pi}^2}{\omega^2} - \frac{\omega_{pe}^2}{\omega^2} \approx 1 - \frac{\omega_{pe}^2}{\omega^2}. \tag{4.68}$$

This is the same as the refractive index of an electromagnetic wave in isotropic plasma (4.47). Its electric field is linearly polarized in the direction of the background magnetic field ($\mathbf{E} \parallel \mathbf{B}_0$). For exactly perpendicular propagation the dispersion equation does not contain the magnetic field. The mode has a cut-off at $\omega = \omega_{pe}$ (Fig. 4.6).

The *extraordinary (X)* mode is the solution of $n_X^2 = RL/S$. With the trivial approximation $\omega_{ce} \gg \omega_{ci}$ two *hybrid resonances* are found (Fig. 4.6). The upper hybrid resonance is

$$\omega_{UHR}^2 \approx \omega_{pe}^2 + \omega_{ce}^2 \tag{4.69}$$

---

[2] The modes are called ordinary and extraordinary for historical reasons although there is nothing particularly ordinary or extraordinary with them.

and the lower, written here for one ion species,

$$\omega_{LHR}^2 \approx \frac{\omega_{ci}^2 + \omega_{pi}^2}{1 + (\omega_{pe}^2/\omega_{ce}^2)} \approx \omega_{ce}\omega_{ci}\left(\frac{\omega_{pe}^2 + \omega_{ce}\omega_{ci}}{\omega_{pe}^2 + \omega_{pi}^2}\right). \tag{4.70}$$

The upper hybrid resonance can be used to determine the plasma density from wave observations if there is an independent way to determine the local magnetic field. The waves propagating close to the lower hybrid resonance frequency are important because they can resonate with both electrons and ions. In low-density plasma ($\omega_{pe} \ll \omega_{ce}$) $\omega_{LHR} \to \omega_{ci}$. When $\omega_{pe} > \omega_{ce}$, as is the case close to the equator within radiation belts, $\omega_{LHR} \approx \sqrt{\omega_{ce}\omega_{ci}}$ is a good approximation.

At the limit of low frequency

$$n_X^2 \to 1 + \frac{\omega_{pi}^2}{\omega_{ci}^2} = 1 + \frac{c^2}{v_A^2}. \tag{4.71}$$

This is the cold plasma representation of the MHD *magnetosonic mode*. In MHD (Sect. 4.4) its dispersion equation is found to be

$$\frac{\omega^2}{k^2} = v_s^2 + v_A^2, \tag{4.72}$$

where v_s is the speed of sound. In cold plasma v_s is neglected ($\to 0$), whereas in MHD the displacement current is neglected corresponding to the limit $c \to \infty$. In tenuous space plasmas v_A can, however, be a considerable fraction of c. Combining finite v_s and the cold plasma solution the dispersion equation is

$$\frac{\omega^2}{k^2} = \frac{v_s^2 + v_A^2}{1 + v_A^2/c^2}. \tag{4.73}$$

For increasing $k_\perp$ the magnetosonic/X-mode branch approaches the lower hybrid resonance.

4.3.4 Propagation at Arbitrary Wave Normal Angles

The propagation of plasma waves at wave normal angles between $0°$ and $90°$ depends on the local plasma parameters. In Chaps. 5 and 6 we present several examples of obliquely propagating whistler- and X-mode waves in observations and numerical analyses. As noted at the end of Sect. 4.2.4 the solutions of the dispersion equation propagating at arbitrary WNAs can be represent as *dispersion surfaces* in three-dimensional $(\omega, k_\parallel, k_\perp)$-space.

Figure 4.3 was an example of the surface containing the parallel and obliquely propagating right-hand polarized whistler mode and the perpendicular propagating linearly polarized magnetosonic/X-mode below the lower hybrid resonance frequency. While both the whistler mode and the X-mode are observable at frequencies below the lower hybrid resonance frequency, they can be distinguished if enough components of the wave fields are measured to determine the wave polarization.

4.4 Magnetohydrodynamic Waves

The ULF Pc4 and Pc5 waves well below the ion gyro frequency in the magnetosphere belong to the family of *magnetohydrodynamic* or *Alfvén waves*. Their wavelengths are comparable to the Earth's radius and thus the dipole geometry constrains the modes that can propagate in the inner magnetosphere. As will be discussed in Chap. 6, these waves play a major role in the diffusive transport of charged particles in the inner magnetosphere.

4.4.1 Dispersion Equation for Alfvén Waves

We start the discussion by introducing the linearized dispersion equation for Alfvén waves in a homogeneous ambient magnetic field. Consider a compressible, non-viscous, perfectly conductive fluid in a magnetic field described by the MHD equations

$$\frac{\partial \rho_m}{\partial t} + \nabla \cdot (\rho_m \mathbf{V}) = 0 \tag{4.74}$$

$$\rho_m \frac{\partial \mathbf{V}}{\partial t} + \rho_m (\mathbf{V} \cdot \nabla)\mathbf{V} = -\nabla P + \mathbf{J} \times \mathbf{B} \tag{4.75}$$

$$\nabla P = v_s^2 \nabla \rho_m \tag{4.76}$$

$$\nabla \times \mathbf{B} = \mu_0 \mathbf{J} \tag{4.77}$$

$$\nabla \times \mathbf{E} = -\frac{\partial \mathbf{B}}{\partial t} \tag{4.78}$$

$$\mathbf{E} + \mathbf{V} \times \mathbf{B} = 0 . \tag{4.79}$$

In Eq. (4.76) we have taken the gradient of the equation of state and introduced the speed of sound

$$v_s = \sqrt{\gamma_p P / \rho_m} = \sqrt{\gamma_p k_B / m} , \tag{4.80}$$

where γ_p is the polytropic index and k_B the Boltzmann constant.

From this set of equations we can eliminate $\mathbf{J}$, $\mathbf{E}$, and P

$$\frac{\partial \rho_m}{\partial t} + \nabla \cdot (\rho_m \mathbf{V}) = 0 \qquad (4.81)$$

$$\rho_m \frac{\partial \mathbf{V}}{\partial t} + \rho_m (\mathbf{V} \cdot \nabla)\mathbf{V} = -v_s^2 \nabla \rho_m + (\nabla \times \mathbf{B}) \times \mathbf{B}/\mu_0 \qquad (4.82)$$

$$\nabla \times (\mathbf{V} \times \mathbf{B}) = \frac{\partial \mathbf{B}}{\partial t} \, . \qquad (4.83)$$

Assume that in equilibrium the density ρ_{m0} is constant and look for the solution in the rest frame of the plasma where $\mathbf{V} = 0$. Furthermore, let the background magnetic field $\mathbf{B}_0$ be uniform. By considering small perturbations to the variables

$$\mathbf{B}(\mathbf{r}, t) = \mathbf{B}_0 + \mathbf{B}_1(\mathbf{r}, t) \qquad (4.84)$$

$$\rho_m(\mathbf{r}, t) = \rho_{m0} + \rho_{m1}(\mathbf{r}, t) \qquad (4.85)$$

$$\mathbf{V}(\mathbf{r}, t) = \mathbf{V}_1(\mathbf{r}, t) \qquad (4.86)$$

we can linearize the equations by picking up the first-order terms

$$\frac{\partial \rho_{m1}}{\partial t} + \rho_{m0}(\nabla \cdot \mathbf{V}_1) = 0 \qquad (4.87)$$

$$\rho_{m0} \frac{\partial \mathbf{V}_1}{\partial t} + v_s^2 \nabla \rho_{m1} + \mathbf{B}_0 \times (\nabla \times \mathbf{B}_1)/\mu_0 = 0 \qquad (4.88)$$

$$\frac{\partial \mathbf{B}_1}{\partial t} - \nabla \times (\mathbf{V}_1 \times \mathbf{B}_0) = 0 \, . \qquad (4.89)$$

From these we find an equation for the velocity perturbation $\mathbf{V}_1$

$$\frac{\partial^2 \mathbf{V}_1}{\partial t^2} - v_s^2 \nabla(\nabla \cdot \mathbf{V}_1) + \mathbf{v}_A \times \{\nabla \times [\nabla \times (\mathbf{V}_1 \times \mathbf{v}_A)]\} = 0 \, , \qquad (4.90)$$

where we have introduced the *Alfvén velocity* as a vector

$$\mathbf{v}_A = \frac{\mathbf{B}_0}{\sqrt{\mu_0 \rho_{m0}}} \, . \qquad (4.91)$$

By looking for plane wave solutions $\mathbf{V}_1(\mathbf{r}, t) = \mathbf{V}_1 \exp[i(\mathbf{k} \cdot \mathbf{r} - \omega t)]$ we get an algebraic equation

$$-\omega^2 \mathbf{V}_1 + v_s^2(\mathbf{k} \cdot \mathbf{V}_1)\mathbf{k} - \mathbf{v}_A \times \{\mathbf{k} \times [\mathbf{k} \times (\mathbf{V}_1 \times \mathbf{v}_A)]\} = 0 \, . \qquad (4.92)$$

After straightforward vector manipulation this leads to the dispersion equation for ideal MHD waves

$$-\omega^2 \mathbf{V}_1 + (v_s^2 + v_A^2)(\mathbf{k} \cdot \mathbf{V}_1)\mathbf{k} +$$
$$+(\mathbf{k} \cdot \mathbf{v}_A)[((\mathbf{k} \cdot \mathbf{v}_A)\mathbf{V}_1 - (\mathbf{v}_A \cdot \mathbf{V}_1)\mathbf{k} - (\mathbf{k} \cdot \mathbf{V}_1)\mathbf{v}_A)] = 0 . \quad (4.93)$$

Parallel Propagation

For $\mathbf{k} \parallel \mathbf{B}_0$, the dispersion equation reduces to

$$(k^2 v_A^2 - \omega^2)\mathbf{V}_1 + \left(\frac{v_s^2}{v_A^2} - 1 \right) k^2 (\mathbf{V}_1 \cdot \mathbf{v}_A)\mathbf{v}_A = 0 . \quad (4.94)$$

This describes two different wave modes. $\mathbf{V}_1 \parallel \mathbf{B}_0 \parallel \mathbf{k}$ yields the *sound wave*

$$\frac{\omega}{k} = v_s . \quad (4.95)$$

The second solution is a linearly polarized transverse wave with $\mathbf{V}_1 \perp \mathbf{B}_0 \parallel \mathbf{k}$. Now $\mathbf{V}_1 \cdot \mathbf{v}_A = 0$ and we find the *Alfvén wave*

$$\frac{\omega}{k} = v_A . \quad (4.96)$$

The magnetic field of the Alfvén wave is

$$\mathbf{B}_1 = -\frac{\mathbf{V}_1}{\omega/k} B_0 . \quad (4.97)$$

The wave magnetic and electric fields are perpendicular to the background field. This mode does not perturb the density or pressure but causes shear stress on the magnetic field ($\nabla \cdot (\mathbf{BB})/\mu_0$). Consequently, it is also called the *shear Alfvén wave*.

Parallel propagating linearly polarized waves can be decomposed to left- and right-handed circularly polarized components. With increasing k the circularly polarized components of the Alfvén wave split to two branches found in cold plasma theory: the left-hand polarized electromagnetic ion cyclotron wave approaching the ion cyclotron frequency from below and the right-hand polarized whistler mode (Fig. 4.4). Physically, this splitting is due to the decoupling of the electron and ion motions through the Hall effect (3.41).

Perpendicular Propagation

Perpendicular propagation ($\mathbf{k} \perp \mathbf{B}_0$) implies $\mathbf{k} \cdot \mathbf{v}_A = 0$, and the dispersion equation (4.93) reduces to

$$\mathbf{V}_1 = (v_s^2 + v_A^2)(\mathbf{k} \cdot \mathbf{V}_1)\mathbf{k}/\omega^2 . \tag{4.98}$$

Clearly $\mathbf{k} \parallel \mathbf{V}_1$, and we have found the *magnetosonic wave* in the MHD approximation.

$$\frac{\omega}{k} = \sqrt{v_s^2 + v_A^2} . \tag{4.99}$$

For a plane wave the linearized convection equation (4.89) becomes

$$\omega \mathbf{B}_1 + \mathbf{k} \times (\mathbf{V}_1 \times \mathbf{B}_0) = 0 , \tag{4.100}$$

which yields the magnetic field of the wave

$$\mathbf{B}_1 = \frac{V_1}{\omega/k} \mathbf{B}_0 . \tag{4.101}$$

The wave magnetic field is in the direction of the background magnetic field $\mathbf{B}_0$. The wave electric field is obtained from the ideal MHD Ohm's law $\mathbf{E}_1 = -\mathbf{V}_1 \times \mathbf{B}_0$ and is perpendicular to $\mathbf{B}_0$ and we have obtained the same polarization as in the cold plasma description. In MHD the wave is known as the *compressional (or fast) Alfvén (or MHD) wave*.

Propagation at Oblique Angles

To find the dispersion equation at arbitrary wave normal angles insert θ into the dot products of the dispersion equation. Selecting the z-axis parallel to $\mathbf{B}_0$ and the x-axis so that $\mathbf{k}$ is in the xz-plane, the components of the dispersion equation are

$$V_{1x}(-\omega^2 + k^2 v_A^2 + k^2 v_s^2 \sin^2 \theta) + V_{1z}(k^2 v_s^2 \sin \theta \cos \theta) = 0 \tag{4.102}$$

$$V_{1y}(-\omega^2 + k^2 v_A^2 \cos^2 \theta) = 0 \tag{4.103}$$

$$V_{1x}(k^2 v_s^2 \sin \theta \cos \theta) + V_{1z}(-\omega^2 + k^2 v_s^2 \cos^2 \theta) = 0 . \tag{4.104}$$

The y-component yields a linearly polarized mode with the phase velocity

$$\frac{\omega}{k} = v_A \cos \theta . \tag{4.105}$$

This is the extension of the shear Alfvén wave to oblique directions. It does not propagate perpendicular to the magnetic field, as there $\omega/k \to 0$.

The non-trivial solutions of the remaining pair of equations are found by setting the determinant of the coefficients of V_{1x} and V_{1z} equal to zero

$$\left(\frac{\omega}{k}\right)^2 = \frac{1}{2}(v_s^2 + v_A^2) \pm \frac{1}{2}[(v_s^2 + v_A^2)^2 - 4v_s^2 v_A^2 \cos^2 \theta]^{1/2} . \tag{4.106}$$

These modes are *compressional*. The plus sign gives the generalization of the *fast* MHD mode. It can propagate to all directions with respect to the background magnetic field. The magnetic field and density compressions of the fast mode oscillate in the same phase. The solution with the minus sign is the *slow* MHD mode. Its density and magnetic perturbations oscillate in opposite phases. The slow mode is strongly damped through the Landau mechanism, the calculation of which requires a kinetic approach.

The discussion above assumes homogeneous magnetic field and isotropic plasma pressure. The simplest extension of MHD into anisotropic plasma is the *double adiabatic theory* (Chew et al. 1956) with separate equations of state for parallel and perpendicular pressures. This leads to the *firehose mode* in the direction parallel and the *mirror mode* perpendicular to the background magnetic field (see, e.g., Koskinen 2011). The mirror mode is of interest in the inner equatorial magnetosphere with anisotropic plasma pressure ($P_\perp > P_\parallel$). Its density and magnetic field oscillate in opposite phases[3] similar to the slow-mode wave. Note that the mirror mode propagates perpendicular to the background field, whereas the phase velocity of the slow mode goes to zero when $\theta \to 90°$. The lowest-frequency long-wavelength ULF oscillations observed in the inhomogeneous magnetosphere can be either slow-mode or mirror-mode waves (e.g., Southwood and Hughes 1983; Chen and Hasegawa 1991).

The compressional MHD waves can steepen to shocks. Fast-mode shocks are ubiquitous in the solar wind. They form when an obstacle moves faster than the local magnetosonic speed, e.g., in front of planetary magnetospheres or when an ICME is fast enough relative to the background flow. Also SIRs gradually develop fast forward and fast reverse shocks, although mostly beyond the Earth orbit. Compressional shocks hitting the magnetopause can launch ULF waves inside the magnetosphere (Sect. 5.4).

Slow-mode shocks are strongly damped and thus difficult to observe. In the magnetosphere they have been found in association with magnetic reconnection where they have an important role decoupling the ion motion from the electron plasma flow and accelerating inflowing ions to the outflow velocities.

---

[3] This motivates the name of the mirror mode: The magnetic oscillation forms local magnetic bottles with increased density in the center of the bottle.

4.4.2 MHD Pc4–Pc5 ULF Waves

The wavelengths of magnetospheric MHD waves with periods in the Pc4–Pc5 range (45–600 s, or 1.7–22 mHz) are very long. For example, assuming equatorial Alfvén speed of $300 \, \text{km s}^{-1}$, a Pc5 wave with $f = 2 \, \text{mHz}$ has the wavelength of about $10 \, R_E$, which is comparable to the size of the inner magnetosphere. In fact, the frequency of about 1 mHz is in practice the lowest for which the oscillation can still be described as a wave in the inner magnetosphere. At such long wavelengths the assumption of a homogeneous background magnetic field $\mathbf{B}_0$ assumed in Sect. 4.4.1 is no more valid, nor can the fluctuations be considered as plane waves. The solutions to the full set of coupled nonlinear hydromagnetic equations must be sought using numerical methods. The boundary conditions are usually given at the magnetopause and in the ionosphere.

The ULF waves in the quasi-dipolar inner magnetosphere retain the mode structure of the MHD waves in a homogeneous magnetic field: the shear Alfvén wave with the wave vector along the background magnetic field and the fast compressional mode wave that can propagate to all directions. Because the Alfvén speed in the inner magnetosphere is much larger than the sound speed, the phase speed of the perpendicular propagating fast mode (4.106) can be approximated by the Alfvén speed $v_A = B/\sqrt{\mu_0 \rho_m}$.

However, the polarization of the ULF waves becomes more complicated and depends on the background field geometry. In the nearly dipolar inner magnetospheric field the electric and magnetic components of the ULF waves are useful to give in *local magnetic field-aligned coordinates*. In the literature several different notations are used. A well-motivated convention is to use the right-handed set of unit vectors $\{\mathbf{e}_\nu, \mathbf{e}_\phi, \mathbf{e}_\mu\}$, where $\mathbf{e}_\mu$ is along the background magnetic field line, $\mathbf{e}_\phi$ is in the azimuthal direction (eastward) and $\mathbf{e}_\nu = \mathbf{e}_\phi \times \mathbf{e}_\mu$ points radially outward in the equatorial plane. The electric and magnetic field components in these directions are often indicated using other subscripts, e.g., $\{r, a, p\}$ for radial, azimuthal and parallel.

The wave electric field $\mathbf{E}_1$ of the MHD waves is always perpendicular to the background magnetic field and can thus have only two components δE_ν and δE_ϕ. The wave magnetic field $\mathbf{B}_1$ is perpendicular to $\mathbf{E}_1$ and can point to all directions. The different polarizations are characterized according to the appearance of the magnetic field fluctuations. The the wave with the magnetic field in the azimuthal direction, i.e., $B_1 = \delta B_\phi$ is called the *toroidal mode* corresponding to the shear Alfvén wave. The associated wave electric field must be in the radial direction $E_1 = \delta E_\nu$. The *poloidal mode*, in turn, refers to the fast mode, which can propagate at all wave normal angles. The perpendicular propagating (compressional) mode has the wave magnetic field $B_1 = \delta B_\mu$ and the parallel propagating mode $B_1 = \delta B_\nu$. The associated wave electric field is in both cases azimuthal $E_1 = \delta E_\phi$. The observed ULF waves usually contain a mixture of the different polarizations.

To keep the discussion simple, let us consider the different polarizations on the dipole equator where $\mathbf{e}_\nu$ is the radial unit vector $\mathbf{e}_r$ and we can use cylindrical coordinates. In the cylindrically symmetric geometry, the total electric field can be expanded in cylindrical harmonics as

$$\mathbf{E}(r, \phi, t) = \mathbf{E}_0(r, \phi) + \sum_{m=0}^{\infty} \delta E_{rm} \sin(m\phi \pm \omega t + \xi_{rm}) \, \mathbf{e}_r$$

$$+ \sum_{m=0}^{\infty} \delta E_{\phi m} \sin(m\phi \pm \omega t + \xi_{\phi m}) \, \mathbf{e}_\phi . \qquad (4.107)$$

Here $\mathbf{E}_0(r, \phi)$ is the time-independent convection electric field, m is the azimuthal mode number, δE_{rm} are the amplitudes of the toroidal modes and $\delta E_{\phi m}$ of the poloidal modes, and ξ_{rm} and $\xi_{\phi m}$ represent their respective phase lags.

Note that the wave field is often expanded in terms of exponential basis functions $\exp(\mathrm{i}(m\phi - \omega t))$. In such an expansion the azimuthal mode number m is an integer from $-\infty$ to ∞. Since ϕ increases eastward, negative m corresponds to a westward and positive m to an eastward propagating wave phase. In the expansion (4.107) the opposite propagation directions correspond to the two signs of $\pm\omega t$.

The terminology varies in the literature. Often only the division to toroidal and poloidal modes according to the electric field components is used and the poloidal mode includes both compressional and non-compressional polarizations. Sometimes the parallel propagating (non-compressional) poloidal mode is called the poloidal Alfvén mode to distinguish it from the compressional oscillation. Furthermore, since the observed oscillation typically has both toroidal and poloidal electric field components and all three magnetic field components simultaneously, the combination of toroidal electric field component δE_ν and compressional magnetic field component δB_μ is sometimes considered as another compressional mode. It is, however, a redundant combination of toroidal and compressional polarizations.

Figure 4.7 shows two examples of THEMIS-A satellite observations of Pc5 waves exhibiting all polarizations simultaneously. In both cases the fluctuations took place following solar wind pressure enhancements. During Event A the spacecraft was in the post-midnight sector (about 03 MLT) at $L \approx 10$, during Event B on the duskside flank (about 18 MLT) at $L \approx 9$. Both events feature clear ULF range fluctuations with all three polarizations superposed. The amplitude of the toroidal mode was the largest during both events. This is consistent with the statistical result of Hudson et al. (2004a), that the Pc5 waves in the dawn and dusk sectors are preferentially toroidal.

ULF waves with small azimuthal mode number m have a predominantly toroidal polarization, whereas waves with large m are primarily poloidal. For limiting cases $m = 0$ and $|m| \to \infty$, the ideal-MHD solution in a dipole field configuration yields purely toroidal and purely poloidal modes, respectively. For finite m, different polarizations are coupled and toroidal and poloidal waves can have both small and large m.

The division of the ULF waves to toroidal, poloidal and compressional waves is further complicated by the fact that the magnetic field and density oscillations are

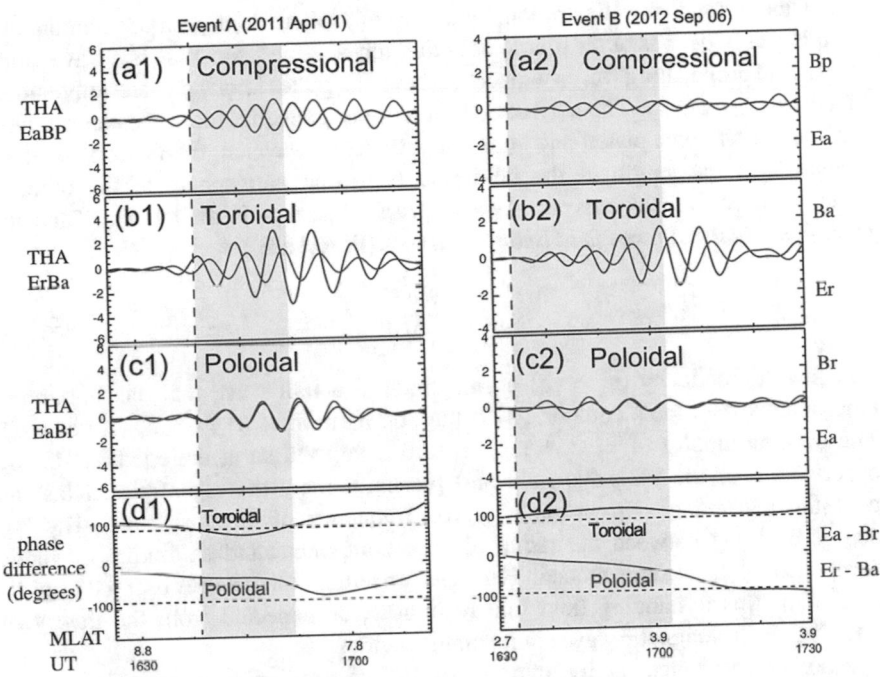

Fig. 4.7 Magnetic (blue) and electric field (red) components in magnetic field-aligned coordinates. The data have been band-pass filtered to match the observed ULF wave frequencies in the Pc5 range, 0.9–2.7 mHz (Event A) and 1.8–2.5 mHz (Event B). Here the components are: p is directed along the background magnetic field, r points (nearly) radially outward and a is directed azimuthally eastward (From Shen et al. 2015, reprinted by permission from American Geophysical Union)

often found to be in opposite phases (e.g., Zhang et al. 2019, and references therein) suggesting that they would be slow-mode waves or, in the case of anisotropic pressure, mirror-mode waves as noted in Sect. 4.4.1. Empirical determination between these is challenging because the direction and velocity of the wave propagation are difficult to observe.

The magnetospheric magnetic field lines are connected to the ionosphere where toroidal Alfvén waves are partially reflected and partially transmitted through the neutral atmosphere to the ground. This allows for remote observations of magnetospheric ULF oscillations using ground-based magnetometers with appropriate sampling rates. The ground-based measurements provide a wider latitudinal and longitudinal coverage of a given wave event than single-point space observations. For example, longitudinally separated magnetometers can be used to determine the azimuthal wave number (m), provided that their distance is smaller than half of the azimuthal wavelength of the oscillation. The waves are, however, attenuated when propagating from the ionosphere to the ground, which obscures their properties.

Let us assume, for illustration, that the ionosphere is a perfectly conductive boundary at both ends of the dipole field flux tubes. Such a flux tube is a wave guide for parallel propagating waves with conductive end plates known in electrodynamics as a *resonance cavity*. The perfect conductivity implies that the wave electric field vanishes at the end plates and thus only selected wavelengths fulfil Maxwell's equations. If the length of the field line from one hemisphere to the other is l, the allowed wavelengths are $\lambda_\parallel = 2l/n$, where n is an integer. Thus the *eigenfrequencies* of these *field line resonances* (FLRs) are

$$f = \frac{n v_A}{2l} . \tag{4.108}$$

The lowest frequency ($n = 1$) corresponds to a half-wave that has maximum amplitude at the dipole equator, as do the odd harmonics ($n = 3, 5, \ldots$) as well. The even harmonics ($n = 2, 4, \ldots$) in turn have minima at the equator. Having observations of the magnetic field and plasma density the eigenfrequencies can be estimated and related to the observed frequency of the oscillation. The 90-degree time lag between the radial electric component and azimuthal magnetic component of the toroidal mode in Fig. 4.7 is an indication that the observed toroidal oscillation was a standing field line resonance, as expected from the theory of standing electromagnetic waves in resonant cavities.

Another resonance cavity may form for perpendicular propagating waves between the dayside magnetopause and the near-equatorial ionosphere leading to standing *cavity mode oscillations* (CMOs), again with vanishing electric field at the boundaries. When compressional waves launched externally at the dayside magnetopause propagate inward, they are attenuated. The cavity modes peak at *resonant L–shells* where the frequency matches with the field line resonance of the toroidal (shear) wave, and the shear mode is amplified at the expense of the compressional mode (Kivelson and Southwood 1986).

4.5   Summary of Wave Modes

To keep track of the multitude of wave modes in the radiation belts, the most important wave modes for our treatise are briefly summarized in Table 4.2.

Table 4.2 Key wave modes in the inner magnetosphere relevant for dynamics of radiation belt particles, their frequencies, polarization and dominant wave normal angles

Wave mode	Frequency	Polarization	Wave normal angle
Whistler-mode chorus	0.5–10 kHz lower band: $0.1 f_{ce}$–$0.5 f_{ce}$ upper band: $0.5 f_{ce}$–$1.0 f_{ce}$	RH circular	$\sim 0°$ near equator, more oblique at higher latitudes
Plasmaspheric hiss	$\lesssim 100$ Hz[a]	RH circular	$\sim 0°$ near equator, more oblique at higher latitudes
Magnetosonic	From a few Hz to a few 100 Hz	Linear X	$\sim 90°$ confined near equator
EMIC	0.1–5 Hz H^+ band: $< f_{cH^+}$ He^+ band: $< f_{cHe^+}$	LH circular	H^+ band: $\sim 0°$ near equator, more oblique at higher latitudes; He^+ band: $\sim 30°$ near equator, more oblique with increasing latitude and L
Pc4 and Pc5 ULF waves	1.7–22 mHz	mixture of toroidal & poloidal	from field-aligned toroidal to perpendicular compressional oscillations

[a] The higher-frequency hiss (from 100 Hz to several kHz) has low intensity and is less relevant for radiation belt dynamics

Open Access This chapter is licensed under the terms of the Creative Commons Attribution 4.0 International License (http://creativecommons.org/licenses/by/4.0/), which permits use, sharing, adaptation, distribution and reproduction in any medium or format, as long as you give appropriate credit to the original author(s) and the source, provide a link to the Creative Commons license and indicate if changes were made.

The images or other third party material in this chapter are included in the chapter's Creative Commons license, unless indicated otherwise in a credit line to the material. If material is not included in the chapter's Creative Commons license and your intended use is not permitted by statutory regulation or exceeds the permitted use, you will need to obtain permission directly from the copyright holder.

Chapter 5
Drivers and Properties of Waves in the Inner Magnetosphere

How different wave modes are driven, is a central issue in space plasma physics. A practical problem is that often only indirect evidence of the driver can be identified in observations. The plasma environment is complex and variable and already a small difference in background or initial conditions may lead to widely different observable outcomes. In this chapter we discuss drivers of waves causing acceleration, transport and loss of radiation belt particles, whereas Chap. 6 discusses these effects in detail. We note that while this division is motivated in a textbook, it is somewhat artificial and the growth of the waves and their consequences often need to be studied together. For example, a whistler-mode wave can grow from thermal fluctuations due to gyro-resonant interactions until a marginally stable state is reached or nonlinear growth takes over. The growing wave starts to interact with different particle populations leading to damping or further growth of the wave. The fluxes of the higher-energy radiation belt particles are, however, small compared to the lower-energy background population, which supports the wave. Thus their effects on the overall wave activity usually remain small, although the waves can have drastic effect on higher-energy populations. Consequently, these two chapters should be studied together.

Our focus is on the waves that are most relevant to the evolution of radiation belts. A reader interested in a more comprehensive discussion of space plasma waves and instabilities is guided to general textbooks (e.g., Treumann and Baumjohann 1997; Koskinen 2011).

5.1 Growth and Damping of Waves

In Sect. 4.2.2 we found that a small-amplitude electrostatic Langmuir wave in a Maxwellian plasma is attenuated by heating the plasma population. We say that such a plasma is *stable* against *small* perturbations in the particle distribution. To

121

© The Author(s) 2022
H. E. J. Koskinen, E. K. J. Kilpua, *Physics of Earth's Radiation Belts*,
Astronomy and Astrophysics Library, https://doi.org/10.1007/978-3-030-82167-8_5

excite a plasma wave in the magnetosphere either an external source, e.g., a VLF transmitter, lightning stroke or an interplanetary shock hitting the magnetopause, or an internal *plasma instability* driven by an unstable particle distribution or magnetic field configuration is required.

To drive plasma unstable the system must contain *free energy* to be transformed to wave energy. The free energy may be stored in the magnetic field configuration such as magnetic tension of a thin current sheet, in an anisotropic plasma pressure, in the streaming of plasma particles with respect to each other, etc. Identification of the free energy source is essential because different sources of free energy can lead to widely different consequences.

The solution to the plasma dispersion equation, $\omega(\mathbf{k}) = \omega_r(\mathbf{k}) + i\omega_i(\mathbf{k})$, where ω_r is the real part and ω_i the imaginary part of the wave frequency, and $\mathbf{k}$ the wave vector, depends on the local plasma parameters. In radiation belts the plasma conditions vary both spatially and temporally, making the wave environment diverse and complex.

In a stable plasma, the perturbations will eventually be damped ($\omega_i < 0$). For a small damping rate ($|\omega_i| \ll \omega_r$) the perturbation is a normal mode of the plasma. Sometimes the damping is so strong that the oscillation is overdamped. The fluctuation is still there but the wave energy is quickly absorbed by plasma particles. A well-known example of this is the damped ion–acoustic mode in the ionosphere, which determines the spectral shape of the received signal of incoherent scatter radars.

If $\omega_i > 0$, the wave grows and we have an instability. Without doing actual calculations it is impossible to say how large the wave amplitude can grow. This is further complicated by the transport of wave energy from the position where the growth rate is calculated and it is necessary to apply *ray-tracing* to follow the spatial evolution of the wave mode (an example of a widely used ray-tracing procedure is described in detail by Horne 1989). The wave growth in the inner magnetosphere is often limited quasi-linearly by acceleration of particles. The growth may, however, continue to a state where nonlinear effects begin to limit the growth rate. If nothing quenches the growth, the system develops toward a major configurational change, large-scale magnetic reconnection being an important example of such.

5.1.1 Macroscopic Instabilities

Macroscopic instabilities in the configuration space may be intuitively more comprehensible than velocity-space microinstabilities. For example, we can imagine the Earth's magnetic field lines as strings of a huge musical instrument. When an external perturbation hits the system, it tries to restore its original configuration launching a compressional magnetosonic mode, which in turn can excite field-aligned shear Alfvén waves at the eigenfrequencies of the field lines known as *field line resonances* (Sect. 4.4.2).

However, a quantitive description of macroscopic instabilities is far from simple. As stated by Krall and Trivelpiece (1973), "the fluid theory, though of great practical use, relies heavily on the cunning of its user". In collisionless space plasmas the truncation of the moment equations leading, e.g., to magnetohydrodynamics, involves several critical approximations (e.g., Koskinen 2011), which may not be valid under unstable plasma conditions. A well-known phenomenon is the magnetic reconnection, which in the magnetospheric context is an instability of a thin current sheet. In collisionless space plasmas reconnection is often characterized as a tearing of the current sheet in almost ideal MHD. However, the cutting and reconnecting of the macroscopic magnetic field lines is essentially a microscopic process allowing the "de-freezing" of the plasma particles and the magnetic field from each other.

Another example is the MHD version of the hydrodynamic *Kelvin–Helmholtz instability* (KHI) occurring on the magnetopause. The instability is driven by the velocity shear between the faster solar wind flow and the slower flow in the magnetospheric side of the boundary leading to surface oscillations similar to those caused by wind blowing over water. These oscillations may lead to a perturbation that can propagate as an MHD wave into the magnetosphere. The macroscopic process on the magnetospheric boundary requires some type of viscosity similar to the drag in the viscous interaction model of magnetospheric convection proposed by Axford and Hines (1961) already mentioned in Sect. 1.4.1. The viscosity under the plasma conditions at the magnetopause cannot be collisional. Instead it must be provided by wave–particle interactions, thus connecting the macroscopic instability to a microscopic process.

5.1.2 Velocity-Space Instabilities

The discussion of velocity-space instabilities is instructive to start within the linearized Vlasov theory of Sect. 4.2 in the electrostatic approximation. It is an easy exercise to show that in this framework any monotonously decreasing ($\partial f/\partial v < 0$) distribution is stable against small perturbations.

Figure 5.1 is a textbook example of a double-peaked distribution function known as a *gentle-bump distribution*. It consists of a Maxwellian background (density n_1, temperature T_1) and a Maxwellian beam (n_2, T_2) moving at velocity V_0 with respect to the background. We again consider electrons only and assume that the ions form a cold background. The electron distribution function is now

$$f_{e0} = \frac{n_1}{n_e}\left(\frac{m_e}{2\pi k_B T_1}\right)^{3/2}\exp\left(-\frac{m_e v^2}{2k_B T_1}\right) + \frac{n_2}{n_e}\delta(v_x)\delta(v_y)\left(\frac{m_e}{2\pi k_B T_2}\right)^{1/2} \times$$

$$\frac{1}{2}\left\{\exp\left(-\frac{m_e(v_z - V_0)^2}{2k_B T_2}\right) + \exp\left(\frac{m_e(v_z + V_0)^2}{2k_B T_2}\right)\right\}. \tag{5.1}$$

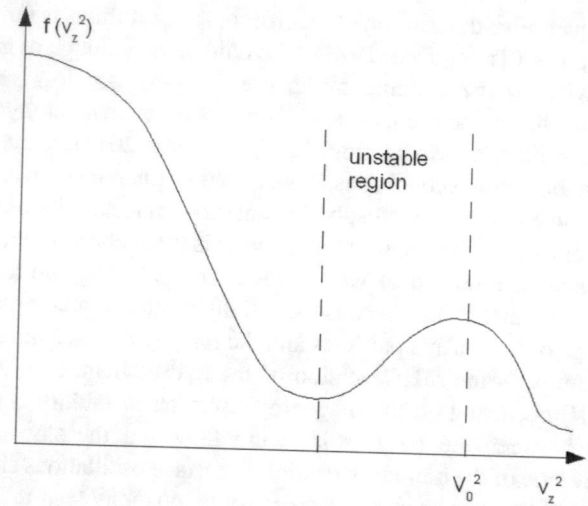

Fig. 5.1 Gentle-bump distribution function with a *potentially* unstable region where $\partial f/\partial v >$ 0. In order to avoid a current driven by the bump the electron distribution has been assumed to be symmetric about $v_z = 0$, thus the velocity axis is v_z^2. This way the example remains strictly electrostatic

We further assume that $n_e = n_1 + n_2 \gg n_2$, $T_2 \ll T_1$, $V_0^2 \gg 2k_B T_1/m_e$, i.e., the density and the temperature of the beam are much smaller than those of the background and the beam is faster than the thermal velocity of the background.

In the absence of the bump the solution is the damped Langmuir wave. Now the calculation of $K(\omega, k)$ is somewhat more tedious but still analytically doable applying the same strategy as in Sect. 4.2.2. Due to the "gentleness" of the bump ($n_1 \gg n_2$ and $T_1 \gg T_2$) the real part of the frequency can be approximated by the frequency of the Langmuir wave

$$\omega_r \approx \omega_{pe}(1 + 3k^2\lambda_{De}^2)^{1/2} \approx \omega_{pe}(1 + \frac{3}{2}k^2\lambda_{De}^2) \ . \tag{5.2}$$

The imaginary part is modified from the Landau solution by a term depending on the relative number densities and temperatures of the bump and the background

$$\omega_i \approx -\sqrt{\frac{\pi}{8}} \frac{\omega_{p1}}{|k^3\lambda_{D1}^3|} \exp\left(-\frac{1}{2k^2\lambda_{D1}^2} - \frac{3}{2}\right) + \tag{5.3}$$

$$+ \frac{n_2}{n_1}\left(\frac{T_1}{T_2}\right)^{3/2} \frac{k^3}{k_z^3}\left(\frac{k_z V_0}{\omega_r} - 1\right) \exp\left\{-\frac{T_1/T_2}{2k^2\lambda_{D1}^2}\left(1 - \frac{k_z V_0}{\omega_r}\right)^2\right\} \ .$$

The first term gives the Landau damping caused by the monotonously decreasing background distribution. The second term is stabilizing (damping) to the right from the bump ($v_z > V_0$), where the distribution is decreasing, but it *may* be unstable ($\omega_i > 0$) to plasma oscillations on the ascending slope to the left from the bump. The wave vector of the growing Langmuir wave is close to the direction of the motion of the beam (z-axis), which in magnetized plasmas is typically along the background magnetic field.

The instability requires that the contribution from the positive gradient due to the beam overcomes the damping by the background. This is known as the *gentle-bump* (or *gentle-beam*) *instability*. If the bump is too gentle, it is not powerful enough to overcome the damping by the background and drive an instability. The only way to find out whether the distribution is stable or unstable is to calculate the imaginary part of the frequency. Recall that even if the wave remains damped ($\omega_i < 0$), the normal mode is there. Its role is to transform kinetic energy of the beam to temperature of the background and beam populations. This gradually leads to filling of the trough between the background and the beam, forming a marginally stable non-Maxwellian distribution.

Two more wave modes, whose growth rate can be found within the electrostatic approximation, are worth of mentioning: the *ion–acoustic wave* (IAC), which is a short-wavelength solution of (4.12), and the *electrostatic ion cyclotron wave* (EIC). They can be driven unstable in a magnetized plasma by a magnetic field aligned current carried by a field-aligned electron beam, e.g., within and above the auroral region. As these electrostatic modes are of lesser importance in radiation belt physics, we refer to Chap. 7 of Koskinen (2011) for further discussion.

Similarly to the discussion of Landau damping in Sect. 4.2.3, the damping or growth can be illustrated with particles gliding down the slopes of the distribution function and either gaining energy from or losing it to the plasma wave. The instability is enhanced if the number of particles in the bump is increased, if the bump becomes narrower (colder) in the velocity-space, or if the speed of the bump (V_0) increases. In the latter two cases the process approaches the *two-stream instability* of cold plasma theory.

In magnetized space plasmas important unstable distribution functions are loss cone, ion ring, and butterfly distributions (Sect. 3.4), which have positive velocity gradients perpendicular to the magnetic field ($\partial f/\partial v_\perp > 0$). A gyrotropic velocity distribution $f(v_\parallel, v_\perp)$ can also be given as a function of kinetic energy and pitch angle $f(W, \alpha)$, and instabilities can be found if $\partial f/\partial W > 0$ or $\partial f/\partial \alpha > 0$. For example, at the edge of a loss cone the distribution function may have a strong positive gradient $\partial f/\partial \alpha$. Figure 5.2 is a sketch of a potentially unstable butterfly distribution as function of pitch angle. Note further that although an anisotropic bi-Maxwellian pancake distribution has $\partial f/\partial v_\perp < 0$, it can be unstable, because $\partial f/\partial \alpha > 0$. This is a key element of the discussion in Sect. 5.2.

Fig. 5.2 A butterfly distribution function as a function of the pitch angle α, where the shading indicates the potentially unstable ($\partial f/\partial \alpha > 0$) pitch-angle regimes. For a time series of observed flux of butterfly-distributed relativistic electrons, see Fig. 3.4

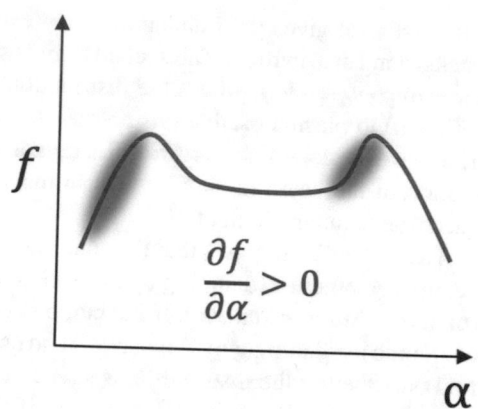

5.1.3 Resonant Wave–Particle Interactions

To illustrate the resonant interactions between radiation belt particles and electro-magnetic plasma waves we investigate the resonance condition (4.38) found from Vlasov theory written in relativistic form as

$$\omega - k_\| v_\| = \frac{n\omega_{c\alpha}}{\gamma}. \tag{5.4}$$

Here $k_\|$ can be written as $k\cos\theta$, where θ is the wave normal angle (WNA), i.e. the angle between the background magnetic field and the wave vector, and γ is the Lorentz factor

$$\gamma = (1 - v^2/c^2)^{-1/2} = (1 - v_\|^2/c^2 - v_\perp^2/c^2)^{-1/2}. \tag{5.5}$$

Note that in the relativistic resonance condition the parallel velocity appears both in the Doppler-shift term ($k_\| v_\|$) of the wave frequency and in the gyro-frequency term through γ.

In (5.4) n is the order of the Bessel functions in the dielectric tensor of the hot magnetized plasma (4.34) and runs from $-\infty$ to $+\infty$. Thus, both right- and left-hand polarized waves can resonate with both positively and negatively charged particles. $n = 0$ corresponds to the Landau resonance $k_\| v_\| = \omega$, whereas $n \neq 0$ give the gyro resonances. Note that for a circularly polarized wave to be in Landau resonance, the wave needs to have a finite WNA because only then the wave has an electric field component parallel to the magnetic field that can accelerate/decelerate the particle. The effect of Landau resonance thus becomes more important the more obliquely the wave propagates.

Equation (5.4) shows that for a gyro resonance to take place the Doppler-shifted frequency of the wave ($\omega - k_\| v_\|$) has to match with the particle's gyro frequency $\omega_{c\alpha}/\gamma$ or its higher harmonic. If a particle is mirroring equatorially ($v_\| = 0$), and/or the wave propagates purely perpendicular to the background magnetic field ($k_\parallel = 0$), the wave frequency needs to match exactly with the particle's gyro frequency.

The importance of the Doppler shift and the Lorentz factor in fulfilling the resonance condition is obvious when we recall that the frequencies of whistler-mode waves are below, and the frequencies of EMIC waves much below the local electron gyro frequency, which in the outer electron belt is in the range 5–10 kHz. Note that the Lorentz factor of ultra-relativistic electrons is of the order of 5, which alone is not sufficient to shift the frequency to fulfil the resonance condition. Thus both the Doppler shift and the Lorentz factor are essential, in particular, in the interaction of electrons with EMIC waves.

We often need to find the velocities at which radiation belt particles can be in resonance with a wave having a particular frequency and phase speed. Consider waves with a fixed ω and $k_\parallel$ that fulfil the dispersion equation of the wave mode in question. The *resonant velocity* in the non-relativistic case is

$$v_{\parallel,res} = (\omega - n\omega_{c\alpha})/k_\parallel . \tag{5.6}$$

Thus the resonance picks up only the velocity component in the parallel direction. The resonance condition does not constrain the perpendicular velocity ($v_\perp$) and the resonant particle can be anywhere on a straight line, called the a *resonant line*, in the ($v_\parallel, v_\perp$)-plane, provided that the resonance condition is consistent with the dispersion equation of the wave in question. Resonance is thus possible over a wide range of particle energies.

For relativistic particles (given here for $n = 1$) the gyro-resonant velocity can be solved from

$$v_{\parallel,res} = -\frac{\omega}{k_\parallel}\left(1 - \frac{\omega_{c\alpha}}{\omega}\frac{1}{c}\sqrt{c^2 - v_{\parallel,res}^2 - v_{\perp,res}^2}\right) \tag{5.7}$$

and the parallel and perpendicular velocities are coupled. Instead of a straight resonant line the relativistic resonance condition (5.7) defines a semi-ellipse in the ($v_\parallel, v_\perp$)-plane, called a *resonant ellipse*, which constrains the resonant energies. For a wave with a given frequency ω there is now a range of parallel resonant velocities instead of a single $v_{\parallel,res}$ in the nonrelativistic case. Because natural waves in radiation belts have finite frequency bandwidths, there is a finite volume of resonant ellipses in the velocity space. In case of Landau interaction the resonance condition is always $v_{\parallel,res} = \omega/k_\parallel$, and the resonance is independent of $v_\perp$ also in case of relativistic particles.

Let us then consider the effect of several small resonant interactions between the wave and particles, which can lead to damping or amplification of the wave. The combined effect of the interactions is called *diffusion* of the particle distribution in the velocity space.

Kennel and Engelmann (1966) introduced a simple graphical illustration of the diffusion process of non-relativistic particles. Let $\Delta W = \hbar\omega$ be a quantum of energy that a particle gains or loses during a brief interaction with the wave. The change of parallel momentum can be written as $m\triangle v_\parallel = \hbar k_\parallel$ yielding $\triangle W = m\triangle v_\parallel \omega/k_\parallel$. One the other hand, assuming that the increment of energy from a single interaction is small compared to the total energy of the particle, the change of energy can be expanded as $\triangle W = m(v_\parallel \triangle v_\parallel + v_\perp \triangle v_\perp)$. Equating these two expressions for energy change leads to

$$v_\parallel \triangle v_\parallel + v_\perp \triangle v_\perp = \triangle v_\parallel\, \omega/k_\parallel \,, \tag{5.8}$$

the integral of which is

$$v_\perp^2 + (v_\parallel - \omega/k_\parallel)^2 = \text{constant} . \tag{5.9}$$

Equation (5.9) defines circles in the $(v_\perp, v_\parallel)$-plane. The circles are centered at $(0, \omega/k_\parallel)$ and have an increasing radius with increasing $v_\perp$ for a given $v_{\parallel,res}$. For relativistic particles the corresponding equation again defines ellipses (for details, see Summers et al. 1998). These circles (or ellipses) are called *single-wave characteristics*. In Landau resonance the single-wave characteristic is a straight line in the parallel direction.

The resonance occurs when the single-wave characteristic crosses the resonant ellipse, or the straight resonant line in the nonrelativistic case. The characteristic defines the direction of the particle's motion at the time of interaction in the $(v_\perp, v_\parallel)$-plane. In Landau resonance the diffusion is in the $\pm v_\parallel$ direction familiar from the electrostatic Vlasov theory. In gyro resonances the particles move randomly in either direction along the single-wave characteristics and the diffusion is in the direction of the tangent of the single-wave characteristics. The resonant interactions can thus change both the pitch angle and energy of the particles. The net flux of particles is toward the direction of decreasing distribution function along the single-wave characteristics. If the flux is toward increasing energy, as is the case, e.g., with a Maxwellian distribution, the particles gain energy and the wave is damped. In the opposite case, such as for the unstable distributions discussed at the end of Sect. 5.1.2, the wave grows at the expense of particle energy.

Recall again that natural waves in radiation belts have finite frequency bandwidths. Thus there is a family of adjacent single-wave characteristics the particle can resonate with even though it moves away from the single-wave characteristic of the previous interaction. This formulation applies to small-amplitude (linear) waves because $\triangle W$ at each interaction was assumed to be small.[1]

Figure 5.3 illustrates the diffusion due to resonant interaction of electrons with whistler-mode waves in the near-equatorial region of the outer radiation belt. The color-coded distribution function represents an anisotropic population injected from the magnetotail. Due to pitch-angle anisotropy the particles on single-wave

---

[1] The diffusion along the single-wave characteristics is analogous to the method of characteristics used to solve the Vlasov equation in a magnetized plasma (Sect. 4.2.4). In both cases the changes (either $\triangle W$ or $\triangle f$) at each step are assumed to be small.

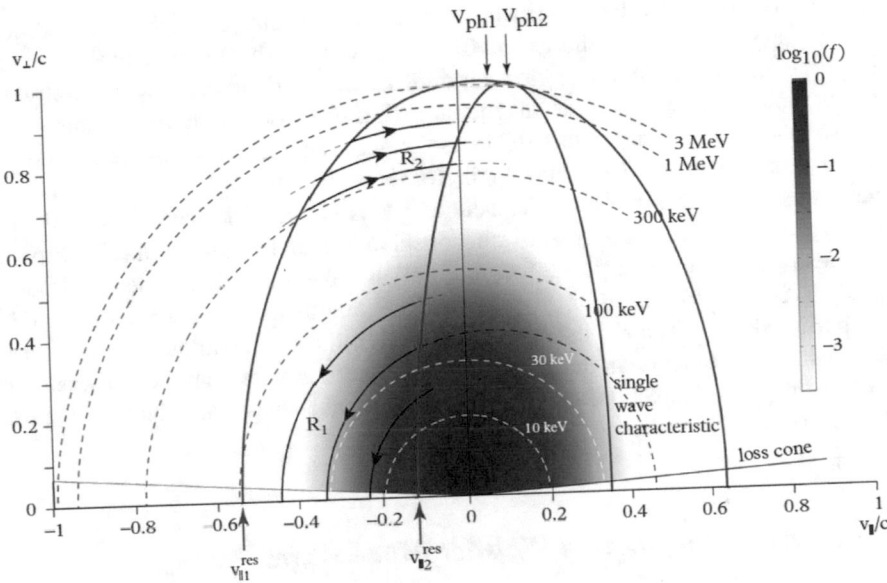

Fig. 5.3 Illustration of resonant ellipses and single-wave characteristics of whistler-mode waves. The dashed lines illustrate constant energy surfaces. The color shading represents a pancake distribution function of thermal/suprathermal electrons. The red lines are resonant ellipses corresponding to 0.1 f_{ce} (the narrower ellipse) and 0.5 f_{ce} (the wider ellipse). The displacement of the ellipses to the right is due to the Doppler shift $k_\parallel v_\parallel$. The black lines are selected single wave characteristics with arrows showing the direction of diffusion. (From Bortnik et al. 2016, reprinted by permission from Oxford University Press)

characteristics labeled R_1 crossing the resonant ellipse, corresponding to resonant velocity $v_{\parallel,2}^{res}$, diffuse toward smaller pitch angles and lose energy. Thus the whistler-mode wave is amplified as discussed in Sect. 5.2 below. At higher energies ($\gtrsim$ 300 keV), the distribution function is no more anisotropic and the particles on characteristics labeled R_2 diffuse toward larger energies, corresponding to whistler-mode acceleration of radiation belt electrons to be discussed in Sect. 6.4.5.

5.2 Drivers of Whistler-Mode and EMIC Waves

The central role of electromagnetic right-hand polarized whistler-mode waves and left-hand polarized EMIC waves driven by anisotropic velocity distribution functions in the inner magnetosphere was understood already during the 1960s (e.g., Kennel 1966; Kennel and Petschek 1966, and references therein). Those days observations were limited but understanding of plasma theory made it possible to reach results that have later been confirmed using much more extensive and detailed observations.

We discuss below how electron and proton anisotropies drive instabilities generating whistler–mode chorus and EMIC waves in the background inner magnetospheric plasma. Anisotropic electron and proton distributions arise naturally as the particles injected from the magnetotail during substorms move adiabatically toward the stronger magnetic field in the inner magnetosphere. Conservation of the magnetic moment (μ) leads to increase of $W_\perp$ through the drift betatron mechanism. At the same time the bounce paths between northern and southern hemispheres become shorter and the conservation of the longitudinal invariant (J) leads to increase of $W_\parallel$ through Fermi acceleration. In the nearly-dipolar field B is proportional to L^{-3} whereas the length of the bounce motion is proportional to L. Thus in the earthward motion the conservation of μ stretches the distribution function in the perpendicular direction more than the conservation of J in the parallel direction, and the result is a pancake-shaped distribution function ($T_\perp > T_\parallel$).

5.2.1   Anisotropy-Driven Whistler Mode Waves

We follow here the classic presentation of Kennel and Petschek (1966). We begin by considering the cold plasma whistler-mode solution, i.e., the parallel propagating R mode in the frequency range $\omega_{ci} \ll \omega < \omega_{ce}$. In this approximation we can neglect the ion contribution to Eq. (4.62). We are mainly interested in the outer radiation belt domain where electron plasma frequency is larger than the cyclotron frequency (e.g., at $L = 5$ the ratio ω_{pe}/ω_{ce} is about 4). The real part of the frequency can under these conditions be approximated as

$$\frac{c^2 k_\parallel^2}{\omega^2} \approx \frac{\omega_{pe}^2}{\omega(\omega_{ce} - \omega)}. \tag{5.10}$$

To make the discussion simple we consider the interaction with the fundamental ($n = 1$) harmonic of the gyro frequency in non-relativistic approximation ($\gamma = 1$). Inserting (5.10) into the equation of resonant velocity $v_{\parallel,res} = (\omega - \omega_{ce})/k_\parallel$ the *resonant energy* becomes

$$W_{e,res} = \frac{1}{2} m_e v_{\parallel,res}^2 = W_B \frac{\omega_{ce}}{\omega} \left(1 - \frac{\omega}{\omega_{ce}} \right)^3, \tag{5.11}$$

where $W_B = B^2/(2\mu_0 n_0)$ is the magnetic energy per particle, i.e., magnetic energy divided by particle number density n_0. Note that, as $\omega < \omega_{ce}$, the wave needs to propagate toward the electrons yielding $k_\parallel v_\parallel < 0$ in order the Doppler shift to increase the wave frequency to match with ω_{ce}. In addition, (5.11) implies that the resonant energies are largest at the lowest frequencies and decrease toward zero when $\omega \to \omega_{ce}$. Under conditions typical in the inner magnetosphere the resonant energies are in the range 1–100 keV.

Let $F_e(v_\parallel, v_\perp)$ be the equilibrium distribution function normalized to 1. With some effort the growth rate of the Vlasov theory solution at the resonant velocity can be written as

$$\omega_i = \pi \omega_{ce} \left(1 - \frac{\omega}{\omega_{ce}}\right)^2 \Delta_e(v_{\parallel,res}) \left(A_e(v_{\parallel,res}) - \frac{1}{(\omega_{ce}/\omega) - 1}\right) \tag{5.12}$$

where

$$\Delta_e(v_{\parallel,res}) = 2\pi \frac{\omega_{ce} - \omega}{k_\parallel} \int_0^\infty v_\perp \, dv_\perp \, F_e(v_\perp, v_\parallel) \Bigg|_{v_\parallel = v_{\parallel,res}} \tag{5.13}$$

and

$$A_e(v_{\parallel,res}) = \frac{\int_0^\infty v_\perp \, dv_\perp \left(v_\parallel \frac{\partial F_e}{\partial v_\perp} - v_\perp \frac{\partial F_e}{\partial v_\parallel}\right) \frac{v_\perp}{v_\parallel}}{2\int_0^\infty v_\perp \, dv_\perp \, F_e} \Bigg|_{v_\parallel = v_{\parallel,res}}$$

$$= \frac{\int_0^\infty v_\perp \, dv_\perp \, \tan\alpha \frac{\partial F_e}{\partial \alpha}}{2\int_0^\infty v_\perp \, dv_\perp \, F_e} \Bigg|_{v_\parallel = v_{\parallel,res}}. \tag{5.14}$$

The factor $\Delta_e(v_{\parallel,res})$ is a measure of the fraction of the total electron distribution close to the resonance. Since $\omega < \omega_{ce}$, Δ_e is always positive. Recall that while the resonance picks up only one velocity component $v_{\parallel,res}$ in the parallel direction, the electron can have any perpendicular velocity along the resonant line in the $(v_\perp, v_\parallel)$-plane. This motivates the integration over all perpendicular velocities above. If the wave has a wide frequency band, as is the case with naturally occurring whistler-mode waves, a large part of the electron distribution can interact with the wave.

A_e is, in turn, a measure of the anisotropy. It depends on the gradient of F_e with respect to the pitch angle at constant energy. For pancake, loss-cone, and butterfly distributions there are domains of α where $\partial F_e/\partial \alpha > 0$.

Whether ω_i is positive (growing) or negative (damping) depends on the sign of the last term in brackets in the RHS of (5.12). The wave grows when

$$A_e > \frac{1}{(\omega_{ce}/\omega) - 1} \tag{5.15}$$

and attenuates otherwise. This condition can also be expressed in terms of the resonant energy

$$W_{e,res} > \frac{W_B}{A_e(A_e + 1)^2}. \tag{5.16}$$

In case of a bi-Maxwellian distribution (3.56) A_e is independent of $v_{\parallel,res}$ and reduces to

$$A_e = \frac{T_\perp - T_\parallel}{T_\parallel}.$$ (5.17)

Assuming a pancake distribution with $T_\perp > T_\parallel$, which is typically observed in the radiation belt region outside of the plasmapause, the whistler mode grows if the anisotropy is large enough and the condition given by (5.15) or (5.16) is met. The minimum resonant energy can be determined from observations. The more anisotropic the population is, the lower is the minimum resonant energy.

The instability condition for the whistler mode depends on the anisotropy A_e only, but the actual growth or damping rate depends on both A_e and the fraction of resonant electrons $\triangle_e$. Furthermore, (5.15) indicates that the closer to the gyro frequency the wave frequency is, the stronger anisotropy is required to drive the wave due to the increasing resonant damping when $\omega \to \omega_{ce}$.

We have here considered only waves that propagate purely parallel to the background magnetic field. As shown, e.g., by Kennel (1966) the growth rate of the wave decreases with increasing wave normal angle. The generation of oblique whistler-mode waves requires thus gyro resonances occurring for a long enough time. At perpendicular propagation (WNA $\approx 90°$, i.e., the magnetosonic / X-mode wave) the resonant energy goes to zero. As discussed in Sect. 5.3.2, the magnetosonic mode can be driven unstable by proton ring distributions with free energy in the perpendicular direction ($\partial f/\partial v_\perp > 0$).

5.2.2 Whistler-Mode Chorus

The observed whistler-mode waves outside the plasmapause are known as *chorus waves*. They are different from the lightning-generated whistlers with decreasing frequency–time spectra (Sect. 4.3.2). The chorus waves are composed of short, mostly rising, right-hand polarized emissions in the kilohertz range. When played through an audio loudspeaker, the signal resembles a "dawn chorus" of a rookery. According to the appendix of Storey (1953), these dawn choruses had already been heard in the 1930s.

By the early twenty-first century the chorus waves have been demonstrated to be able to accelerate radiation belt electrons to relativistic energies and, on the other hand, to scatter a fraction of the population to the atmospheric loss cone. The acceleration and diffusion processes are dealt in Chap. 6. Here we discuss the main observational features of the waves by walking through Fig. 5.4 reproduced from Bortnik et al. (2016).

The sketch on the top left (Fig. 5.4a) indicates that the chorus emissions are preferentially observed close to the equator outside the plasmasphere from around midnight through the dawn sector to noon. This is consistent with the concept

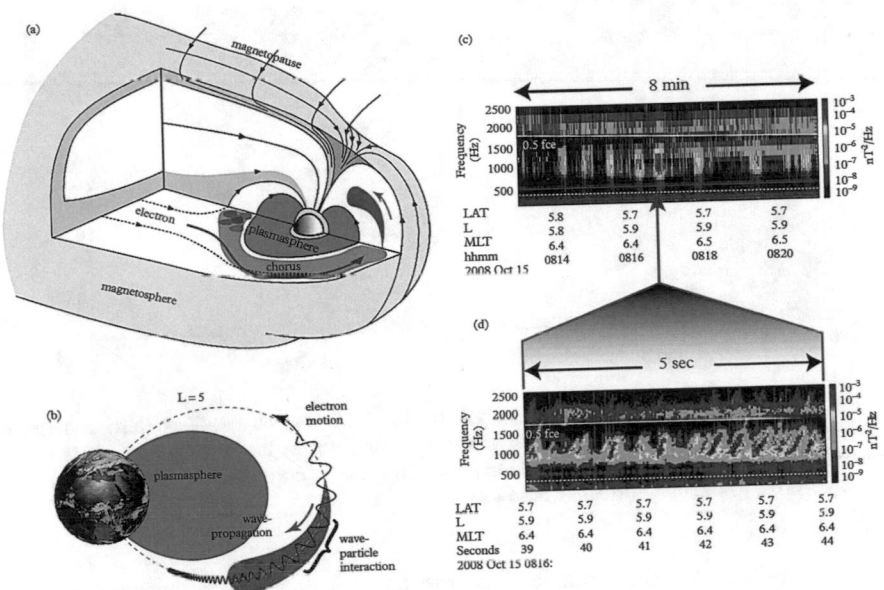

Fig. 5.4 Main observational features of chorus waves. (**a**) The waves appear predominantly close to the equator from midnight through dawn to noon. (**b**) The waves propagate away from the equator, but are attenuated before being reflected from the ionosphere. (**c**) The waves appear as brief bursts in two distinct bands with a cap around 0.5 f_{ce}. (**d**) The individual bursts are composed of short rising tones giving the emissions their chirping characteristics (From Bortnik et al. 2016, reprinted by permission from Oxford University Press)

that they are driven by anisotropic electron populations in the energy range 1–100 keV injected from the magnetotail, since these electrons drift eastward around the plasmapause.

The waves have been found to propagate away from the equator (Fig. 5.4b), but not back to the equator. This suggests that, unlike shear Alfvén waves, the chorus waves do not reflect back from the ionosphere but are attenuated by wave–particle interactions not too far from the equator.

The damping and growth of the waves depend on the wave normal angle. Figure 5.5 based on *Cluster* observations (left) and ray-tracing analysis (right) shows that near the equator the chorus waves propagate parallel or nearly parallel to the background magnetic field, while the obliquity increases with increasing geomagnetic latitude. The generation region close to the equatorial plane can be understood since most of the free energy is concentrated close to the equator, where $\sim 90°$ electrons are trapped and the anisotropy is strongest, which maximizes the gyro-resonant growth of parallel propagating waves. The main attenuation mechanism is likely Landau damping by suprathermal electrons around 1 keV as demonstrated in the ray-tracing study by Bortnik et al. (2007). The attenuation

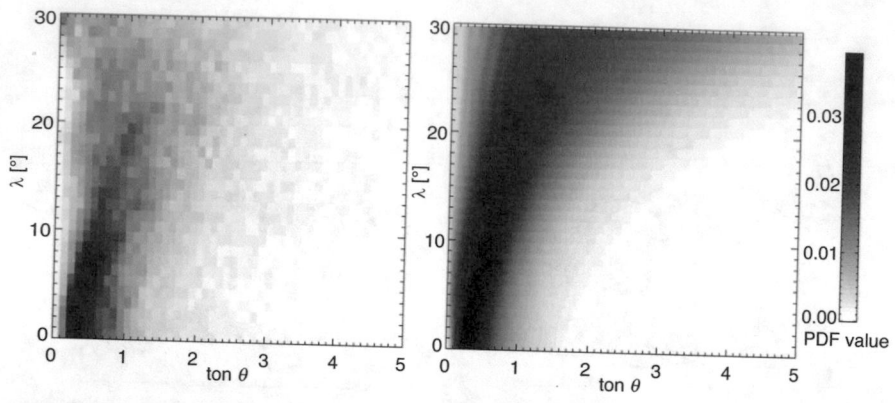

Fig. 5.5 Probability distribution functions (PDF) from *Cluster* observations (left) and three-dimensional ray-tracing (right) of chorus waves as a function of the wave normal angle θ and geomagnetic latitude λ. The observed WNAs are mostly $< 45°$ ($\tan\theta < 1$). (From Breuillard et al. 2012, Creative Commons Attribution 3.0 License)

increases with increasing latitude since the Landau damping is more effective at higher obliquity.

The spectrograms (c) and (d) of Fig. 5.4 illustrate the observed characteristics of the chorus emissions. There are two distinct frequency bands: the lower band in the range $0.1\,f_{ce} < f < 0.5\,f_{ce}$ and the less intense higher-frequency band $0.5\,f_{ce} < f < f_{ce}$. The upper spectrogram indicates that the emissions appear in short bursts of about 10–20 s, whereas the lower spectrogram shows how a single burst is composed of upward chirping signals shorter than a second, which give the chorus-like tone to the emission.

5.2.3 Two-Band Structure of the Chorus

The splitting of the chorus emission to two frequency bands (Fig. 5.4) has been a longstanding problem since the OGO 1 and OGO 3 satellite observations during the second half of the 1960s (Burtis and Helliwell 1969, 1976). Several explanations have been proposed ranging from different drivers for each band to nonlinear wave–particle or wave–wave coupling phenomena (see Li et al. 2019, and references therein). The growth of two-band whistler-mode wave with a gap in amplitude at $0.5\,f_{ce}$ has been demonstrated using particle-in-cell plasma simulations assuming the presence of two different anisotropic hot electron distributions (Ratcliffe and Watt 2017), but it is not clear how these two populations would form.

A possible and quite simple scenario to generate electron anisotropy in two distinct energy domains is based on the interplay of the first order gyro resonance $\omega - k_{\parallel}v_c = \omega_{ce}$ and the Landau resonance $\omega - k_{\parallel}v_L = 0$, where v_c and v_L represent the gyro- and Landau-resonant parallel velocities. At $\omega = 0.5\,\omega_{ce}$ both conditions

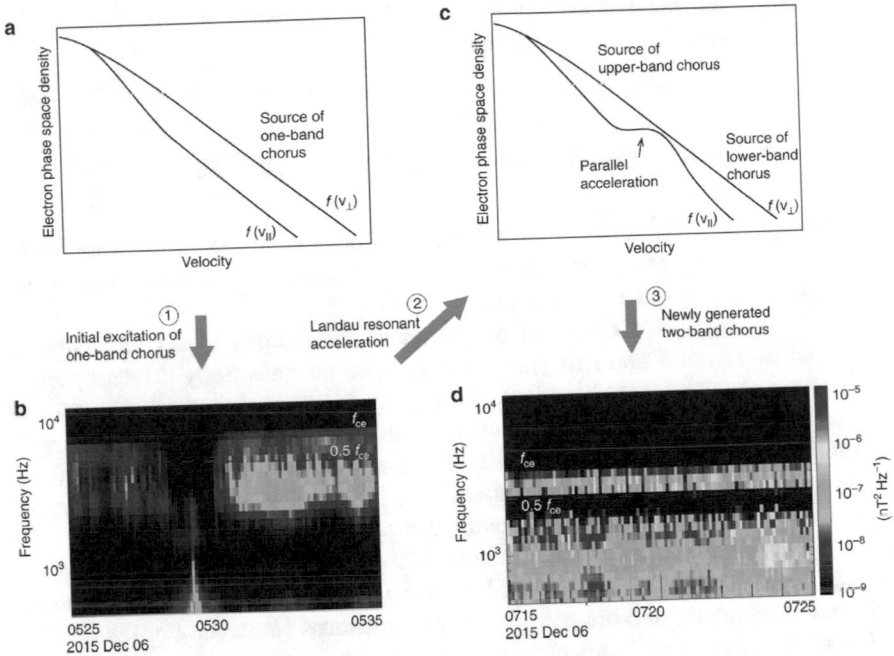

Fig. 5.6 A scenario of the excitation of lower and upper band chorus waves based on Van Allen Probes measurements and numerical simulations. Note that the spectrograms in (**b**) and (**d**) were taken at different periods separated by about 2 hours, when the spacecraft was moving toward a larger distance from the Earth. (From Li et al. 2019, Creative Commons Attribution 4.0 International License)

are fulfilled if v_c and v_L have the same magnitude but are in opposite directions. Thus electrons that should drive the whistler-mode growth actually Landau damp the waves at $\omega = 0.5\,\omega_{ce}$. The damping accelerates electrons in the direction of the background magnetic field reducing the electron anisotropy around the Landau-resonant energy.

Li et al. (2019) studied this mechanism using numerical simulations consistent with *Van Allen Probes* electron and high-resolution wave observations. During the investigated event the electron data indicated two anisotropic electron distributions: one in the range 0.05–2 keV and another > 10 keV. The initial source of free energy in the simulation was an unstable pancake electron distribution injected from the plasma sheet (Fig. 5.6a). First a single-band whistler mode starts growing (Fig. 5.6b) but the Landau damping at $\omega = 0.5\,\omega_{ce}$ quickly sets in and two different anisotropic distributions start forming (Fig. 5.6c). In the simulation the gyro resonance occurred in the energy range 0.22–24 keV, while the Landau-resonant energy was in the range 1.3–2.1 keV. After their formation the two anisotropic populations act separately: the lower energy electrons drive the upper band whistler mode and the upper energy population the lower band (Fig. 5.6d) consistent with the relation of resonant energy and frequency (5.11). Two separate processes are further supported by the common observation of independent evolution of the upper and lower bands, in particular the different appearance of the rising chirps.

5.2.4 Formation and Nonlinear Growth of the Chirps

The linear theory discussed in Sect. 5.2.1 does not explain the formation of the characteristic rising-tone chirps of the chorus emissions. Their short time scales and large amplitudes point to a nonlinear process.

Omura et al. (2013) reviewed theories and simulations based on nonlinear formation of *electron holes* in phase space. Once an anisotropy-driven coherent wave grows to a finite amplitude, the wave potential around the resonant velocity is able to trap a fraction of the resonant electrons and distort the trajectories of non-trapped resonant electrons. Consequently a hole or a hill forms in the $(\mathbf{r}, \mathbf{v})$ phase space (Fig. 5.7). The deformed electron trajectories correspond to resonant currents that modify the wave field with components: J_E in the direction of the wave electric field and J_B in the direction of the wave magnetic field. It turns out that J_E is responsible for the growth of the wave amplitude whereas J_B leads to the drift in the frequency and the rising tones of the chorus elements. For the detailed calculations we refer to Omura et al. (2013) and references therein.

The continuous filter bank data from the EFW instrument of *Van Allen Probes* has offered the opportunity to reconstruct the frequency and amplitude of large-amplitude whistler-mode waves. Tyler et al. (2019) performed the first statistical analysis based on 5 years of *Van Allen Probes* data. They looked for amplitudes $> 5\ \mathrm{mV\,m^{-1}}$, which are 1–2 orders of magnitude larger than average chorus wave amplitudes. This threshold avoided the risk of contaminating the data set with much smaller-amplitude plasmaspheric hiss emissions. Large-amplitude whistler waves exceeding this level were observed to occur 1–4% of the time from pre-midnight through dawn to noon, mostly between 0–7 MLT, typically above $L = 3.5$. This distribution of the observed wave-packets is consistent with the assumption that they grow from initially anisotropy-driven linear whistler-mode waves.

The nonlinear growth of the chorus elements can continue to very large amplitudes. For example, exceptionally strong electric fields of whistler-mode emissions, about $240\ \mathrm{mV\,m^{-1}}$, were observed by the S/WAVES instruments on the STEREO spacecraft when they passed through the radiation belts on their way to their final orbits (Cattell et al. 2008).

Detailed investigation of the large-amplitude wave packets requires high sampling rate in the time domain and transmission of the waveform to the ground. Thus the relevant observations in any particular region of the magnetosphere have been sparse. Figure 5.8 shows another example of large whistler wave packets in the inner magnetosphere, captured by the Time Domain Sampler onboard the *Wind* spacecraft. The waveform illustrates that the amplitude grows and damps within tens of milliseconds, which makes direct comparisons with local electron data difficult because the particle instruments usually do not have so good time resolution.

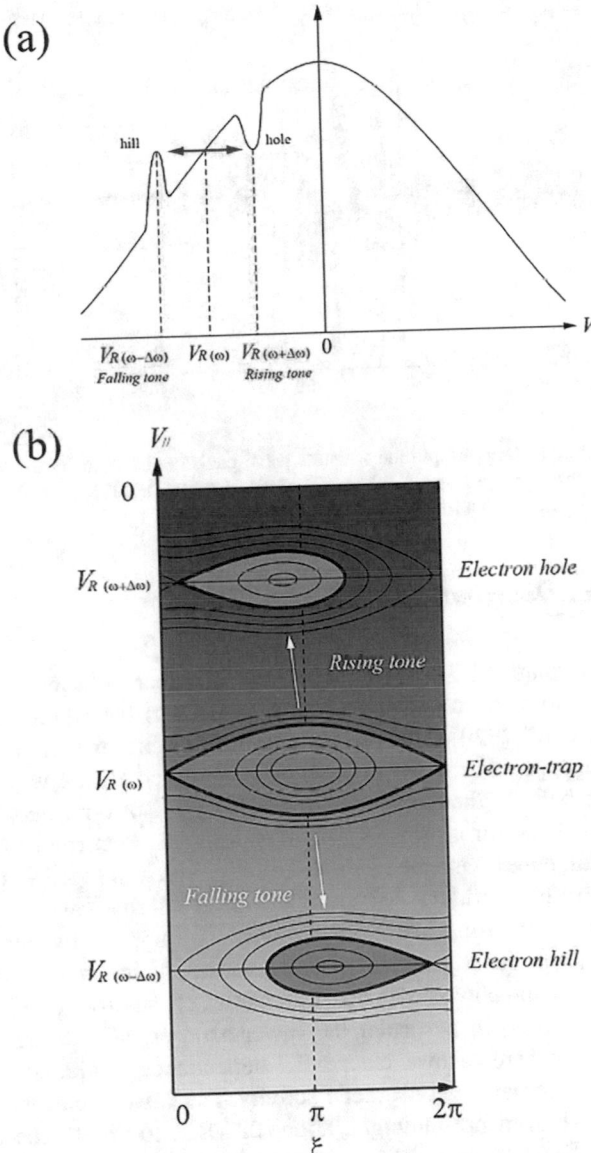

Fig. 5.7 Sketch of the formation of electron holes and hills (**a**) in one-dimensional distribution function $F(v_\parallel)$ and (**b**) in the phase space $(v_\parallel, \xi)$, where ξ is the angle between the perpendicular velocity of a resonant electron and the wave magnetic field (From Omura et al. 2015, reprinted by permission from American Geophysical Union)

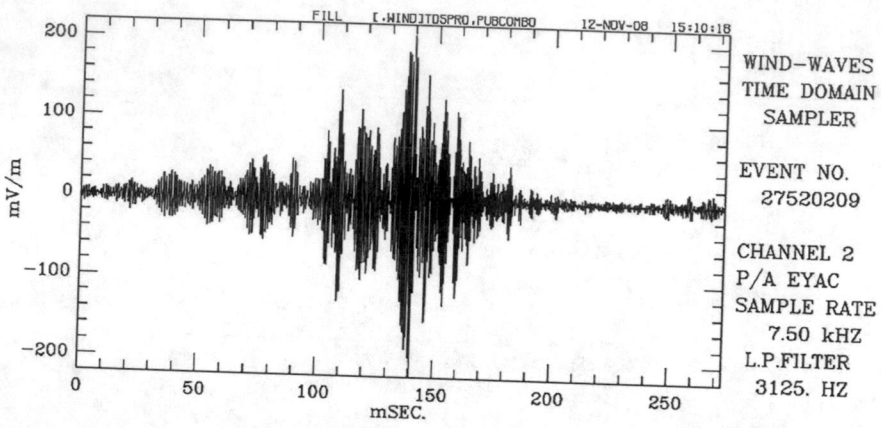

Fig. 5.8 An example of large-amplitude whistler wave packets observed by the Wind spacecraft during its passage through the magnetosphere. (From Kellogg et al. 2011, reprinted by permission from American Geophysical Union)

5.2.5   Spatial Distribution of Chorus Waves

Figure 5.9 shows maps of the average intensity of upper- and lower-band chorus waves compiled from several satellite data sets close to the equator at $L^* \leq 10$ (Meredith et al. 2012, 2020). The low-altitude limit of chorus emissions coincides with the plasmapause and the waves have been observed all the way to the outer rims of the outer belt. As the emissions are driven by anisotropic electrons injected from the plasma sheet during storms and substorms, the occurrence and intensity have strong dependence on magnetospheric activity, which in Fig. 5.9 is represented by the AE index. It is evident that the upper-band chorus waves are limited to a narrower L-range than the lower-band waves. The upper-band emissions also have, on average, significantly smaller peak intensities, typically a few hundred pT^2 compared to lower band chorus with peak intensities of the order of $2000\,pT^2$.

While chorus waves, in particular the lower-band emissions, are observed at all MLTs, their intensity shows clear MLT-dependence, which becomes more pronounced with increasing geomagnetic activity. The wave occurrence and intensities are strongest from pre-midnight, about 23 MLT, to noon. The database of Meredith et al. (2020) also demonstrates that chorus waves occur considerably more frequently and have larger intensities close to the equator than at higher magnetic latitudes. The trend is particularly clear for the upper-band chorus waves that are rarely detected at higher latitudes. Observations and ray-tracing studies (Bortnik et al. 2007) also show that at the dawn sector chorus waves can propagate to higher latitudes, reaching in the dayside 25°–30°, or above, compared to only 10°–15° in the nightside.

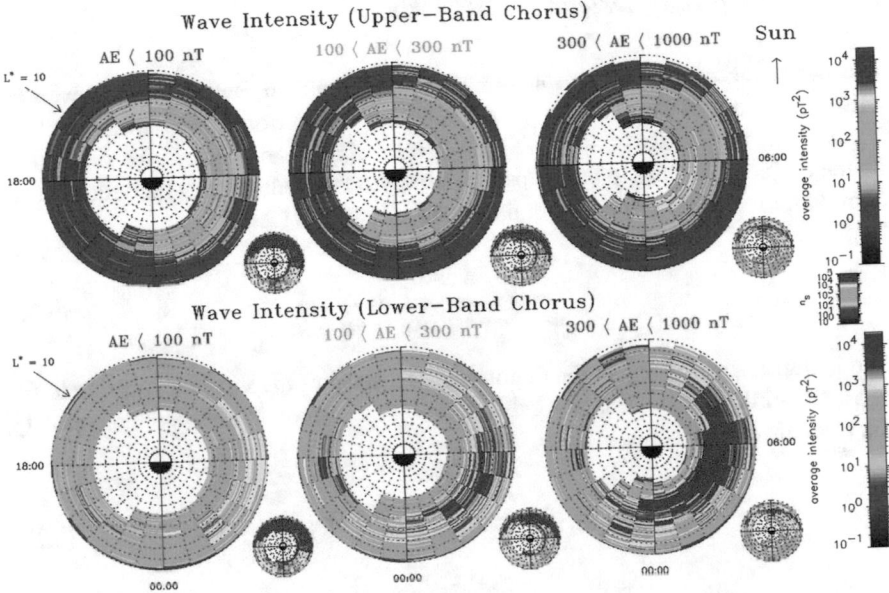

Fig. 5.9 Global maps of the average intensity of upper- and lower-band chorus waves close to the equator $|\lambda_m| < 6°$ as a function L^* and MLT. The small maps show the sampling distributions. The database from which the maps were calculated is a combination of observations from *Dynamics Explorer* 1, CRRES, *Cluster*, *Double Star* TC1, THEMIS and *Van Allen Probes* (From Meredith et al. 2020, reprinted by permission from American Geophysical Union)

These features can be explained by combination of Landau damping by electrons in the keV range and gyro-resonant amplification, which is dominated by tens of keV electrons. The intensification of chorus with geomagnetic activity close to midnight is explained by enhanced substorm injections and strengthening of the earthward convection, both key mechanisms creating the anisotropic electron population to excite chorus waves. The magnetospheric convection also transports the electrons responsible for the Landau damping restricting the propagation of chorus to higher magnetic latitudes. As these electrons drift toward the dayside, they get scattered by wave–particle interactions and the lower-energy electrons scatter faster than higher-energy electrons. As a consequence, closer to noon, chorus waves can propagate to higher latitudes due to smaller amount of Landau-resonant electrons.

A fraction of high-latitude chorus may also be generated locally, e.g., by wave–particle interaction processes with low-energy electrons (from a few hundred eV to a few keV), electron beams or nonlinear wave–wave coupling processes. Furthermore, the scattering of electrons to the atmospheric loss cone, during the drift from midnight toward the noon, can increase the anisotropy and lead to local generation of dayside chorus.

5.2.6   Anisotropy-Driven EMIC Waves

The derivation of the growth rate for left-hand polarized EMIC waves driven by anisotropic ions is similar to that of anisotropic electron-driven whistler-mode waves discussed in Sect. 5.2.1. We start with a single ion species. We focus here on the frequencies close to but below the ion gyro frequency ω_{ci}, which requires an approximation of the dispersion equation different from (5.10)

$$\frac{c^2 k_\parallel^2}{\omega^2} \approx \frac{\omega_{pi}^2}{\omega_{ci}(\omega_{ci} - \omega)} . \tag{5.18}$$

The multiplication by ω_{ci} in the denominator instead of ω as in (5.10) is due to the approximation $\omega \approx \omega_{ci}$. Inserting this again in the expression of resonant velocity, the resonant energy is found to be

$$W_{i,res} = \frac{1}{2} m_i v_{\parallel,res}^2 = W_B \left(\frac{\omega_{ci}}{\omega}\right)^2 \left(1 - \frac{\omega}{\omega_{ci}}\right)^3 \tag{5.19}$$

and the growth rate is

$$\omega_i = \frac{\pi \omega_{ci}}{2} \left(\frac{\omega_{ci}}{\omega}\right) \frac{(1 - \omega/\omega_{ci})^2}{(1 - \omega/(2\omega_{ci}))} \Delta_i(v_{\parallel,res}) \left(A_i(v_{\parallel,res}) - \frac{1}{(\omega_{ci}/\omega) - 1}\right), \tag{5.20}$$

where Δ_i and A_i are defined in the same way as in Sect. 5.2.1. The main difference to the electron case is the narrower frequency range close to the ion gyro frequency where the ions can cause significant wave growth. The threshold resonant energy for the EMIC wave is

$$W_{i,res} > \frac{W_B}{A_i^2(A_i + 1)} . \tag{5.21}$$

In the inner magnetosphere the frequencies of the chorus waves are a few kHz and of EMIC waves $\lesssim 1$ Hz. Once generated by the anisotropic suprathermal anisotropic electrons the chorus waves can be in resonance with radiation belt electrons of energies $\gtrsim 30$ keV, whereas gyro-resonant interaction of electrons with EMIC waves requires MeV energies. These interactions can lead to both damping or further growth of the waves depending on the actual shape of the distribution function of the high-energy population as discussed in Chap. 6.

5.2.7 Multiple-Ion Species and EMIC Waves

The inner magnetospheric plasma contains a variable mixture of protons and He^+ and O^+ ions. The multi-ion dispersion equation has resonances at the gyro frequencies of each species. Consequently, the EMIC waves appear in separate frequency bands: Hydrogen band emission occurs between helium and proton gyro frequencies and the helium band between oxygen and helium gyro frequencies (Fig. 4.5). Sometimes an oxygen band below the oxygen gyro frequency is also observed.

The presence of cold ions of ionospheric origin lowers the threshold for the excitation of EMIC waves and enhances the wave growth. Cold ions are indeed sometimes referred to as "generation catalyst" for EMIC waves (Young et al. 1981). EMIC waves are excited in the regions of minimum magnetic field of a given magnetic flux tube close to the equator, where the hot anisotropic ions are concentrated. After generation the waves propagate along the magnetic field toward the increasing field. Ray-tracing studies further show that the growth rates of EMIC waves are considerably larger outside than inside the plasmasphere.

Keika et al. (2013) conducted a comprehensive statistical analysis of EMIC wave observations by AMPTE/CCE from years 1984–1989. As shown in Fig. 5.10, EMIC waves were observed mostly beyond $L = 4$ at all local times and preferentially in the noon–afternoon sector beyond $L = 6$. Similar results were obtained by Meredith et al. (2003) using CRRES observations and by Chen et al. (2019) who analyzed 64 months of Van Allen Probes observations, the latter of which do not reach as far out as AMPTE/CCE but allow for more detailed analysis of wave properties.

Keika et al. (2013) found that, similar to chorus waves, the MLT distribution of EMIC waves depends on geomagnetic activity. During quiet times EMIC waves are distributed more symmetrically and their occurrence peaks close to noon. During geomagnetically active conditions the occurrence is most frequent near noon and in the dusk sector, where waves in the He^+ band have the strongest concentration. The plasmaspheric plume with cold ions at afternoon hours (Sect. 1.3.2) overlaps with the region of anisotropic hotter ring current ions. In other words, in the afternoon

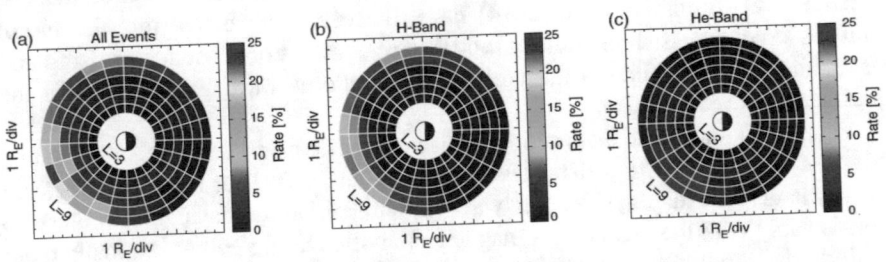

Fig. 5.10 Occurrence rates of all EMIC wave events (left), hydrogen band events (middle) and helium band events during 4.5 years of AMPTE/CCE data. Note that different to Fig. 5.9 noon is to the left. (From Keika et al. 2013, reprinted by permission from American Geophysical Union)

sector hot and cold ion populations coexist. The hot population provides free energy to excite the waves and the cold population enhances the growth. The amplitude of EMIC waves also increases with geomagnetic activity and shows a similar MLT trend as the occurrence rate. The wave amplitudes are strongest in the He^+ band.

Based on their statistical results Keika et al. (2013) suggested that the helium band would be more sensitive to ion injections, whereas the hydrogen band would benefit from solar wind compression of the magnetosphere. The compression enhances drift shell splitting and formation of Shabansky orbits (Sect. 2.6.2), which both can lead to temperature anisotropy ($T_\perp > T_\parallel$) and drive EMIC waves locally on the dayside (e.g., Usanova and Mann 2016, and references therein).

5.3   Plasmaspheric Hiss and Magnetosonic Noise

Inside the plasmasphere the main wave modes affecting electron dynamics are *plasmaspheric hiss* and *equatorial magnetosonic noise*. Hiss is of key importance to scattering electrons to the atmospheric loss cone at wide range of energies and forming the slot region, while magnetosonic waves can resonate with energetic electrons and transfer energy from ring current protons to radiation belt electrons. While the hiss is confined within the plasmasphere, the magnetosonic noise can occur both inside and outside the plasmapause.

5.3.1   Driving of Plasmaspheric Hiss

The plasmaspheric hiss is a whistler-mode emission that derives its name from the early observations of structureless spectral properties that resemble audible hiss found at all magnetic local times in the plasmasphere (Thorne et al. 1973). The frequencies of the emissions extend from a few tens of Hz to a few kHz, which is well below the local electron gyro frequency of more than 10 kHz. The high time-resolution observations with the EMFISIS instrument of the *Van Allen Probes* (Summers et al. 2014) have, however, shown that the hiss is not quite as structureless as previously thought but contains quasi-coherent rising and descending tones similar to the whistler-mode chorus wave packets outside the plasmapause (Sect. 5.2.2).

Figure 5.11 shows the distribution of plasmaspheric hiss based on more than two years of *Van Allen Probes* data. Hiss occurs at all MLTs, but the amplitudes are clearly largest on the dayside. The wave amplitudes on the dayside also increase considerably with the level of geomagnetic activity. During geomagnetically quiet periods the amplitudes range from a few pT to a few tens of pT, whereas during magnetic storms the amplitudes increase to 100–300 pT. Figure 5.11 also demonstrates that hiss waves propagate predominantly parallel to the magnetic field at low magnetic latitudes becoming more oblique at higher latitudes.

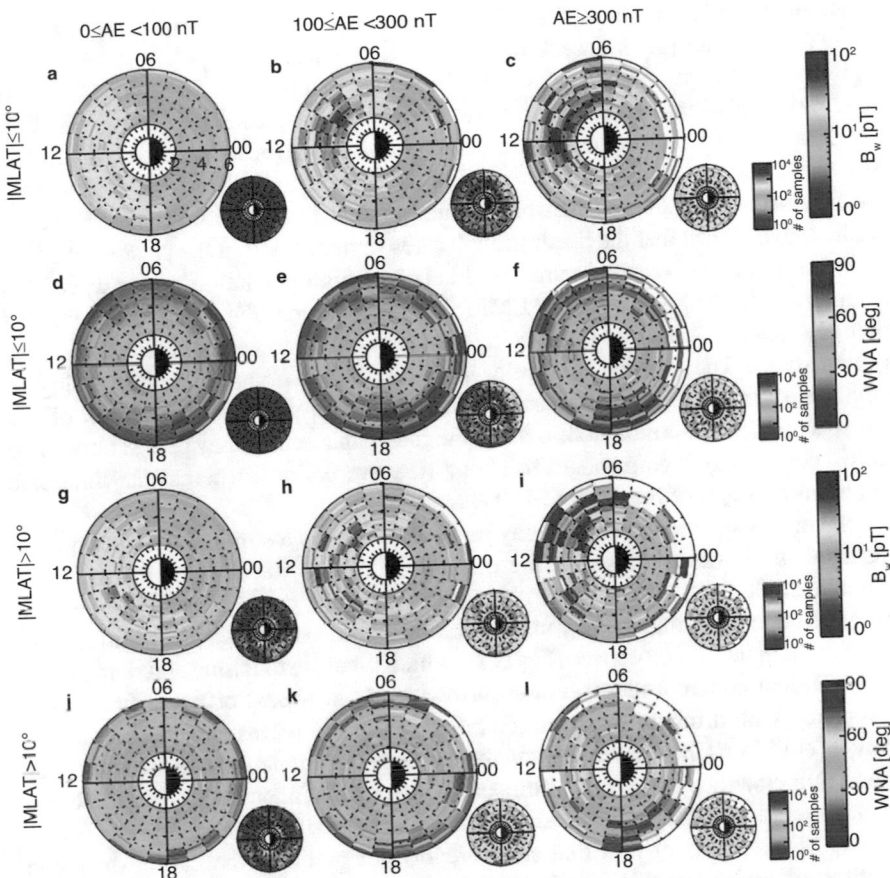

Fig. 5.11 Distribution of hiss amplitude and wave normal angles as observed by *Van Allen Probes* in the range from 10 Hz to 12 kHz for different geomagnetic activity conditions in terms of the *AE* index. The uppermost row indicates the median wave magnetic field amplitude in picoteslas and the second row the median wave normal angle at magnetic latitudes close to the equator ($\lambda \leq 10°$). The third and fourth rows are the corresponding quantities at higher magnetic latitudes. The small maps give the distribution of the samples in each picture. (From Yu et al. 2017, reprinted by permission from American Geophysical Union)

The origin of plasmaspheric hiss remains unclear. Suggested generation mechanisms include triggering through terrestrial lightning strikes, local generation within the plasmasphere, and penetration of chorus waves into the plasmasphere.

The occurrence and geographic distribution of the higher-frequency part of hiss (1–5 kHz) correlate with lightning strikes (Meredith et al. 2006), while at lower frequencies (0.1–1 kHz) no such correlation has been found. Furthermore, lightning is not related to magnetospheric activity whereas plasmaspheric hiss is. The wave power of the higher-frequency hiss is about an order of magnitude smaller than the power of the lower-frequency hiss. Consequently, lightning-induced hiss emissions likely are of minor importance to radiation belt dynamics.

Because early ray-tracing studies had indicated that there could be no significant penetration of whistler-mode chorus waves through the plasmapause, Thorne et al. (1973) advocated local generation through a similar gyro-resonant instability as the growth of whistler-mode emissions outside the plasmapause (Sect. 5.2.1). However, the anisotropic suprathermal plasma does not penetrate to the plasmasphere and later studies have shown that the linear growth rates remain small in the plasmasphere.

While the linear growth seems less likely, the high-resolution vector waveform samples of the *Van Allen Probes* EMFISIS observations (Fig. 5.12) have revealed that hiss features complex quasi-coherent fine-structures with discrete rising and falling tones. The spectral intensities peak at lowest frequencies, decreasing in amplitude with increasing frequency. The structures resemble the chirps of the whistler-mode chorus emissions outside the plasmapause, but they persist only a few milliseconds, which corresponds to about 10 wave periods, whereas the timescale of chorus chirps is of the order of 100 wave periods.

The quasi-coherent structures may be explainable by a similar nonlinear growth mechanism as was discussed in the context of chorus wave chirps in Sect. 5.2.4 (Omura et al. 2015; Nakamura et al. 2016). When a critical wave amplitude is exceeded due to resonant electrons' interaction with the waves, electron hills and holes form in the velocity space (Fig. 5.7), which give rise to falling and rising tones. The different coherent tones in hiss correspond to waves at different frequencies associated with different resonant velocities. The seed waves subject to nonlinear growth could originate from any of the sources mentioned above or from local thermal fluctuations. Since the nonlinear growth rate is much larger than the linear, local generation of hiss in the plasmasphere may be a viable option.

The observations of hiss fine structure do not preclude the chorus–hiss connection, of which there is circumstantial evidence in satellite observations (e.g., Santolík et al. 2006; Bortnik et al. 2008b, and references therein). The hypothesis of chorus wave penetration was revived by ray-tracing studies of Bortnik et al. (2008b). Figure 5.13 illustrates an example of tracing of waves launched at $L = 5$ at the equator in the lower end of the chorus frequency range. The initial wave normal angles extended from $-70°$ to $+20°$ with negative angles corresponding to earthward inclination. The waves with negative WNAs of a few tens of degrees on the nightside were found to be able to penetrate to the plasmasphere, while on the dayside this may occur over a wider range of WNAs, from approximately $-60°$ to $-30°$. This is due to significantly smaller dayside fluxes of about 1 keV-electrons that would Landau-damp the waves. Inside the plasmasphere the waves reflect at higher latitudes ($\lambda \approx 30$–$40°$) closer to the Earth, where their frequency becomes less than the local cut-off frequency.

The chorus waves have specific entry points into the plasmasphere but they become quickly randomized after a few cycles of reflections. This is consistent with the unstructured appearance of hiss in low-resolution observations as illustrated in the bottom panels of Fig. 5.13. The figure also shows that chorus waves outside the plasmapause have larger wave power than hiss in the plasmasphere.

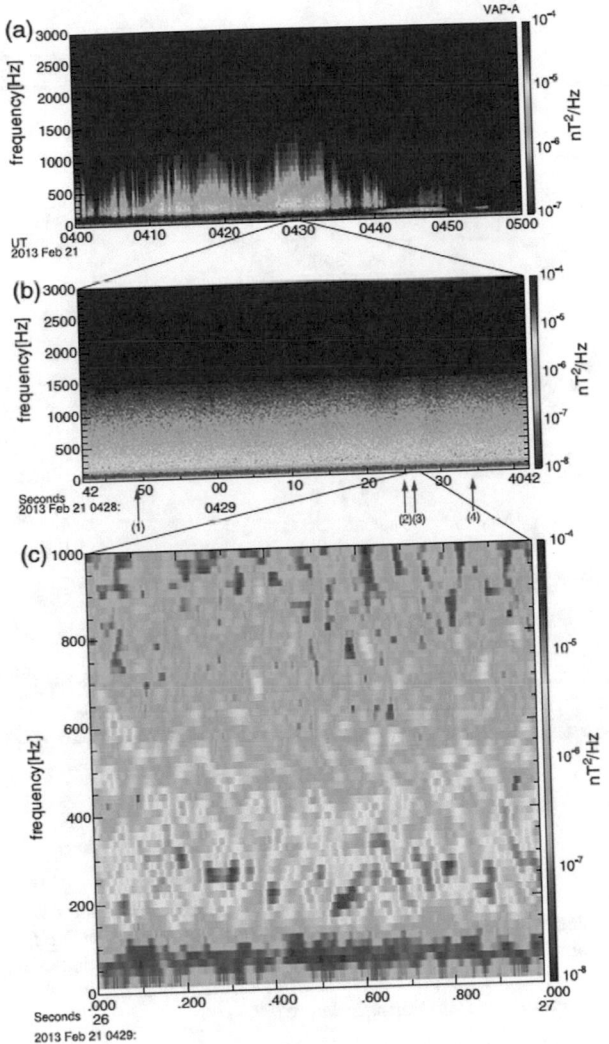

Fig. 5.12 An example of plasmaspheric hiss consisting of coherent rising and falling tones (From Summers et al. 2014, reprinted by permission from American Geophysical Union)

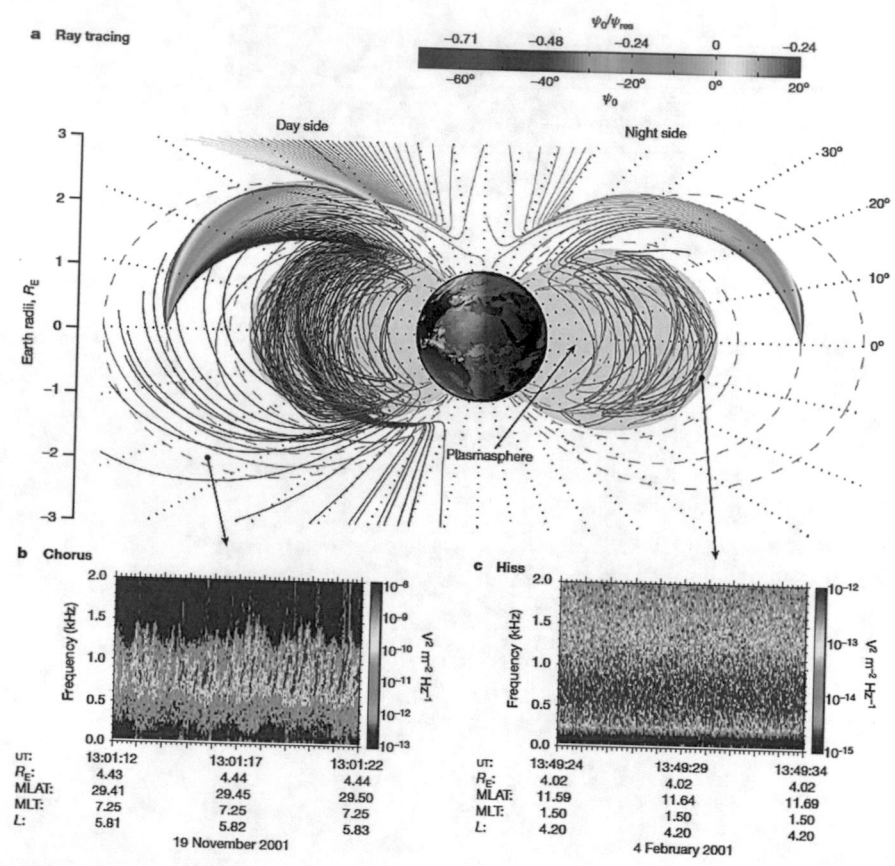

Fig. 5.13 (**a**) Ray-tracing results of chorus waves launched from the equator with different wave normal angles ψ_0 (negative values correspond to earthward propagation) at $0.1\ f_{ce}$. Bottom panels show the typical examples of intensities of (**b**) chorus on the dayside outside the plasmasphere and (**c**) hiss on the nightside inside the plasmasphere from *Cluster* observations. (From Bortnik et al. 2008b, reprinted by permission from SpringerNature)

   Once the waves have penetrated to the plasmasphere, they refract from the large density gradient at the plasmapause and become trapped inside the plasmasphere. In contrast to chorus waves outside the plasmapause, Landau damping of the hiss waves in the plasmasphere is weak due to the high number density of cold electrons and small flux of Landau-resonant electrons. The waves can thus propagate over long time periods passing repeatedly through the equatorial region. The waves become more field-aligned, as they approach the equator, where they can get gradually amplified by gyro resonance with electrons.

   The day–night asymmetry of hiss amplitudes (Fig. 5.11) agrees with the ray-tracing results and the chorus penetration hypothesis as the origin of hiss. This is also consistent with hiss intensifying with geomagnetic activity, similarly to chorus waves. Furthermore, the compression of the dayside magnetosphere creates local magnetic field minima away from the equator (Fig. 2.9) where the growth rate of chorus waves increases and from where they have a shorter distance to propagate to the plasmapause, minimizing the Landau damping before entering the plasmasphere (Tsurutani et al. 2019).

The penetration of chorus waves through the plasmapause depends on both the WNA and the inclination of the wave vector, of which the latter must naturally be toward the Earth. Using *Van Allen Probes* EMFISIS data Hartley et al. (2019) investigated how often the observed wave vectors of chorus emissions are in favorable direction for penetration. They found that the inclination is actually predominantly oriented in the anti-earthward direction. Their ray-tracing computations indicated that only a very small fraction of wave power, typically less than 1%, would propagate to the plasmasphere. The only exception were waves that were emitted very close to the morning sector plasmaspheric plume, located in their model at $L = 5$ and MLT = 14. In the plume region about 90% of the lower-band chorus power was found to propagate into the plasmasphere. In another study Kim and Shprits (2019) showed, based on four years of *Van Allen Probes* observations, that similar to hiss in the plasmasphere proper, the hiss in the plume has amplitudes from a few pT to more than 100 pT. In fact, the plume may provide an efficient entrance for the hiss to penetrate to the plasmasphere, but whether it is enough, remains unclear.

Further indication of chorus–hiss connection was found by Agapitov et al. (2018) in an extensive correlation analysis of THEMIS observations of lower-band chorus and hiss waves during 2007–2017 at times when one of the spacecraft was in the plasmasphere and another outside the plasmapause. They considered 2-min intervals of events where the wave amplitude was required to be larger than 1 pT and the distance between the spacecraft more than $2 R_E$ but less than 3 h in MLT. The correlations were calculated when chorus waves were observed within 10 s before hiss or hiss during 10 s following the chorus observation. They found 71,000 time intervals when the correlation coefficient between chorus and hiss wave power dynamics was larger than 0.5, often larger than 0.7. The best correlations were in the noon to afternoon sector consistent with favorable penetration of chorus waves to the plasmasphere in the sector of the plasmaspheric plume. Even if the amount of penetrating wave energy may be small, Agapitov et al. (2018) argued that it may form an embryonic source for local amplification by, e.g., the above mentioned nonlinear mechanism.

5.3.2 Equatorial Magnetosonic Noise

Perpendicular propagating waves in the frequency range between the proton gyro frequency (ω_{cp}) and the lower hybrid resonance frequency ($\omega_{LHR} \approx \sqrt{\omega_{ce}\omega_{cp}}$) confined within a few degrees from the Earth's magnetic equator were first identified in the OGO 3 satellite observations and were named *equatorial noise* (Russell et al.

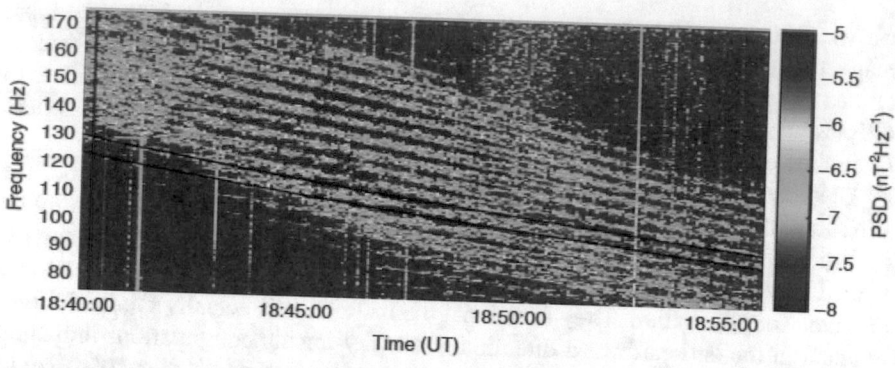

Fig. 5.14 Example of fine-structured magnetosonic waves from *Cluster* observations. The frequency bands are separated by the local proton gyro frequency. The descending trend if the stripes is due to the motion of the spacecraft in the direction of decreasing magnetic field (From Balikhin et al. 2015, Creative Commons Attribution 4.0 International License)

1970). This emission features distinct bands of ion Bernstein modes organized by multiples of the proton gyro frequency (Fig. 5.14). The wave mode is the hot plasma equivalent of the linearly polarized cold plasma X-mode that propagates almost perpendicular (WNA $\approx 89°$) to the background magnetic field. The mode is an extension of the MHD fast magnetosonic mode above ω_{cp} (Fig. 4.6) and the observed emission is commonly called *equatorial magnetosonic noise*. While these waves evidently are related to ion dynamics, they are also of significant interest for radiation belt electrons, as they can resonate with energetic electrons through Landau, gyro and bounce resonances as discussed in Chap. 6.

Let us briefly discuss the generation of magnetosonic waves following Horne et al. (2000) and Chen et al. (2010). The instability is expected to be caused by the *proton ring distribution* (e.g., Thomsen et al. 2017, and references therein) with a positive slope perpendicular to the magnetic field around 10 keV (Fig. 5.15). In the linear regime the growth rate is proportional to the sum of all harmonic resonant interactions between the wave and the protons

$$\omega_i \propto \sum_n \int_0^\infty \left(J_n^2(x) \frac{\partial f(v_\parallel, v_\perp)}{\partial v_\perp} \right) \Bigg|_{v_\parallel = v_{\parallel res}} dv_\perp, \tag{5.22}$$

where $J_n(x)$ are the Bessel functions of order n, $x = k_\perp v_\perp / \omega_{cp}$, and $f(v_\parallel, v_\perp)$ is the proton distribution function (for a detailed calculation, see Chen et al. 2010). The integral is evaluated at the resonant velocity $v_{\parallel res}$ given by

$$v_{\parallel res} = \frac{\omega}{k_\parallel} \left(1 - \frac{n\omega_{cp}}{\omega} \right). \tag{5.23}$$

The instability requires that $\partial f / \partial v_\perp > 0$. The positive gradient maximizes when $v_\parallel$ is small because f decreases with increasing $v_\parallel$. The corresponding

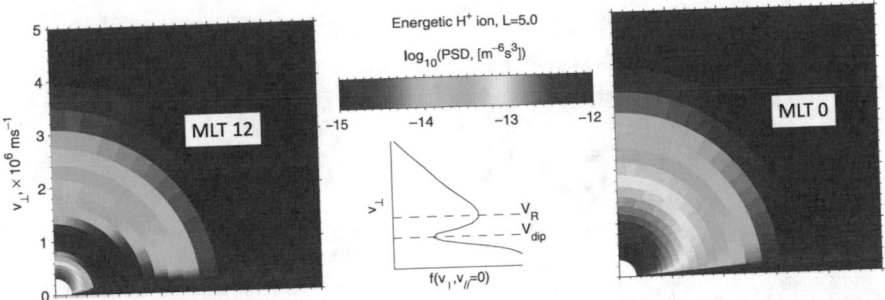

Fig. 5.15 Proton ring distribution. The velocity scales on all axes are the same as on the left vertical axis. The gradient $(\partial f/\partial v_\perp)$ of the phase space density of the ring is positive on the dayside (on the left and in the middle), whereas on the nightside (right) the gradient is positive only at the edge of the atmospheric loss cone (From Chen et al. 2010, reprinted by permission from American Geophysical Union)

perpendicular velocity is called the *ring velocity*. Magnetosonic waves have very small $k_\parallel$ and, in order to have small enough resonant velocities, dominant resonances occur at high multiples of proton gyrofrequency $\omega \approx n\omega_{cp}$.

With increasing $k_\perp$ the magnetosonic mode approaches the lower hybrid resonance frequency where $\omega/k_\perp$ decreases. The effective growth of the magnetosonic wave further requires that the Bessel function J_n maximizes in the region where $\partial f/\partial v_\perp$ is positive. The argument of J_n can be written as

$$x = \frac{\omega}{\omega_{cp}} \frac{v_\perp}{v_A}, \tag{5.24}$$

where v_A is the local Alfvén velocity. At high harmonics ($n \gtrsim 10$) J_n maximizes when $x \approx n$ corresponding to the perpendicular velocity close to the Alfvén velocity. If the ring velocity is larger than v_A, the wave can grow. For smaller n the Bessel function peaks instead in a region where $\partial f/\partial v_\perp < 0$ and the wave is damped. For a growing solution at smaller harmonics the ring velocity must exceed the Alfvén velocity with a larger margin. An important factor controlling the growth of the magnetosonic wave is thus the ratio of the ring velocity to the Alfvén velocity.

The energetic proton phase space densities shown in Fig. 5.15 are simulation results of the main phase of the geomagnetic storm on 22 April 2001 calculated at $L = 5$. On the nightside (right) the distribution is bi-Maxwellian with a loss cone. On the dayside (left) the distribution illustrates a ring with a clear peak at about 20 keV. A schematic of typical ring-like phase space density as a function of the perpendicular velocity, indicating velocities at which the phase space density peaks (the ring velocity) and has a minimum (dip velocity), is shown in the middle.

The formation of ion ring distribution can be understood as follows: Lower-energy protons in the nightside plasma sheet E×B drift as shown in Fig. 2.3. Higher-energy protons ($\gtrsim 10$ keV) are affected by the gradient and curvature effects and drift from the tail predominantly around the dusk toward the dayside. The

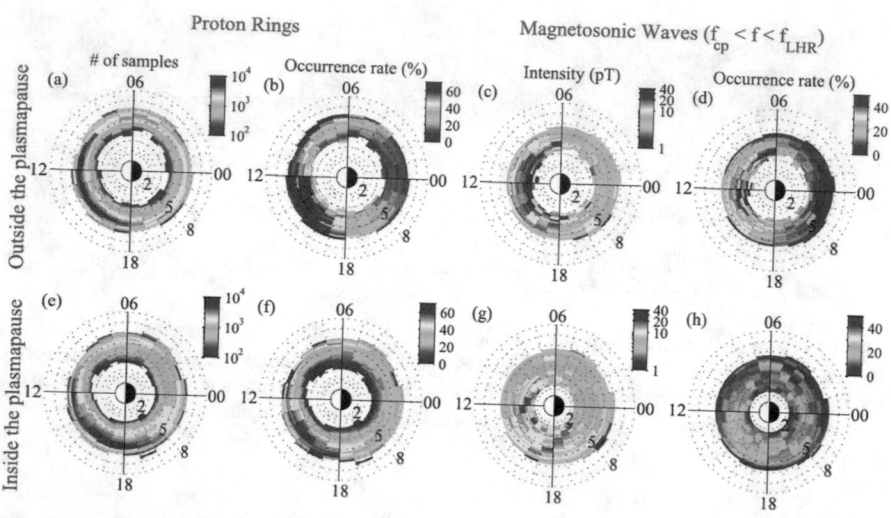

Fig. 5.16 Global distribution of occurrence of proton rings, and occurrence and intensity of magnetosonic waves inside and outside the plasmasphere over 3 years of Van Allen Probes observations. (From Kim and Shprits 2018, reprinted by permission from American Geophysical Union)

gradient and curvature drift rates are proportional to the proton energy, enhancing at dusk and noon the phase space density of higher-energy protons compared to lower energies. In addition, lower-energy protons are subject to charge exchange collisions with exospheric neutrals depleting the core of the proton distribution. The sketch in the middle of Fig. 5.15 demonstrates that below the ring velocity $\partial f / \partial v_\perp$ is positive. Thus there is free energy available for the excitation of magnetosonic waves. The negative gradients above the ring velocity and below the velocity at the trough of the distribution function can contribute to the damping of the waves. The natural place for the generation of magnetosonic waves is thus close to the magnetic equator where pitch angles are close to $90°$ and thus $v_\parallel$ is small to maximize the positive gradient of f.

Figure 5.16 shows global spatial distributions of the occurrence of proton rings, and the occurrence and intensity (wave amplitude) of magnetosonic waves inside and outside the plasmapause in *Van Allen Probes* observations. Proton rings and magnetosonic waves are observed over a relatively wide L-range throughout the dayside magnetosphere, both in and outside the plasmasphere. The occurrence of rings and most intense waves is strongest and the L-coverage widest from noon to dusk hours. An exception is outside the plasmasphere in the pre-noon sector at low L-values. The waves are likely generated outside the plasmapause, since in the plasmasphere the Alfvén velocity is well below the ring velocity.

5.4 Drivers of ULF Pc4–Pc5 Waves

While the microscopic instabilities driving whistler-mode, EMIC, and X-mode waves can often be attributed to specific properties of the particle distribution functions, the question of driving ULF waves in the Pc4–Pc5 range is more complicated.

5.4.1 External and Internal Drivers

Magnetospheric ULF waves can be generated both externally by solar wind–magnetopause interactions and internally inside the magnetosphere over a wide range of frequencies. The frequencies of Pc5 oscillations correspond to the longest wavelengths that can be described as propagating or standing waves in the quasi-dipolar domain of the inner magnetosphere. Excitation mechanisms affect the polarization, azimuthal mode number (m), amplitude and frequency of the ULF waves. Azimuthally large-scale (small m) waves are thought to arise primarily from external sources, while azimuthally smaller-scale (large m) waves are more likely excited by internal mechanisms. This division does, however, not always apply (e.g., James et al. 2016, and references therein).

A thorough discussion of external drivers is beyond the scope of this book (for a review, see Hwang and Sibeck 2016, and the extensive set of references therein). In fact, many different perturbations in the upstream solar wind and in the magnetosheath can shake the magnetospheric magnetic field leading to propagating or standing ULF oscillations in the magnetosphere. Obvious candidates are solar wind pressure pulses hitting the magnetopause, the Kelvin–Helmholtz instability (KHI) caused by large enough velocity shear across the magnetopause at the flanks of the magnetosphere (Chen and Hasegawa 1974), and Flux Transfer Events (FTE) through the dayside magnetopause (Russell and Elphic 1979). Furthermore, the magnetosheath and the foreshock region upstream of the magnetosheath host several plasma instabilities from ion gyro-scale kinetic to large-scale mirror-mode instabilities, which may lead to waves penetrating to the magnetosphere.

Correlating different solar wind perturbations with magnetospheric fluctuations is obscured by the interdependence between upstream parameters (e.g., Bentley et al. 2018). Moreover, the dayside magnetopause acts as a low-pass filter suppressing large-amplitude transient pressure pulses in the magnetosheath in timescales shorter a few minutes (Archer et al. 2013). Thus several different effects can result in similar toroidal, poloidal, or compressional ULF waves launched from the magnetopause inward (Fig. 5.17).

A basic scenario of how a perturbation proceeds from the dayside magnetopause inward, is sketched in Fig. 5.18. The inward propagating fast compressional magnetosonic wave launched by an upstream solar wind perturbation encounters increasing Alfvén speed. When the frequency of the wave matches the eigenfrequency

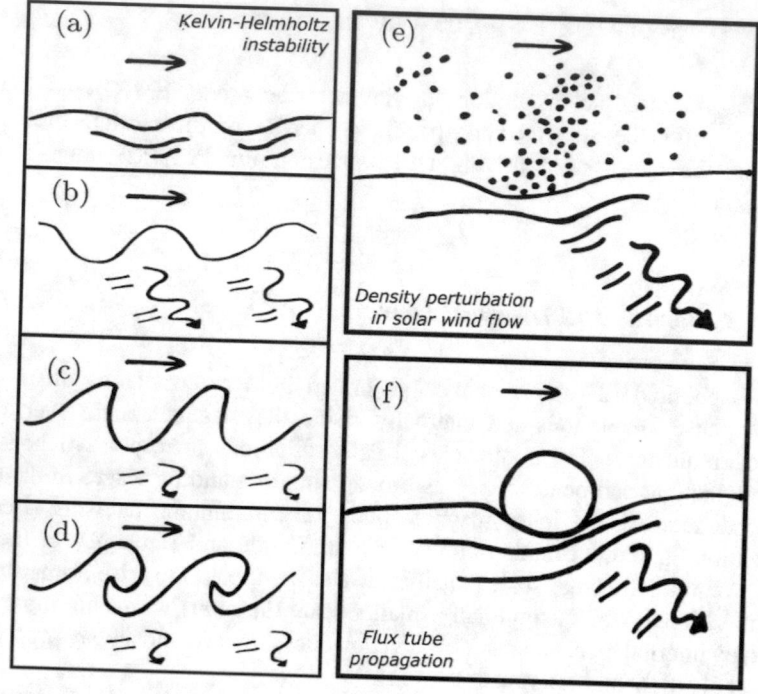

Fig. 5.17 A sketch of different mechanisms on the magnetopause that can drive global magnetospheric ULF oscillations in the Pc4–Pc5 range. Panels (**a**)–(**d**) illustrate the deepening of Kelvin–Helmholtz surface waves to nonlinear vortices, panel (**e**) a solar wind pressure pulse and (**f**) a flux tube pressing the magnetopause (From Bentley et al. 2018, reprinted by permission from American Geophysical Union)

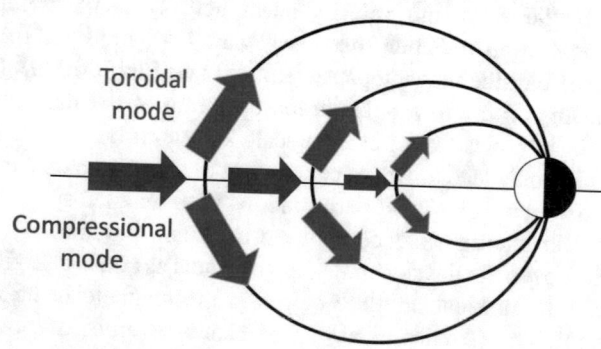

Fig. 5.18 A sketch of the coupling of a fast compressional magnetosonic wave launched at the dayside magnetopause to a toroidal shear Alfvén wave forming field line resonances

$f = nv_A/(2l)$ (Eq. 4.108) of the field line, the wave couples to a toroidal-mode shear Alfvén wave. This leads to a field line resonance (FLR) in the wave guide with reflecting boundaries in the northern and southern ionospheres (Sect. 4.4.2). As the length of the field line (l) is proportional to L, the eigenfrequencies increase with decreasing L-shells. When the compressional mode feeds the toroidal oscillation it gets gradually damped. A similar scenario also applies to the dawn and dusk sectors where the initial perturbation is more likely due to KHI, which may also drive the toroidal mode directly.

The solar wind perturbations can also result in cavity mode oscillations (CMO) of compressional waves in the radial direction (Sect. 4.4.2), although the eigenfrequencies of the cavity modes and their harmonics are rather in the Pc3–Pc4 frequency range than at Pc5 frequencies (e.g., Takahashi et al. 2018).

The largest amplitude ULF waves are associated with the strongest interplanetary shocks (e.g., Hao et al. 2014). The waves can occur over a wide range of L-shells and frequencies and thus affect radiation belt electrons of various energies. The shock-induced ULF waves, however, are quickly damped, most likely via Landau damping with ions of energies of a few keV (e.g., Wang et al. 2015).

Solar wind dynamic pressure oscillations can also directly drive oscillations in the Earth's magnetosphere. The magnetopause responds by contracting and relaxing which leads to compressional magnetic field oscillations in the magnetosphere. This scenario is supported by correlations between fluctuation power in the ULF Pc5 range in the solar wind density/dynamic pressure and in magnetospheric magnetic field found in several experimental studies as well as in simulations showing FLRs when the frequency of solar wind dynamic pressure oscillations matches the local eigenfrequency of the geomagnetic field line (e.g., Claudepierre et al. 2010). If the period of the upstream oscillation is longer than the Alfvén wave travel time through the inner magnetosphere (about 3 min) and the time it takes for a pressure disturbance to propagate past Earth (about 5 min), the perturbation inside the magnetosphere is quasi-static resulting in a phenomenon called *forced breathing of the magnetosphere* (Kepko and Viall 2019).

Internally driven Pc5 waves are commonly ascribed to instabilities due to westward drifting ring current ions and ions injected by substorm dipolarizations or bursty bulk flows from the magnetotail. A possible instability mechanism is the *bounce–drift resonance* between the ions and the ULF wave mode

$$\omega - l\omega_{bi} - m\omega_{di} = 0, \qquad (5.25)$$

where ω_{bi} and ω_{di} are the bounce and drift frequencies of the ions, l the longitudinal and m the azimuthal mode number of the wave (Southwood et al. 1969).[2]

---

[2] In Chap. 6 we discuss the bounce and drift resonances in the context of electron diffusion and transport.

Note that the suprathermal ion populations that drive Pc5 waves are the same ions as those driving EMIC waves and may have substantial anisotropies ($T_\perp > T_\parallel$ or $P_\perp > P_\parallel$), being capable to drive mirror mode waves with magnetic field and density oscillations in opposite phases to each other (Sect. 4.4.1). The threshold for the mirror mode instability must be calculated from kinetic theory resulting in

$$\sum_\alpha \frac{\beta_{\alpha\perp}^2}{\beta_{\alpha\parallel}} > 1 + \sum_\alpha \beta_{\alpha\perp} \,. \tag{5.26}$$

where β_α are the beta parameters, i.e., the ratios of the plasma and magnetic pressures, for each plasma species. Chen and Hasegawa (1991) conducted a theoretical kinetic treatment in realistic magnetospheric plasma conditions assuming a core ($\sim 100\,eV$) and energetic ($\sim 10\,keV$) components and concluded that mirror instability is an important internal mechanism to drive ULF waves. Consequently, the local instability can support a slow mode ULF oscillation against its damping through the Landau mechanism.

Figure 5.19 shows an example of an observed ULF wave event during a weak storm on 6 July 2013 investigated by Xia et al. (2016). At the time indicated by the vertical dashed lines in the figure the *Van Allen Probe* B was in the evening sector (MLT $\approx$ 21:40) at $L \approx 5.5$ close to magnetic equator. At this time there was a clear pressure anisotropy ($P_\perp > P_\parallel$) and the total magnetic field and the plasma density oscillated in opposite phases. As usual the wave had mixed polarization. The parallel

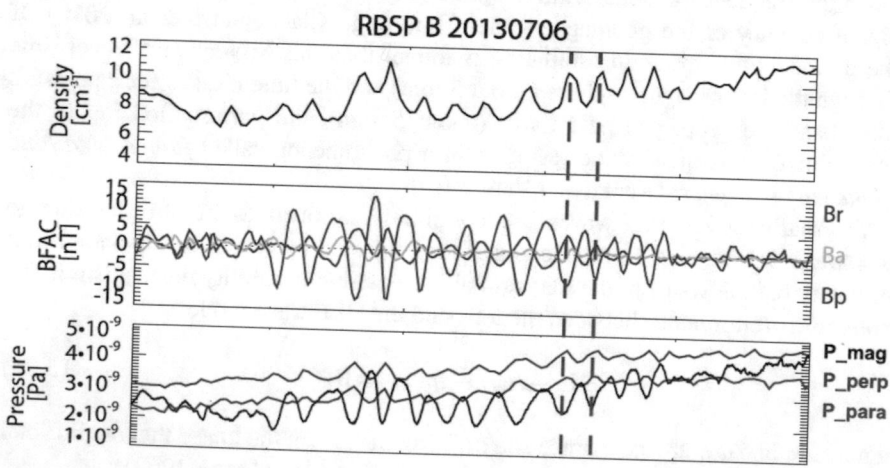

Fig. 5.19 *Van Allen Probe* B observation of a poloidal mode ULF wave with magnetic and density fluctuations in opposite phases. The top panel shows the density fluctuation, the center panel the magnetic field components in field-aligned coordinates (blue: radial, green: azimuthal, red: parallel). In the bottom panel the black line shows the magnetic pressure which is clearly in opposite phase to parallel (blue) and perpendicular (red) plasma pressures. the dashed vertical lines are there to guide the eye (From Xia et al. 2016, reprinted by permission from American Geophysical Union)

(poloidal) component was larger than the radial (compressional) component and the azimuthal component of the waves was smallest. Thus the wave was a dominantly poloidal mode with mirror-mode type compression consistent with the anisotropy-driven drift-mirror instability.

James et al. (2016) investigated substorm-associated ULF waves using observations of the far-ultraviolet imager of the IMAGE satellite, several ground-based magnetometer arrays and the SuperDARN network of coherent ionospheric radars. The advantage of the radars is that they measure large-scale plasma fluctuations in the ionosphere whereas only a fraction of the wave power is transmitted through the ionosphere to ground. In the three events studied in detail by James et al. (2016) the properties of the waves varied widely both within and between the events, being different at different distances from the ionospheric location of the substorm. The waves were found to be poloidal modes with azimuthal wave numbers from -9 to -44, indicating that the phase of the waves propagated westward, which is consistent with westward drifting protons as wave drivers in the magnetosphere. The energies of the protons were estimated from the resonance condition to be in the range 2–66 keV.

The determination of the azimuthal mode number from spacecraft observations is notoriously difficult. The measurements need to be performed at the same L-shell by at least two spacecraft close enough to each other in order to avoid the 2π ambiguity (aliasing) in the calculation of the phase difference between the observed oscillations. Murphy et al. (2018) performed a detailed analysis of ULF Pc4–Pc5 wave observations of the closely-spaced MMS spacecraft during a period of solar wind high-speed stream between 25 September and 10 October 2016. Due to the highly-elliptical orbit the velocity of the satellites close to the perigee of the constellation (1.2 R_E) was so high that the analysis was limited outside 4 R_E but reaching up to the magnetopause in the evening sector where the apogee was at the time of the observations. The ULF wave power peaked close to the magnetopause and in the inner magnetosphere at equatorial distances 6–8 R_E.

Murphy et al. (2018) calculated the azimuthal mode numbers for discrete ULF waves observed during the investigated period. The distribution of m at distances 4–14 R_E is shown in Fig. 5.20. The mode numbers were found to be both positive (indicating eastward propagation of the wave) and negative (indicating westward propagation of the wave) reaching up to ±100 but preferentially within $|m| < 20$. At equatorial distances up to 8 R_E the mode numbers were predominantly positive and < 20, between 8 and 11 R_E predominantly between -5 and -40 and close to the magnetopause beyond 13 R_E again mostly positive and < 20. The positive mode numbers were interpreted to indicate an external driver and the negative ones an internal driver. The latter is consistent with the internal driver being protons passing the Earth on the evening side and driving westward propagating ULF waves similarly to the above mentioned results of James et al. (2016).

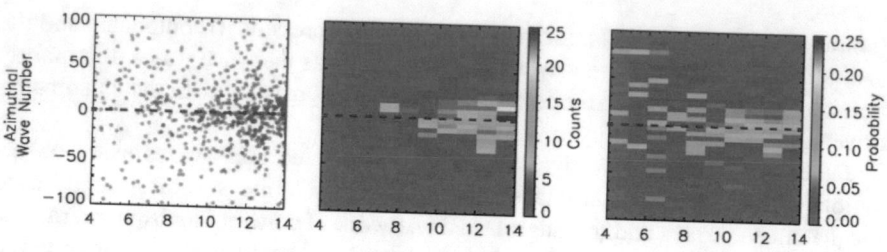

Fig. 5.20 Left: the distribution of azimuthal mode numbers; middle: the histogram of the distribution; right: the probability distribution. The horizontal axis is the Earth-centered distance in R_E (From Murphy et al. 2018, Creative Commons Attribution License)

5.4.2 Spatial Distribution of ULF Waves

The wide spatial range of ULF oscillations from the magnetopause to the ionosphere and the great variability of the wave properties under different solar wind conditions and magnetospheric activity pose challenges to the production of comprehensive maps of the distribution of the waves. Furthermore, a reliable determination of the polarization requires simultaneous observation of a sufficient number of electric and magnetic field components, preferentially all of them. Measuring the magnetic field components often is sufficient, but the polarization of the electric field is needed, in particular, close to the equator where the magnetic field of the fundamental ($n = 1$) FLR has a node yielding a weak magnetic field signature in the observation. Consequently, different studies based on different satellite and ground-based observations have led to different, sometimes contradictory, conclusions.

Hudson et al. (2004a) investigated ULF oscillations in the range $L = 4 - 9$ over 14 months of CRRES observations, which took place close to the maximum of Solar Cycle 22. Toroidal Pc5 oscillations were found on the dusk and dawn flanks of the magnetosphere inside $L = 8$, preferentially at the higher end of the L-shells. Based on the observed local plasma frequency the waves were found to be standing FLRs at the fundamental frequency $f = v_A/(2l)$. Poloidal (including compressional) modes were found to occur in the dusk-to-midnight sector mostly from $L = 5$ to $L = 8$. This is consistent with above discussed instability driven by ions injected from the magnetotail. The orbit of CRRES did not allow sufficient sampling on the dayside and thus the important dayside compressional modes were not covered in the study.

The most comprehensive picture of Pc4–Pc5 oscillations has been obtained from the THEMIS mission. As already illustrated in Fig. 4.7, the magnetometer and the electric field instrument of THEMIS have made possible a complete characterization of different polarization components. Furthermore, after October 2007 the orbital configuration of the spacecraft allowed a full nearly-equatorial coverage of all local times in 13 months reaching out to about 10 R_E.

Liu et al. (2009) performed a statistical study of Pc4–Pc5 events observed from November 2007 to December 2008. The total observation time was more than 3000 h. Of the identified wave events 9805 were in the Pc4 range and 50,184 in

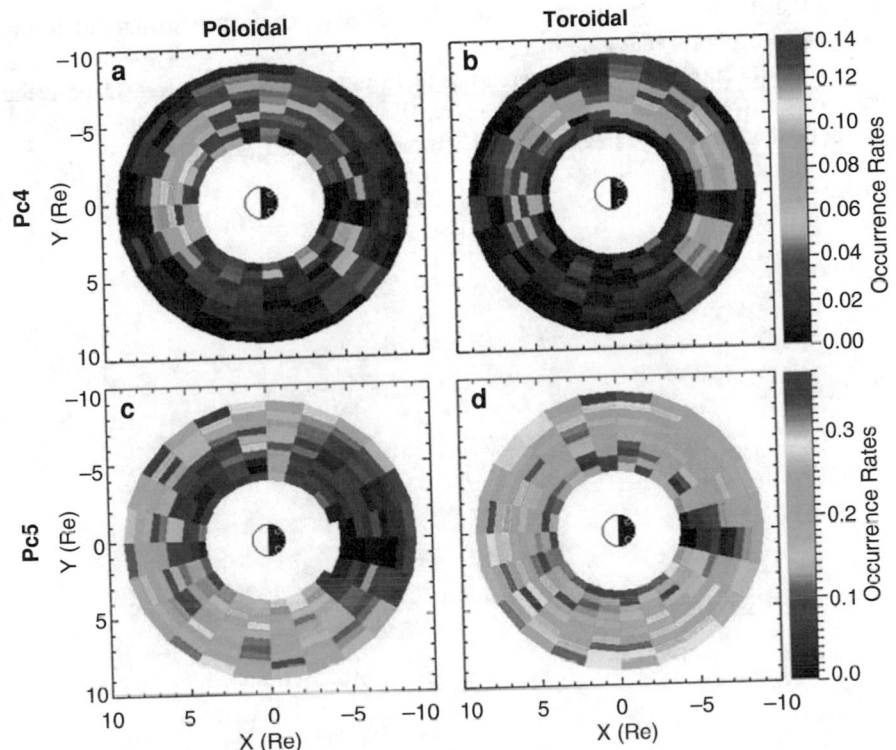

Fig. 5.21 Spatial distribution of the occurrence rates of poloidal and toroidal ULF Pc4 and Pc5 events. The bins are 0.5 R_E wide in the radial direction from 4 to 9 R_E and 15 min in the local time. (From Liu et al. 2009, reprinted by permission from American Geophysical Union)

the Pc5 range. In both frequency ranges the number of toroidal events was a little larger than the number of poloidal events (Pc4: 51%, Pc5: 59%).

The spatial distribution of the occurrence rates of poloidal (including the compressional) and toroidal modes are shown in Fig. 5.21. Pc4 events were most frequent at radial distances 5–6 R_E form the post-midnight (mostly toroidal) to noon (mostly poloidal), whereas Pc5 events were most frequent at distances 7–9 R_E. That poloidal events occur mostly on the dayside is consistent with upstream solar wind perturbations. The enhancement of poloidal Pc5 events on the dusk flank is, in turn, consistent with the internal driving through bounce–drift resonance with ring current ions and/or ions freshly injected from the tail.

The different radial distributions of the toroidal Pc4 and Pc5 modes are likely related to the inverse dependence of the FLR frequency on the length of the field line. The high occurrence rates of toroidal Pc5 events close to the dawn and dusk sectors is an indication that the waves may be driven directly by the Kelvin–Helmholtz instability on the flanks of the magnetosphere. Liu et al. (2009) suggested that the relatively low occurrence rates of both poloidal and toroidal Pc4 events in the dusk sector would be due to the extension of the plasmapause further out during low magnetospheric activity.

Liu et al. (2009) also investigated the distribution of wave power. Overall the wave power was higher in the Pc5 than the Pc4 band. In both frequency bands the power decreased with decreasing radial distance from the Earth.

Open Access This chapter is licensed under the terms of the Creative Commons Attribution 4.0 International License (http://creativecommons.org/licenses/by/4.0/), which permits use, sharing, adaptation, distribution and reproduction in any medium or format, as long as you give appropriate credit to the original author(s) and the source, provide a link to the Creative Commons license and indicate if changes were made.

The images or other third party material in this chapter are included in the chapter's Creative Commons license, unless indicated otherwise in a credit line to the material. If material is not included in the chapter's Creative Commons license and your intended use is not permitted by statutory regulation or exceeds the permitted use, you will need to obtain permission directly from the copyright holder.

Chapter 6
Particle Source and Loss Processes

The main sources of charged particles in the Earth's inner magnetosphere are the Sun and the Earth's ionosphere. Furthermore, the Galactic cosmic radiation is an important source of protons in the inner radiation belt, and roughly every 13 years, when the Earth and Jupiter are connected via the interplanetary magnetic field, a small number of electrons originating from the magnetosphere of Jupiter are observed in the near-Earth space. The energies of solar wind and ionospheric plasma particles are much smaller than the particle energies in radiation belts. A major scientific task is to understand the transport and acceleration processes leading to the observed populations up to relativistic energies. Equally important is to understand the losses of the charged particles. The great variability of the outer electron belt is a manifestation of the continuously changing balance between source and loss mechanisms, whereas the inner belt is much more stable.

In the preceding chapters we already have encountered various aspects of acceleration and loss of charged particles, e.g., betatron and Fermi acceleration (Sect. 2.4.4), bounce and drift loss cones (Sect. 2.6.1), magnetopause shadowing (Sect. 2.6.2), as well as the basics of growth and damping of waves in Chaps. 4 and 5. In this chapter we present the general framework of quasi-linear theory of diffusion and transport, and discuss the sources and losses of different particle species in more detail. At the end of the chapter (Sect. 6.7) we point out that the different mechanisms do not only affect the radiation belt particles additively but also synergistically, e.g., through nonlinear modulation of whistler-mode or EMIC waves by large-amplitude ULF waves.

© The Author(s) 2022
H. E. J. Koskinen, E. K. J. Kilpua, *Physics of Earth's Radiation Belts*, Astronomy and Astrophysics Library, https://doi.org/10.1007/978-3-030-82167-8_6

6.1 Particle Scattering and Diffusion

The response of charged particles to temporally and spatially variable electric and magnetic fields is deterministic and, according to Liouville's theorem of statistical physics, in absence of external sources and losses the phase space density is conserved along the dynamical trajectories of the particles. However, the empirical determination of temporal evolution of the phase space density $\partial f / \partial t$ in any given location is limited by imperfect observations of electromagnetic fields and waves and the particle populations. The finite angular, energy, and temporal resolution of particle instruments makes them insensitive to *phase mixing*.[1] Consequently, we are not able to observationally distinguish individual particles with different phases in their gyro, bounce, or drift motions and the empirical information is in most cases limited to phase-averaged description of the radiation belts. On the other hand, the phase mixing makes the theoretical description much more tractable allowing the use of diffusion formalism to describe the time evolution of particle distributions. As stated by Schulz and Lanzerotti (1974): "Thus, the ultimate inability to distinguish particle phases by observations is a simplifying virtue".

Diffusion is a statistical concept to describe the evolution of the phase space density. It was already encountered in Sect. 5.1.3 as random walk of particles along single-wave characteristics. Although it is customary to talk about particle diffusion, individual particles actually do not diffuse. They are scattered in the phase space by spatial and temporal inhomogeneities, wave–particle interactions and collisions. In wave–particle interactions the resonant scattering is the most efficient, but not the only, cause of diffusion.

Wave–particle interactions can act both as sources and losses of particles in the belts. For example, acceleration of lower-energy electrons can be considered as a source of higher-energy electrons. The losses through wave–particle interactions are due to lowering the particles' pitch angles small enough to precipitate into the atmosphere.

In Chap. 5 we have discussed the growth and decay of waves in Landau and gyro resonances $\omega - k_\parallel v_\parallel = n\omega_{c\alpha}/\gamma$ with the particles. The Landau resonance ($n = 0$) either increases or decreases the parallel energy of the particles depending on the shape of the particle distribution function close to the resonant velocity $v_{\parallel,res} = \omega/k_\parallel$, which leads to energy and pitch-angle diffusion of the phase space density. The perpendicular momentum does not change in Landau resonance and thus the first adiabatic invariant $\mu = p_\perp^2/(2mB)$ is conserved but not the second $J = \oint p_\parallel \, ds$.

The gyro resonance ($n \neq 0$) breaks the invariance of μ and, consequently, the invariance of J and Φ, and leads again to pitch-angle and energy diffusion. Since the gyro resonance takes place in much smaller temporal and spatial scales than the azimuthal drift, its effect on Φ (or on L^*) of the particle remains practically

---

[1] Phase mixing is analogous to the hiding of the initial perturbation in the ballistic term in the Landau damping process (Sect. 4.2.3).

negligible, and thus does not need to be taken into account in pitch-angle diffusion calculations.

In radiation belts gyro-resonant interactions with whistler-mode chorus waves and plasmaspheric hiss at kHz frequencies and with EMIC waves around 1 Hz are the most efficient mechanisms to scatter charged particles toward the atmospheric loss cone. A single wave–particle interaction does not change much the pitch angle and energy, unless the wave amplitude grows to nonlinear regime (see the discussion in Sect. 6.4.4, Fig. 6.4). As the width of the equatorial loss cone in radiation belts is only a few degrees (Fig. 2.6), particles at large equatorial pitch angles must scatter numerous times before they approach the edge of the loss cone and can precipitate to the atmosphere. Consequently, the pitch-angle scattering often is a slow process depleting the radiation belts in timescales of days to hundreds of days, depending on the wave mode and particle energy. Furthermore, for the whistler-mode waves with $\omega < \omega_{ce}$ the resonance vanishes when $v_\parallel \to 0$. Thus, in order to limit a larger than observed excess of close-to-equator mirroring electrons additional mechanisms are needed to scatter the electrons to smaller pitch angles. The same is true for scattering of radiation belt and ring current ions with EMIC waves.

In theoretical investigations it is common to consider particles mirroring exactly at the equator ($\alpha = 90°$), as is frequently done also in this book. For such particles there is no bounce motion and $J = 0$.[2] This is a bit of a singular special case because the inner magnetosphere never is so symmetric that the motion of the particles would remain strictly perpendicular. Due to finite temperature there are thermal velocities and fluctuations in the parallel direction, and once a particle gets parallel momentum, it will be affected by the mirror force. However, this does not solve the problem of finding *efficient enough* scattering mechanism for almost equatorially mirroring particles.

A mechanism that has been invoked to scatter electrons from nearly 90° pitch angles toward larger parallel velocity, where the whistler-mode scattering can take over, is resonance between the electron bounce motion and the equatorial magnetosonic noise introduced in Sect. 5.3.2. The bounce motion requires, by definition, that μ is conserved and the bounce frequency ω_b must be lower than ω_{ce}. If the wave frequency matches with a multiple of ω_b, the resonance can break the invariance of J and lead to scattering in pitch angle and energy.

The bounce-resonant scattering can be investigated by supplementing the parallel equation of motion with a time-dependent force field $F_\parallel(s, t)$, where s is the coordinate along the magnetic field line

$$\frac{dp_\parallel}{dt} = -\frac{\mu}{\gamma}\frac{\partial B(s)}{\partial s} + F_\parallel(s, t), \tag{6.1}$$

---

[2] Recall that the bounce time is well-defined and finite also for "equatorially mirroring" particles, as discussed in Sect. 2.4.2.

where μ is the relativistic adiabatic invariant $p_\perp^2/(2mB)$ (e.g., Shprits 2016, and references therein). The parallel force $F_\parallel$ can be due to an electrostatic wave or, in the case of equatorial magnetosonic noise, due to the parallel electric field component of an oblique (WNA $\approx 89°$) X-mode wave.

Another mechanism to break the invariance of J may arise from compressional ULF fluctuations that affect the length of the bounce path. The net parallel acceleration can in this case be described as Fermi acceleration due to the mirror force oscillating in resonance with the bounce motion. As pointed out by Dungey (1965), the bounce motion is associated with the azimuthal drift. Expanding the azimuthal fluctuation as $\exp(-i\omega t + im\phi)$ the resonance condition can be expressed in terms of bounce and drift frequencies as

$$\omega - l\omega_b - m\omega_d = 0, \tag{6.2}$$

where l is the longitudinal mode number and m the azimuthal mode number of the wave.

We can consider the bounce-resonance associated with a ULF wave mode of a given m. Let ϑ be the co-latitude of spherical coordinates, which at the equator is by definition 90°. Defining the azimuthal wave number as $k_\phi = m/(r \sin \vartheta)$ and bounce-averaged drift velocity as $v_\phi = \omega_d r \sin \vartheta$ at given radial distance r and co-latitude, the resonance condition can be rewritten analogous to the gyro resonance as

$$\omega - k_\phi v_\phi = l\omega_b, \tag{6.3}$$

where $k_\phi v_\phi$ takes the role of the Doppler shift $k_\parallel v_\parallel$ in gyro-resonant interactions. As co-latitude cancels from the equation, the resonance condition is independent of latitude.

Finally, the third invariant Φ, which is inversely proportional to L^*, is violated by the *bounce-averaged drift resonance*

$$\omega - m\omega_d = 0. \tag{6.4}$$

The drift resonance is associated with cross-field motion of trapped particles when the first two adiabatic invariants are conserved. This leads to changes in particle energy and to spreading of the particle distribution to different drift shells, which is commonly referred to as *radial diffusion* (for a review, see Lejosne and Kollmann 2020, and references therein). The concept of radial diffusion was introduced already during the early days of radiation belt research to explain the existence of the outer radiation belt (e.g., Parker 1960). The attribute "radial" is a bit misleading because L^* is not a spatial coordinate but inversely proportional to the magnetic flux enclosed by the particle's drift shell, corresponding to the equatorial radial distance LR_E in a purely dipolar field only. The drift resonance needs low enough wave frequency to match with the timescale of electron's drift motion.

Table 6.1 Summary of different resonances between the waves in the inner magnetosphere and radiation belt particles

Resonance	Description
Gyro resonance (Eq. (5.4)), $n \neq 0$	Resonance with the particle's gyro motion changes the particle's momentum along the single-wave characteristic.
	Nonrelativistic: resonance depends on the parallel velocity of the particle only; straight resonant line in $(v_{\parallel}, v_{\perp})$-space
	Relativistic: resonance depends on both parallel and perpendicular velocities through the Lorentz factor; resonant ellipse in $(v_{\parallel}, v_{\perp})$-space
Landau resonance (Eq. (5.4)), $n = 0$	Electric field parallel to the magnetic field accelerates/decelerates a particle and increases/decreases its *parallel* energy/velocity.
	Requires a finite wave normal angle, and becomes more effective with increasing wave obliquity.
Bounce resonance (Eq. (6.3))	Resonance with multiples of the particle's bounce frequency.
Drift resonance (Eq. (6.4))	Resonance with multiples of particle's drift frequency around the Earth.

Comparison of Tables 2.2 and 4.1 indicates that ULF Pc4–Pc5 waves can interact resonantly with electrons from about 1 MeV upward. Resonance with lower energy electron populations is also possible, but requires high azimuthal wave numbers. This is the case usually with externally excited poloidal waves (Sect. 5.4.1). We discuss the radial diffusion further in Sect. 6.4, including other ways to transport particles across the drift shells, e.g., inductive electric fields related to shock driven compressions of the dayside magnetopause or to substorm expansive phases.

Table 6.1 summarizes the main features how different waves discussed in Chap. 5, can be in resonance with radiation belt charged particles. These waves often co-exist in the inner magnetosphere and affect the particle populations in multiple ways. For example, some of the resonances increase/decrease the perpendicular energy of the particle (gyro and drift resonances), while others (Landau and bounce resonances) increase/decrease the parallel energy and velocity of electrons. To have higher order ($|n| > 1$) gyro resonances the Doppler-shifted wave frequency $\omega - k_{\parallel} v_{\parallel}$ must not become so large that the dispersion equation would no more be fulfilled in the plasma conditions of the inner magnetosphere.

6.2 Quasi-Linear Theory of Wave–Particle Interactions

The time evolution of the phase space density is commonly described using the *diffusion equation* that can also include effects beyond wave–particle interactions or Coulomb collisions. In this Section we introduce the diffusion equation in the framework of *quasi-linear theory*, which is a standard approach in numerical simulation and modeling studies addressing the wave–particle interactions in the inner magnetosphere. An important element is to find the properties of the waves related to wave–particle interactions (e.g. amplitude, wave normal angle, intensity and MLT distribution). These must often be estimated empirically from various observations.

Quasi-linear theory is a theoretical framework in the domain between the linearized Vlasov theory (Sect. 4.2) and nonlinear plasma physics of shocks, large-amplitude waves, wave–wave couplings, strong plasma turbulence, etc. In quasi-linear theory the wave modes are those of the linear plasma theory, but the slow temporal evolution of particle distribution functions is taken into account. The restriction to linear waves is an evident limitation and cannot rigorously address the problem of plasma perturbations growing to large amplitudes. The practical limits of the validity of quasi-linear computations are difficult to assess. For example, it is not clear what is the effect of small-scale nonlinear chorus elements of whistler-mode waves on the larger-scale diffusion process.

As always, there are no general methods to deal with nonlinear plasma equations and nonlinear processes must in practice be considered on a case-by-case basis. Often the best that can be done is to compute the orbits of a large number of randomly launched charged particles in the presence of prescribed nonlinear fluctuations. If it is possible to determine the diffusion coefficients from, e.g., numerical simulations or empirically from observations, they can be inserted in the diffusion equation and used in computation of the temporal evolution of the phase space density, even if the underlying particle scattering would be due to nonlinear interactions.

6.2.1 Elements of Fokker–Planck Theory

The fundamental task is to find a description for the temporal evolution of the charged particle distribution function $\partial f/\partial t$ at a given location in the phase space in the presence of plasma waves, including inter-particle collisions when needed. While the inner magnetospheric plasma is almost collisionless, in addition to various wave–particle interactions, Coulomb and charge-exchange collisions often need to be included in computations of the ring current and radiation belt dynamics. This can formally be done by introducing a collision term $(\partial f/\partial t)_c$ on the right-hand side of the Vlasov equation and rewriting it as the Boltzmann equation (3.17). The *Fokker–Planck approach* is a common, although not the only, method to determine the frictional and diffusive effects arising from the RHS of the Boltzmann equation and it can also be applied to "collisions" between plasma waves and charged particles.

To formally introduce the Fokker–Planck approach let us consider the function $\psi(\mathbf{v}, \triangle\mathbf{v})$ that gives the probability that a particle's velocity $\mathbf{v}$ is deflected, or scattered, by a small increment $\triangle\mathbf{v}$ due to a collision or to an interaction with a wave electric field. Integrating over all possible deflections that may occur during a period $\triangle t$ before the time t gives the distribution function

$$f(\mathbf{r}, \mathbf{v}, t) = \int f(\mathbf{r}, \mathbf{v} - \triangle\mathbf{v}, t - \triangle t)\, \psi(\mathbf{v} - \triangle\mathbf{v}, \triangle\mathbf{v})\, d(\triangle\mathbf{v})\,. \tag{6.5}$$

In the Fokker–Planck approach ψ is assumed to be independent of t. Thus, the scattering process has no memory of earlier deflections and the process can be characterized as a *Markovian random walk* in the phase space.

Next Taylor expand the integral in (6.5) in powers of the small velocity changes $\Delta \mathbf{v}$

$$f(\mathbf{r}, \mathbf{v}, t) =$$
$$\int d(\Delta \mathbf{v}) \left[f(\mathbf{r}, \mathbf{v}, t - \Delta t)\, \psi(\mathbf{v}, \Delta \mathbf{v}) - \Delta \mathbf{v} \cdot \frac{\partial}{\partial \mathbf{v}} (f(\mathbf{r}, \mathbf{v}, t - \Delta t)\, \psi(\mathbf{v}, \Delta \mathbf{v})) \right.$$
$$\left. + \frac{1}{2} \Delta \mathbf{v} \Delta \mathbf{v} : \frac{\partial^2}{\partial \mathbf{v} \partial \mathbf{v}} (f(\mathbf{r}, \mathbf{v}, t - \Delta t)\, \psi(\mathbf{v}, \Delta \mathbf{v})) + \cdots \right], \tag{6.6}$$

where : indicates scalar product of two dyadic tensors $\mathbf{aa} : \mathbf{bb} = \sum_{ij} a_i a_j b_i b_j$. The total probability of all deflections is unity $\int \psi(\Delta \mathbf{v})\, d(\Delta \mathbf{v}) = 1$ and the rate of change of f due to collisions is

$$\left(\frac{\partial f}{\partial t} \right)_c \equiv \frac{f(\mathbf{r}, \mathbf{v}, t) - f(\mathbf{r}, \mathbf{v}, t - \Delta t)}{\Delta t} \tag{6.7}$$
$$\approx -\frac{\partial}{\partial \mathbf{v}} \cdot \left(\frac{\langle \Delta \mathbf{v} \rangle}{\Delta t} f(\mathbf{r}, \mathbf{v}, t) \right) + \frac{1}{2} \frac{\partial^2}{\partial \mathbf{v} \partial \mathbf{v}} : \left(\frac{\langle \Delta \mathbf{v} \Delta \mathbf{v} \rangle}{\Delta t} f(\mathbf{r}, \mathbf{v}, t) \right),$$

where the averages $\langle \Delta \mathbf{v} \rangle$ and $\langle \Delta \mathbf{v} \Delta \mathbf{v} \rangle$ are defined as

$$\langle \ldots \rangle = \int \psi(\mathbf{v}, \Delta \mathbf{v})(\ldots) d(\Delta \mathbf{v}) \tag{6.8}$$

and the terms of the second and higher orders in Δt have been dropped. Note that the denominator in both terms on the RHS of (6.7) is Δt. In random walk the mean square displacements increase linearly with time.

By inserting (6.7) as the collision term to the Boltzmann equation we have arrived to the *Fokker–Planck equation*. The first term on the RHS of (6.7) describes the acceleration/deceleration ($\propto \langle \Delta \mathbf{v} \rangle / \Delta t$) of particles due to collisions, which in classical resistive media corresponds to dynamical friction. The second term is the diffusion term, containing the *diffusion coefficient* $D_{\mathbf{vv}} \propto \langle \Delta \mathbf{v} \Delta \mathbf{v} \rangle / \Delta t$. Note that diffusion can change both the absolute value and the direction of the velocity of the particles. The former corresponds to *energy diffusion*, the latter, in magnetized plasma, to *pitch-angle diffusion*.

The SI units of $D_{\mathbf{vv}}$ are $\mathrm{m}^2\,\mathrm{s}^{-3}$ because the diffusion takes place in the velocity space. In radiation belt diffusion studies the mostly used coordinates are the drift shell, momentum and pitch angle, which have different units. The diffusion equations are commonly normalized so that all diffusion coefficients are given in same units, e.g., $\mathrm{momentum}^2\,\mathrm{s}^{-1}$ or s^{-1}.

Thus far we have nothing more than a formal equation and the hard task is to determine the correct form of the probability function ψ. The diffusion through Coulomb collisions is treated in several advanced plasma physics textbooks (e.g., Boyd and Sanderson 2003) and we skip the technical details. Our focus is on the diffusion resulting from wave–particle interactions and large-scale inhomogeneities of the magnetic field.

6.2.2   Vlasov Equation in Quasi-Linear Theory

Although the Fokker–Planck theory is fundamentally a collisional theory, also wave–particle interactions can be cast to the same formulation within the framework of the quasi-linear approach. The method is to consider the slowly evolving and fluctuating parts of the distribution function separately.

Diffusion Equation in Electrostatic Approximation

The critical assumption of quasi-linear theory is that the temporal evolution of the distribution function $f(\mathbf{r}, \mathbf{v}, t)$ takes place much more slowly than the oscillations of the waves interacting with the particles. The separation is most transparent for electrostatic waves in non-magnetized plasma familiar from the Landau solution of the Vlasov equation in Sect. 4.2.

Let us consider f as a sum of a slowly varying part f_0, which is the average of f over the fluctuations, and of a fluctuating part f_1. For simplicity, we further assume that f_0 is spatially uniform and write

$$f(\mathbf{r}, \mathbf{v}, t) = f_0(\mathbf{v}, t) + f_1(\mathbf{r}, \mathbf{v}, t) . \tag{6.9}$$

Now the Vlasov equation is

$$\frac{\partial f_0}{\partial t} + \frac{\partial f_1}{\partial t} + \mathbf{v} \cdot \frac{\partial f_1}{\partial \mathbf{r}} - \frac{e}{m} \mathbf{E} \cdot \frac{\partial f_0}{\partial \mathbf{v}} - \frac{e}{m} \mathbf{E} \cdot \frac{\partial f_1}{\partial \mathbf{v}} = 0 , \tag{6.10}$$

where the charge density fluctuations are related to the fluctuating electric field through the Maxwell equation

$$\nabla \cdot \mathbf{E} = -\frac{e}{\epsilon_0} \int f_1 \, \mathrm{d}^3 v . \tag{6.11}$$

Assuming that the fluctuations in f_1 and $\mathbf{E}$ are nearly sinusoidal waves, the averages of functions linear in f_1, including $\mathbf{E}$, over the fluctuation period vanish. The average of (6.10) denoted by $\langle \dots \rangle$ is thus

$$\frac{\partial f_0}{\partial t} = \frac{e}{m} \left\langle \mathbf{E} \cdot \frac{\partial f_1}{\partial \mathbf{v}} \right\rangle . \tag{6.12}$$

This equation describes the temporal evolution of f_0.

By subtracting (6.12) from (6.10) we get an equation for the rapid variations of f_1

$$\frac{\partial f_1}{\partial t} + \mathbf{v} \cdot \frac{\partial f_1}{\partial \mathbf{r}} - \frac{e}{m}\mathbf{E} \cdot \frac{\partial f_0}{\partial \mathbf{v}} = \frac{e}{m}\left(\mathbf{E} \cdot \frac{\partial f_1}{\partial \mathbf{v}} - \left\langle \mathbf{E} \cdot \frac{\partial f_1}{\partial \mathbf{v}}\right\rangle\right) . \tag{6.13}$$

In the quasi-linear approximation we neglect the second order nonlinear terms on the RHS as smaller than the linear terms on the LHS, which leads to

$$\frac{\partial f_1}{\partial t} + \mathbf{v} \cdot \frac{\partial f_1}{\partial \mathbf{r}} - \frac{e}{m}\mathbf{E} \cdot \frac{\partial f_0}{\partial \mathbf{v}} = 0 . \tag{6.14}$$

This is formally the same as the linearized Vlasov equation (4.1) with the exception that now, according to (6.12), f_0 is *time-dependent*.

From here on we continue in the same way as in the derivation of the Landau solution. Assuming, for simplicity, that there is only one pole in the complex Laplace-transformed time domain, corresponding to the complex frequency ω_0, we find the fluctuating part of the distribution function in the **k**-space

$$f_1(\mathbf{k}, \mathbf{v}, t) - \frac{ie\mathbf{E}(\mathbf{k}, t)}{m(\omega_0 - \mathbf{k} \cdot \mathbf{v})} \cdot \frac{\partial f_0}{\partial \mathbf{v}} , \tag{6.15}$$

where

$$\mathbf{E}(\mathbf{k}, t) = \frac{ie\,\mathbf{k}\,\exp(-i\omega_0 t)}{\epsilon_0 k^2 (\partial K(\mathbf{k}, \omega)/\partial \omega)|_{\omega_0}} \int \frac{f_1(\mathbf{k}, \mathbf{v}, 0)}{(\omega_0 - \mathbf{k} \cdot \mathbf{v})}\,d^3 v . \tag{6.16}$$

In this expression $K(\mathbf{k}, \omega)$ is the dielectric function of Vlasov theory (4.4).

Finally, by substituting (6.15) and (6.16) to (6.12) and making the inverse Fourier transformation back to the **r**-space the temporal evolution of f_0 is found to be given by the *diffusion equation*

$$\frac{\partial f_0}{\partial t} = \frac{\partial}{\partial v_i}D_{ij}\frac{\partial f_0}{\partial v_j} , \tag{6.17}$$

where the subscripts $\{i, j\}$ refer to the cartesian components of the velocity vector and to the elements of the *diffusion tensor* **D**. Here summing over repeated indices is assumed. The tensor elements D_{ij} are the *diffusion coefficients*

$$D_{ij} = \lim_{\mathscr{V} \to \infty} \frac{ie^2}{m^2 \mathscr{V}} \int \frac{\langle E_i(-\mathbf{k}, t)E_j(\mathbf{k}, t)\rangle}{(\omega_0 - \mathbf{k} \cdot \mathbf{v})}\,d^3 k , \tag{6.18}$$

where $\langle E_i(-\mathbf{k}, t)E_j(\mathbf{k}, t)\rangle/\mathscr{V}$ is the spectral energy density of the electrostatic field and $\mathscr{V}$ denotes the volume of the plasma.

Note that the components of the electric field in (6.18) are given in Fourier-transformed configuration space. In the following we often express the phase space density in other than cartesian velocity coordinates, e.g., as $f(p, \alpha)$ or $f(\mu, K, L^*)$. In practice the diffusion coefficients must be calculated from the observed or modeled amplitude and polarization of the electric field in the $\mathbf{r}$-space and transform thereafter the diffusion equation into the appropriate coordinate system.

Now we have a recipe to calculate the diffusion of the distribution function f_0 in the velocity space *if we can determine the spectrum of electric field fluctuations* for a given wave mode $(\omega_0, \mathbf{k})$.

Diffusion Equation for Magnetized Plasma

The inner magnetospheric plasma is embedded in a magnetic field and the fluctuations are electromagnetic, which makes the treatment of the diffusion equation technically more complicated than in the electrostatic case. The fundamental quasi-linear theory of velocity space diffusion due to small-amplitude waves in a magnetized plasma was presented by Kennel and Engelmann (1966) and has been discussed thoroughly in the monographs by Schulz and Lanzerotti (1974) and Lyons and Williams (1984).

Kennel and Engelmann (1966) derived the diffusion equation for f_0 due to electromagnetic waves into the form

$$\frac{\partial f_0}{\partial t} = \frac{\partial}{\partial \mathbf{v}} \cdot \left(\mathsf{D} \cdot \frac{\partial f_0}{\partial \mathbf{v}} \right) , \tag{6.19}$$

where the *diffusion tensor* D is defined as

$$\mathsf{D} = \lim_{\mathscr{V} \to \infty} \frac{1}{(2\pi)^3 \mathscr{V}} \sum_{n=-\infty}^{\infty} \frac{q^2}{m^2} \int \mathrm{d}^3 k \, \frac{i}{\omega_{\mathbf{k}} - k_\| v_\| - n\omega_c} (\mathbf{a}_{n,\mathbf{k}})^* (\mathbf{a}_{n,\mathbf{k}}) . \tag{6.20}$$

Here $\mathscr{V}$ is the volume of the plasma, the sum is over all harmonic numbers, the vectors $\mathbf{a}_{n,\mathbf{k}}$ contain information on the amplitude and polarization of the wave electric field, the asterisk indicates the complex conjugate, $\omega_{\mathbf{k}}$ is the complex frequency corresponding to the wave vector $\mathbf{k}$, and $\|$ refers to the direction of the background magnetic field. It is evident that accurate empirical determination of the amplitude and polarization is essential for successful numerical computation of the components of the diffusion tensor.

Kennel and Engelmann (1966) further showed that diffusion brings the plasma to a marginally stable state for all wave modes. In the proof no assumption of a small growth rate was made. Thus the conclusion applies to both *non-resonant adiabatic diffusion*, e.g., large-scale fluctuations of the magnetospheric magnetic field, and to *resonant diffusion* at the limit where the imaginary part of the frequency $\omega_{\mathbf{k}i} \to 0$.

At the limit of resonant diffusion the singularity in the denominator of (6.20) is replaced by Dirac's delta that picks up the waves for which

$$\omega_{\mathbf{kr}} - k_{\parallel} v_{\parallel} - n\omega_c = 0 \qquad (6.21)$$

for an integer n. This is the resonance condition familiar from Chap. 5. The theory describes the diffusion resulting from both Landau ($n = 0$) and gyro-harmonic ($n \neq 0$) resonances assuming that the conditions of quasi-linear approach are met.

In radiation belts the particle distribution functions are safe to assume gyrotropic, which motivates formulation of the quasi-linear theory in two-dimensional ($v_{\perp}, v_{\parallel}$) velocity space. It is a straightforward exercise in coordinate transformations (e.g., Chap. 5 of Lyons and Williams 1984) to write the diffusion equation in the (v, α)-space as

$$
\begin{aligned}
\frac{\partial f}{\partial t} &= \nabla \cdot (\mathbf{D} \cdot \nabla f) \\
&= \frac{1}{v \sin \alpha} \frac{\partial}{\partial \alpha} \sin \alpha \left(D_{\alpha\alpha} \frac{1}{v} \frac{\partial f}{\partial \alpha} + D_{\alpha v} \frac{\partial f}{\partial v} \right) \\
&+ \frac{1}{v^2} \frac{\partial}{\partial v} v^2 \left(D_{v\alpha} \frac{1}{v} \frac{\partial f}{\partial \alpha} + D_{vv} \frac{\partial f}{\partial v} \right),
\end{aligned}
\qquad (6.22)
$$

where the subscript $_0$ has been dropped and the slowly[3] evolving velocity distribution function is denoted by f. Note that here the diffusion equation is written in a form where all diffusion coefficients are given in units of velocity^2 s^{-1}. The non-relativistic equation (6.22) can be formulated relativistically by replacing v with $p = |\mathbf{p}| = \gamma m v$. Formally the Lorentz factor only appears as a relativistic correction to the gyro frequency in the calculation of the diffusion coefficients. The relativistic calculations are, however, more complicated because the resonant lines become resonant ellipses as discussed in Sect. 5.1.3.

The diffusion equation (6.22) expresses the already familiar fact that wave–particle interactions can cause diffusion both in the absolute value of the velocity (or kinetic energy $W = mv^2/2$) and in pitch angle. Kennel and Engelmann (1966) pointed out that the particles scatter primarily in pitch angle. Only for particles whose velocities are of the order of or slightly below the wave phase velocity is the energy scattering rate comparable to the rate of pitch-angle scattering.

The direction of the diffusion depends on the shape of the particle distribution function close to the velocity of the particle. For example, when the anisotropy of suprathermal electron distribution amplifies whistler-mode waves in the outer radiation belt (Sect. 5.2.1), the suprathermal electrons scatter toward smaller pitch angles and lower energy. On the other hand the velocity distribution of radiation belt

---

[3] Recall that "slowly" refers to slow compared to the wave oscillation.

electrons (100 keV and above) is more isotropic with $\partial f / \partial W < 0$ and the scattering in energy leads to electron acceleration at the expense of wave power.

6.2.3 Diffusion Equation in Different Coordinates

In radiation belt physics the phase space density in six-dimensional phase space $f(\mathbf{r}, \mathbf{p}, t)$ is often given as a function of the action integrals $\{J_i\} = \{\mu, J, \Phi\}$ and the corresponding gyro-, bounce- and drift-phase angles $\{\varphi_i\}$. If an action integral is an adiabatic invariant, the corresponding phase angle is a cyclic coordinate and the phase space density is independent of that angle. In a fully adiabatic case, where all action integrals are conserved, the phase space is three-dimensional and $f(\mu, J, \Phi)$.

When the adiabatic invariance of one or several action integrals is broken, particles with different phase angles respond differently to the perturbation. For example, gyro-resonant electrons in the same phase as the electric field of a whistler-mode wave are scattered most efficiently leading to gyro-phase bunching of scattered electrons. However, within the quasi-linear approximation, the random walk of the particles in the phase space leads to phase mixing and after a few oscillation periods the individual phases are no more possible to distinguish in observational data. In the case of whistler-mode waves the phase mixing randomizes the phase angles within a few milliseconds, which is well below the temporal resolution of most particle instruments.

The phase mixing is much slower in the drift motion around the Earth. For example, substorm related particle injections from the magnetotail into the inner magnetosphere and abrupt energization due to interplanetary shocks hitting the dayside magnetopause take place much faster than the drifts around the Earth and break the third adiabatic invariant. The drift periods are from a few minutes to several hours (Table 2.2) and the bunches of energetic particles are readily observable in particle spectra as *drift echoes*. Figures 7.6 and 7.8 in Chap. 7 are two illustrative examples of drift echoes after shock-driven acceleration.

The phase mixing facilitates the use of *phase-averaged phase space density* $\overline{f}(\{J_i\}, t)$ in diffusion studies. As the phase information is lost in the averaging, $\overline{f}$ is not consistent with the Liouville theorem in case of broken adiabatic invariance. However, the Fokker–Planck equation can still be applied in the quasi-linear *approximation*. Now the kinetic equation, supplemented with external sources and losses, can be written as

$$\frac{\partial \overline{f}}{\partial t} + \sum_i \frac{\partial}{\partial J_i} \left[\left\langle \frac{dJ_i}{dt} \right\rangle_v \overline{f} \right] = \sum_{ij} \frac{\partial}{\partial J_i} \left[D_{ij} \frac{\partial \overline{f}}{\partial J_i} \right] - \frac{\overline{f}}{\tau_q} + \overline{S} , \qquad (6.23)$$

where $\langle dJ_i/dt \rangle_v$ are the frictional transport coefficients and D_{ij} the elements of the diffusion tensor. The loss and source terms ($\overline{f}/\tau_q$ and $\overline{S}$) represent the average lifetime of immediate loss processes (e.g., magnetopause shadowing or charge exchange) and the drift-averaged external sources of $\overline{f}$. From here on we simplify the notation by dropping the bars above f and S.

It is customary to write the kinetic equation in some other coordinates $\{Q_i\}$ than the basic action integrals $\{J_i\}$. In radiation belt studies J is frequently replaced by K and Φ by L (or L^*). The general coordinate transformation of the kinetic equation is

$$\frac{\partial f}{\partial t} + \frac{1}{\mathscr{J}} \sum_i \frac{\partial}{\partial Q_i} \left[\mathscr{J} \left\langle \frac{dQ_i}{dt} \right\rangle_v f \right] = \frac{1}{\mathscr{J}} \sum_{ij} \frac{\partial}{\partial Q_i} \left[\mathscr{J} \tilde{D}_{ij} \frac{\partial f}{\partial Q_j} \right] - \frac{f}{\tau_q} + S,$$

(6.24)

where $\mathscr{J} = \det\{\partial J_k/\partial Q_l\}$ is the *Jacobian determinant* of the transformation from coordinates $\{J_k\}$ to coordinates $\{Q_l\}$ and $\tilde{D}_{ij}$ denotes the transformed diffusion coefficients. For example, the Jacobian for the transformation from $\{J_i\} = \{\mu, J, \Phi\}$ to $\{Q_i\} = \{\mu, K, L\}$ is $\mathscr{J} = (8m\mu)^{1/2}(2\pi B_E R_E^2/L^2)$, where B_E is the equatorial magnetic field on the surface of the Earth.

Let us neglect the frictional term. Assuming that μ and K are constant, $\mathscr{J} \propto L^{-2}$. This way we obtain the important *radial diffusion equation*

$$\frac{\partial f}{\partial t} = L^2 \frac{\partial}{\partial L} \left(\frac{D_{LL}}{L^2} \frac{\partial f}{\partial L} \right) + S - \frac{f}{\tau_q}.$$

(6.25)

Radial diffusion refers in this context to the statistical effect of the motion of radiation belt particles across the drift shells while conserving the first two adiabatic invariants.

In radiation belt studies the phase space density is often considered as a function of pitch angle, momentum and drift shell. In this case the detailed formulation of the diffusion equation is a bit more complicated (e.g., Schulz and Lanzerotti 1974)

$$\frac{\partial f}{\partial t} = L^2 \frac{\partial}{\partial L}\bigg|_{\alpha,p} \left(\frac{D_{LL}}{L^2} \frac{\partial f}{\partial L}\bigg|_{\alpha,p} \right)$$

$$+ \frac{1}{G(\alpha)} \frac{\partial}{\partial \alpha}\bigg|_{p,L} G(\alpha) \left(D_{\alpha\alpha} \frac{\partial f}{\partial \alpha}\bigg|_{p,L} + p\, D_{\alpha p} \frac{\partial f}{\partial p}\bigg|_{\alpha,L} \right)$$

(6.26)

$$+ \frac{1}{G(\alpha)} \frac{\partial}{\partial p}\bigg|_{\alpha,L} G(\alpha) \left(p\, D_{\alpha p} \frac{\partial f}{\partial \alpha}\bigg|_{p,L} + p^2 D_{pp} \frac{\partial f}{\partial p}\bigg|_{\alpha,L} \right) + S - \frac{f}{\tau_q}.$$

Here α is the pitch angle at equator, $G = p^2 T(\alpha) \sin\alpha \cos\alpha$, and $T(\alpha)$ the bounce function (Eq. (2.76) or (2.77)). D_{LL} is the diffusion coefficient in L^* (the asterisk has been dropped for clarity). $D_{\alpha\alpha}$, $D_{\alpha p}$ and D_{pp} are the diffusion coefficients in pitch angle, mixed pitch angle–momentum, and momentum. In (6.26) all diffusion coefficients are given in units of s^{-1}. Due to vastly different temporal and spatial scales the cross diffusion between L and (α, p) has been neglected. As the gyro-

and Landau-resonant processes scatter particles both in momentum and pitch angle, there is no reason to diagonalize the diffusion coefficients in the (α, p)-space.

The determination of the diffusion coefficients is the most critical part in studies of transport, acceleration, and loss of radiation belt particles. We discuss the procedures in the context of radiation belt electrons in Sect. 6.4.

6.3 Ring Current and Radiation Belt Ions

There are two partially spatially overlapping energetic ion populations in the inner magnetosphere. The strongly time-variable ring current is carried primarily by westward drifting ions in the energy range 10–200 keV, peaking at geocentric distances 3–4 R_E and reaching roughly to 8 R_E. The much less variable proton population of the inner radiation belt is located earthward of 3 R_E. It consists mainly of 0.1–40-MeV protons with a high-energy tail up to relativistic energies of 1–2 GeV.

Although the ring current is not the main focus of our book, the basic dynamics of the current-carrying ions is similar to the dynamics of radiation belt particles. Thus we start this Section with a brief review of the characteristics of the ring current. As noted in Chap. 1, the ring current is the main, but not the only, cause of temporal perturbations in the north component of the equatorial magnetic field on ground. These perturbations are used to calculate the *Dst* and *SYM-H* indices, which, in turn, are commonly used measures of the strength of the storms in the magnetosphere. As the energy density of ions is much larger than that of electrons, the net current is carried mostly by the westward drifting ions. The variability of *Dst* and *SYM-H* during magnetospheric storms is a signature of the variability in the energy density of the current carriers.

6.3.1 Sources of Ring Current Ions

The ultimate sources of the ring current are the ionosphere and the solar wind. The main carriers of the current are energetic protons and O^+ ions. While singly-charged oxygen must be of ionospheric origin, the protons may come from both sources. Table 6.2 summarizes the relative abundances of ring current ions during quiet and storm-time conditions based on AMPTE/CCE and CRRES satellite observations. The data were gathered during a relatively small number of storm-time observations and the numbers shall be taken as indicative. As always, individual storms exhibit large deviations from typical values.

The ion energies in the ionosphere and the solar wind are smaller than the energies of the ring current carriers. While the solar wind proton population already is in the keV-range, the ionospheric plasma has to be accelerated all the way from a few eV. The main ion outflow from the ionosphere takes place in auroral and

Table 6.2 Relative abundances of different ion species and total ion energy density at $L = 5$ in the ring current during quiet times and under different levels of storm activity based on AMPTE/CCE and CRRES observations (Daglis et al. 1999)

Source and species		Quiet times	Small & medium storms	Intense storms
Solar wind H$^+$	(%)	$\gtrsim 60$	~ 50	$\lesssim 20$
Solar wind He^{++}	(%)	~ 2	$\lesssim 5$	$\gtrsim 10$
Ionospheric H$^+$	(%)	$\gtrsim 30$	~ 20	$\lesssim 10$
Ionospheric O$^+$	(%)	$\lesssim 5$	~ 30	$\gtrsim 60$
Solar wind total	(%)	~ 65	~ 50	~ 30
Ionosphere total	(%)	~ 35	~ 50	~ 70
Total energy density	(keV cm^{-3})	~ 10	$\gtrsim 50$	$\gtrsim 100$

polar cap latitudes. The ions are first transported to the magnetospheric tail and only thereafter to the inner magnetosphere, being meanwhile gradually energized.

The acceleration and heating of the outflowing ionospheric plasma takes place in several steps (for a review, see Chap. 2 of Hultqvist et al. 1999). Enhanced O$^+$ ion outflows are observed during substorm growth and expansion phases in the ionosphere by ground-based radars and by satellites traversing the auroral field lines. Thus the observed large storm-time fluxes of O$^+$ in Table 6.2 are not surprising. Some amount of heating by fluctuating electric fields already takes place in the ionosphere. The more energy the ions gain, the more efficiently the mirror force pushes them up. Further acceleration is provided by the magnetic field-aligned electric potential structures of the order of 1–10 kV, which accelerate auroral electrons downward and ions upward. As different particle species move up and down along the magnetic field and drift across the field, the regions above the auroral zones host a large variety of plasma waves, many of which can contribute to the energization of the ionospheric plasma to the keV-range.

In the magnetotail current sheet $\mathbf{J} \cdot \mathbf{E} > 0$, which according to Poynting's theorem of electrodynamics implies energy transfer from the electromagnetic field to the charged particles. Ions crossing the current sheet with a finite but small magnetic field component normal to the sheet (B_n) are transported for a short while in the direction of the electric field and gain energy (see, e.g., Lyons and Speiser 1982). This is an example of *non-resonant diffusion* in pitch angle and energy, which is due to the breaking of the first adiabatic invariant and results in chaotization of particle motion (e.g., Chen and Palmadesso 1986; Büchner and Zelenyi 1989).

Figure 6.1 illustrates how a low-energy ion entering the nightside magnetosphere from the high-latitude mantle is transported first to the distant tail and from there earthward with the large-scale convection while bouncing between the magnetic mirrors. The closer to the Earth the particle comes, the more frequently it crosses the current sheet. Because the stretching is strongest in the distant tail, the particle motion is most chaotic there and, consequently, the acceleration is most efficient for particles entering the plasma sheet far in the tail. Numerical test-particle simulations

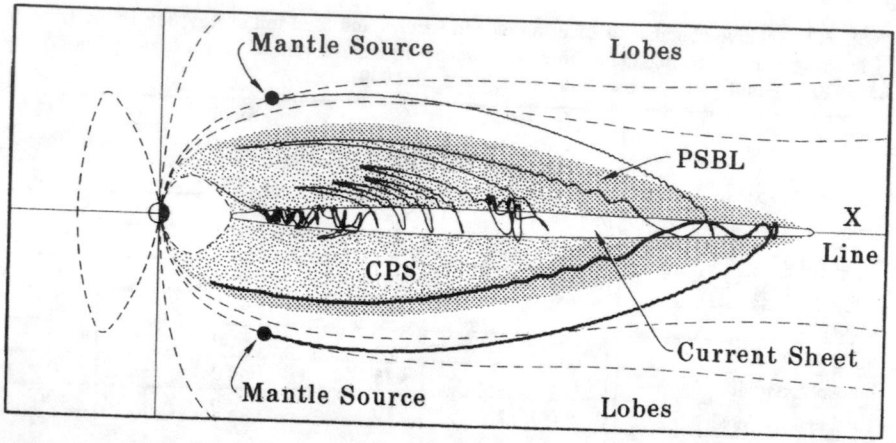

Fig. 6.1 Schematic picture of transport of two solar wind particles with slightly different initial conditions entering to the magnetosphere through the high-latitude mantle to the inner plasma sheet. The difference in the particle trajectories illustrates the sensitivity to the initial conditions, which is characteristic to chaotic motion. The X line is the distant reconnection line of the Dungey cycle (Sect. 1.4.1). The abbreviations CPS and PSBL refer to central plasma sheet and plasma sheet boundary layer (From Ashour-Abdalla et al. 1993, reprinted by permission from American Geophysical Union)

by Ashour-Abdalla et al. (1993) indicated that a particle encountering the current sheet for the first time beyond $80\,R_E$ in the tail with the energy of 0.3 keV can gain energy by a factor of 50 through this process alone.

The north component of the magnetic field in the current sheet increases toward the Earth and the current sheet heating becomes less efficient in the near-Earth space. The particles advecting adiabatically to the inner magnetosphere also gain energy through the drift betatron mechanism (Sect. 2.4.4). However, this is not sufficient to account for ion energies above 100 keV, and wave–particle interactions are called for. Advanced diffusion codes must deal with both resonant and non-resonant source and loss processes (e.g., Jordanova et al. 2010, and references therein).

Substorm dipolarizations can also contribute to ion acceleration through transient inductive electric fields, whose role in reaching 100-keV energies may be important (e.g., Pellinen and Heikkila 1984; Ganushkina et al. 2005). The inductive electric fields can lead to preferential acceleration of O^+ over H^+ because all adiabatic invariants of O^+ can be violated while the magnetic moment of H^+ remains conserved as has been demonstrated in test-particle simulations (e.g., Delcourt et al. 1990).

6.3.2 Loss of Ring Current Ions

The intensity of the ring current is determined by the balance between the sources and losses of the current carriers. The losses are taking place all the time, but during the storm main phase they are overshadowed by the injection of new current carriers. The enhancement of the ring current is a relatively fast process, whereas the losses take more time. This is evident in the rapid negative evolution of the *Dst* index during the storm main phase and much slower decay of the current during the recovery phase (Fig. 1.7).

The main loss of energetic ions, originally suggested by Dessler and Parker (1959), is due to *charge-exchange* collisions between the ring current ions and the neutral hydrogen atoms in the extension of the Earth's collisionless exosphere known as *geocorona*. A typical charge-exchange process is a collision between a positively charged ion and a neutral atom, in which the ion captures an electron from the atom. After the process the charge state of the ion is reduced by one and the neutral particle becomes positively charged

$$X^{n+} + Y \rightarrow X^{(n-1)+} + Y^+ . \tag{6.27}$$

At ring current altitudes the geocorona consists almost purely of hydrogen atoms, but for ions mirroring at low altitudes also the charge exchange with heavier atoms needs to be included in detailed calculations.

The temperature of the neutral geocorona is of the order of 0.1 eV. Thus after a charge exchange with a ring current ion, the emerging particles are an ion of very low energy and an *energetic neutral atom* (ENA). The ENA moves to the direction of the incident ion at the time of the collision and leaves the ring current region. The charge exchange does not directly decrease the number of current carriers, but transfers the charge from fast to very slowly drifting ions. These ions are no more efficient current carriers, instead they become a part of thermal background plasma.

The efficiency of charge exchange as a loss mechanism depends on the lifetimes of the current carriers, which are inversely proportional to the charge-exchange cross sections. The cross sections cannot be calculated theoretically and their empirical determination is difficult because the exosphere is a much better vacuum than can be created in laboratories. Furthermore, the density profile of the geocorona as well as the *L*-shells and pitch angles of the incident ions need to be taken into account because ions mirroring at different altitudes encounter different exospheric densities.

Also Coulomb collisions and wave–particle interactions have a role in removing ring current carriers. The Coulomb collisions are most efficient at lower energies (<10 keV). However, charge exchange and Coulomb collisions jointly do not remove enough ions with energies larger than a few tens of keV, and above 100 keV they lead to flatter pitch-angle distributions (smaller loss cones) than observed (Fok et al. 1996). On the other hand the ring current is embedded in a domain populated by EMIC waves, plasmaspheric hiss and equatorial magnetosonic waves, which can scatter the higher-energy ring current ions to the atmospheric loss cone.

A challenge in inclusion of wave–particle interactions in numerical ring current models is that both the growth and decay of the waves must be modeled self-consistently with the evolution of the particle populations. For example, the growth rate of EMIC waves needs to be calculated solving the hot plasma dispersion equation simultaneously with the kinetic equation. From the growth rate the wave amplitudes are estimated using empirical relations. The effect of wave–particle interactions on the ions is thereafter treated as a diffusion process where the diffusion coefficients are determined using the calculated wave amplitudes (e.g., Jordanova et al. 2010, and references therein).

6.3.3   Sources and Losses of Radiation Belt Ions

The inner radiation belt is relatively stable against short timescale perturbations. The energetic particle content of the inner belt is dominated by protons at MeV energies extending up to a few GeV. For higher energies the gyro radii become comparable to the curvature radius of the background magnetic field and particles cannot any more be trapped in the magnetic bottle. The residence times of protons are long, from years close to the atmospheric loss cone (large adiabatic index K) to thousands of years for equatorially mirroring particles ($K \approx 0$). The particle spectra display variations in decadal (solar cycle) to centennial (secular variation of the geomagnetic field) timescales (e.g., Selesnick et al. 2007).

While the spectrum of trapped ions at energies larger than 100 keV appears to turn quite smoothly from ring current to radiation belt energies, the histories of the ions are different. Ring current carriers up to energies of 100–200 keV originate from the much lower-energy ionosphere and the solar wind being accelerated and transported by various magnetospheric processes. Different mechanisms are needed to produce the radiation belt ions up to tens or hundreds of MeV.

The two main sources of inner radiation belt protons are *solar energetic particle* (SEP) events and the *cosmic ray albedo neutron decay* (CRAND) mechanism. Below energies of 100 MeV and for $L \gtrsim 1.3$ the solar source dominates, whereas below altitudes of 2000 km and at higher energies CRAND is the dominant source.

The solar flares and CMEs produce large fluxes of energetic protons, of which most are shielded beyond $L \approx 4$ by the geomagnetic field. SEPs arriving to the magnetosphere with pitch angles already within the atmospheric loss cone are lost directly, whereas most ions are just deflected by the magnetic field and escape from the near-Earth space. During strong solar particle events solar protons and heavier ions have been found to be injected to L-shells 2–2.5 (e.g. Hudson et al. 2004b, and references therein), from where they are transported inward through radial diffusion. However, the trapped orbits in the innermost magnetosphere are equally difficult to enter into as to escape from, and only a small fraction of incoming protons become trapped. At $L = 2$ the trapping efficiency of 10-MeV protons has been estimated to be of the order of 10^{-4} and of 100-MeV protons only 10^{-7} (Selesnick et al. 2007).

Solar storms have a twofold role in the radiation belt ion dynamics. They provide intermittent source populations and drive perturbations in the magnetosphere that are necessary for particle trapping. However, the energetic solar particles from solar eruptions arrive to the Earth faster than the associated ICME. Thus the trapping is more efficient if the magnetosphere at the time of SEP arrival is perturbed, e.g., by a former ICME or a fast solar wind stream.

Decadal to centennial time series of fresh solar proton injections have been derived from enhancements of the NO_3-rich layers in Arctic and Antarctic ice. Protons with energies larger than 30 MeV penetrate to the Earth's atmosphere and enhance the production of odd nitrates (including NO_3) in the troposphere. The molecules thereafter precipitate to ground and become archived in the polar ice (McCracken et al. 2001, and references therein).

The CRAND mechanism as a source of inner belt protons was suggested by Singer (1958) soon after the early observations of trapped radiation. It results from Galactic cosmic ray bombardment in the atmosphere, which produces neutrons that move to all directions. While the average neutron lifetime is 14 min 38 s, during which a multi-MeV neutron hits the ground or escapes far from the Earth, a small fraction of the neutrons decay to protons while still inside the magnetosphere. Because the Galactic cosmic ray spectrum is hard and temporally constant, the CRAND mechanism produces a hard and stable spectrum.

At energies below 50 MeV the observed proton spectra are too intense and variable to be explained by the CRAND mechanism. Figure 6.2 illustrates results of a model study by Selesnick et al. (2007), in which the main source and loss terms were integrated over 1000-year timescales. The proton fluxes at $L = 1.2$ and $L = 1.7$ are computed for several values of the adiabatic invariant K. The fluxes are highest for the protons mirroring closest to the equator (smallest K), where the trapping times are longest. At $L = 1.2$ the fluxes are several orders of magnitude smaller at all energies than at $L = 1.7$ and the spectrum is completely dominated by CRAND-produced protons. At $L = 1.7$ the softer solar proton spectrum becomes visible below 100 MeV.

The main energy loss mechanisms of the proton belt are the charge exchange and inelastic nuclear reactions with neutral exospheric atoms as well as Coulomb collisions with ionospheric and plasmaspheric charged particles. Furthermore, although small, the adiabatic compression or expansion of the drift shells related to the solar cycle and the secular variation of the geomagnetic field affect the proton energies during their long residence in the inner belt.

The charge exchange cross sections decrease rapidly at energies above 100 keV and the mechanism is a much slower loss process in the inner proton belt than in the ring current. Coulomb collisions and nuclear reactions slowly decrease the ion energy from hundreds of MeV to levels where the ENA production finally can take over the role as a loss process.

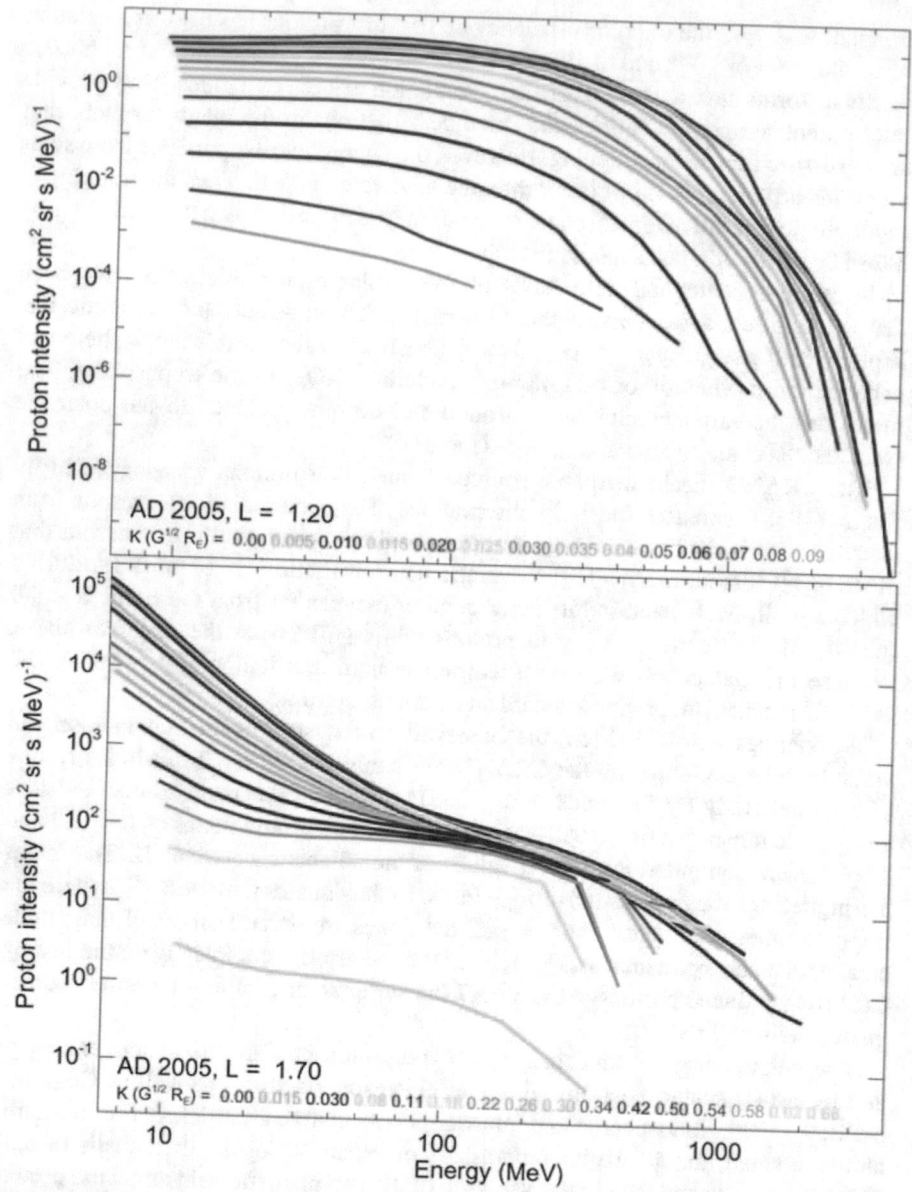

Fig. 6.2 Model calculations of energetic proton spectra at $L = 1.2$ (top) and $L = 1.7$ (bottom). The colors correspond to different values of the adiabatic index K. The uppermost curves are for equatorially mirroring particles ($K \approx 0$). The largest K in the upper picture is $0.09\,G^{1/2}R_E$, in the lower $0.58\,G^{1/2}R_E$, each corresponding to the value for which the mirror points of the entire drift shell are above the Earth's atmosphere. Note different scales on vertical axes in the panels (From Selesnick et al. 2007, reprinted by permission from American Geophysical Union)

6.4 Transport and Acceleration of Electrons

During and after the *Van Allen Probes* mission the dynamical evolution of the energetic electron populations, in particular of the relativistic and ultra-relativistic electrons, has been in the focus of radiation belt research. This has a strong practical motivation because some of the most serious spacecraft anomalies have been addressed to large fluxes of relativistic "killer electrons". In this section we discuss the physical mechanisms of electron acceleration and transport. Electron losses are the topic of Sect. 6.5.

6.4.1 Radial Diffusion by ULF Waves

The traditional theory of the electron belt formation, introduced in the 1960s, is based on inward radial diffusion due to low-frequency electromagnetic fluctuations in asymmetric quasi-dipolar magnetic field. The fluctuations are assumed to conserve the first and second adiabatic invariants but break the third, which in radiation belt studies is usually represented by L^*. In the following discussion we drop the asterisk for clarity and write the diffusion equation (6.25) without external source and loss terms

$$\frac{\partial f}{\partial t} = L^2 \frac{\partial}{\partial L} \left(\frac{D_{LL}}{L^2} \frac{\partial f}{\partial L} \right) , \qquad (6.28)$$

where electromagnetic fluctuations determine the radial diffusion coefficient D_{LL}. When the seed population is transported from the tail toward larger magnetic field, the particles gain energy due to the conservation of the magnetic moment $\mu = p_\perp^2/(2m_e B)$ and, in the presence of ULF waves, by resonant interactions between the waves and the azimuthal drift motion of the electrons.

The practical challenge is to determine the diffusion coefficient D_{LL}. In theoretical analysis one has to make quite a few simplifying assumptions and approximations. Already a slightly distorted dipole field geometry together with standard convection electric field models leads to complications. Furthermore, the intensity of the electromagnetic fluctuations is different at different magnetic local times and a function of magnetospheric activity. On the other hand, the empirical determination of the diffusion coefficients is severely constrained by available observations and different studies have led to different, sometimes contradictory, results (e.g., Ali et al. 2016, and references therein).

Based on purely theoretical arguments Fälthammar (1965) demonstrated that the diffusion coefficient from magnetic field perturbations for equatorially mirroring particles $D_{LL,eq}^{em}$ is proportional to L^{10}. He considered small time-dependent perturbations of the magnetic field assuming that the spatial asymmetry of the perturbation was a stationary stochastic process. Lejosne (2019) re-derived the coefficient in the form

$$D_{LL,eq}^{em} = \frac{1}{8}\left(\frac{5}{7}\right)^2 \left(\frac{R_E}{B_E}\right) L^{10} \omega_d^2 P_A(\omega_d),\qquad(6.29)$$

where B_E is the terrestrial magnetic field at equator, $P_A(\omega_d)$ the power spectral density of the asymmetric compressional magnetic fluctuation and ω_d the angular drift frequency. D_{LL} decreases with decreasing equatorial pitch angle and is at the edge of the atmospheric loss cone reduced to about 10% of the coefficient for equatorially mirroring particles.

For compressional perturbations the diffusion coefficient is determined by the azimuthal component of the inductive electric field ($\nabla \times \mathbf{E} = -\partial \mathbf{B}/\partial t$). Fälthammar (1965) considered electrostatic ($\nabla \times \mathbf{E} = 0$) perturbations separately, for which he found the diffusion coefficient

$$D_{LL}^{es} = \frac{1}{8 R_E^2 B_E^2} L^6 \sum_n P_{E,n}(n\omega_d),\qquad(6.30)$$

where $P_E(n\omega_d)$ is the power spectral density of the nth harmonic of the electric field fluctuation at the drift resonant frequency $\omega = n\omega_d$. In the electrostatic approximation the magnetic field lines are electric equipotentials and the expression is valid for all pitch angles. The SI unit of the ratio of power spectral densities P_E/P_A is that of velocity squared ($m^2\, s^{-2}$). Thus both expressions (6.29) and (6.30) have the same physical dimension (SI unit s^{-1}).

The division of electromagnetic fluctuations to inductive and electrostatic disturbances can be theoretically justified due to their different sources. However, these are difficult to distinguish in satellite observations. Another approach is to calculate a "pure" magnetic diffusion coefficient D_{LL}^b and combine the electrostatic and inductive electric fields into an electric diffusion coefficient D_{LL}^e. This approach was taken by Fei et al. (2006), who developed further the earlier calculations of Elkington et al. (2003). They assumed the electric and magnetic perturbations to be those of compressional Pc5 ULF waves in an asymmetric quasi-dipolar magnetic field in the equatorial plane

$$B(r, \phi) = \frac{B_0 R_E^3}{r^3} + b_1(1 + b_2 \cos\phi),\qquad(6.31)$$

where b_1 describes the global compression of the dipole field and b_2 is the azimuthal perturbation.[4]

---

[4] This model is a simplification of the Mead (1964) model (1.15), as there is no radial dependence in the non-dipolar terms of the magnetic field.

The calculation of Fei et al. (2006) was relativistic, which is important because radial diffusion is often applied to relativistic electrons. They found the diffusion coefficients

$$D_{LL}^b = \frac{\mu^2}{8q^2\gamma^2 B_E^2 R_E^2} L^4 \sum_m m^2 P_{B,m}(m\omega_d) \qquad (6.32)$$

$$D_{LL}^e = \frac{1}{8 B_E^2 R_E^2} L^6 \sum_m P_{E,m}(m\omega_d), \qquad (6.33)$$

where m is the azimuthal mode number and $P_{B,m}$ and $P_{E,m}$ are the power spectral densities of the compressional component of the magnetic field and the azimuthal component of the electric field. D_{LL}^e has the same form as Fälthammar's D_{LL}^{es} (6.30) but the power spectral densities are different. Here the $P_{E,m}$ includes the spectral power of the entire electric field, whereas in (6.30) $P_{E,n}$ represents the electrostatic fluctuations only.

Fei et al. (2006) claimed that their coefficients reduce to those of Fälthammar (1965) in the nonrelativistic limit and taking the different treatment of the electric field into account. However, as pointed out by Ali et al. (2016) and Lejosne (2019) the sum of D_{LL}^e and D_{LL}^b is about a factor of 2 smaller than D_{LL}^{em}. The reason is the assumption of Fei et al. (2006) that the electric and magnetic field perturbations are independent of each other. As demonstrated by Perry et al. (2005), the azimuthal component of the electric field E_ϕ and the time derivative of the poloidal component of the magnetic field $\partial B_\theta / \partial t$ are anticorrelated in the model magnetic field (6.31), as they should be according to Faraday's law ($\nabla \times \mathbf{E} = -\partial \mathbf{B}/\partial t$).

The factor of 2 difference in Fei's and Fälthammar's diffusion coefficients may be a somewhat academic problem in practical diffusion studies, which often involve magnetospheric storms. In such cases the magnetic field model (6.31) is too simple and very different empirically determined diffusion coefficients have been found in different studies. In addition to the compression and stretching of the magnetic field, the time evolving ring current affects the electromagnetic field. For application of different empirically derived diffusion coefficients we refer to the investigation by Ozeke et al. (2020) of the two Saint Patrick's Day (March 17) storms in 2013 and 2015. An example of the practical difficulties is that while most of the time $D_{LL}^b \ll D_{LL}^e$, during the storm main phase this relationship can be the reverse.

Due to the great variability of inner magnetospheric conditions the empirical determination of the diffusion coefficients on the case by case basis may be the only way of finding diffusion rates consistent with particle observations. For a given D_{LL} the diffusion equation (6.28) is fast to compute numerically, which makes it possible to look for an optimal coefficient as a function of appropriate magnetospheric parameters. A widely-used parameterization is that of Brautigam and Albert (2000) based on observations of the magnetospheric storm on 9 October 1990. They used the Kp-index as a parameter and found the diffusion coefficient

$$D_{LL}(L,t) = a L^b 10^{cKp(t)} \qquad (6.34)$$

with coefficients $a = 4.73 \times 10^{-10}$, $b = 10$, and $c = 0.506$. More empirical event-based parameterizations can be found, e.g., in the above cited publications by Elkington et al. (2003), Ali et al. (2016) and Ozeke et al. (2020), and in articles cited therein. It is clear that event-based derivations using different parameters lead to different results but the critical L-dependence is in most cases close to Fälthammar's original L^{10}.

6.4.2 Electron Acceleration by ULF Waves

We illustrate the drift resonant acceleration of electrons by discrete ULF wave modes following Elkington et al. (2003). They considered equatorial electrons in the model magnetic field of Eq. (6.31). The drift contours (2D drift shells) are determined by the constant magnetic field strength, where the L-parameter is replaced by

$$\mathcal{L} = \left(\frac{R_E^3}{r^3} + \frac{b_1 b_2}{B_0} \cos\phi \right)^{-1/3}. \tag{6.35}$$

For small perturbations ($b_1 \ll B_0$), $\mathcal{L} \approx L$ within the radiation belts.

The electric field of the ULF waves in the equatorial plane was given in Eq. (4.107) as

$$\mathbf{E}(r, \phi, t) = \mathbf{E}_0(r, \phi) + \sum_{m=0}^{\infty} \delta E_{rm} \sin(m\phi \pm \omega t + \xi_{rm}) \, \mathbf{e}_r +$$

$$+ \sum_{m=0}^{\infty} \delta E_{\phi m} \sin(m\phi \pm \omega t + \xi_{\phi m}) \, \mathbf{e}_\phi .$$

Here $\mathbf{E}_0(r, \phi)$ is the time-independent convection electric field. δE_{rm} are the electric field amplitudes of the *toroidal* modes and $\delta E_{\phi m}$ of the *poloidal* modes, and ξ_{rm} and $\xi_{\phi m}$ represent their phase lags.

For radial diffusion to be efficient the fluctuations should be global and resonate with a multiple of the angular drift frequency of the electrons

$$\omega - (m \pm 1) \omega_d = 0. \tag{6.36}$$

In the outer radiation belt the $m = 2$ mode fulfils the resonance condition at the fundamental drift frequency ($\omega = \omega_d$) of relativistic electrons. At $L = 6$ the drift periods of 1–5-MeV electrons are 2.7–12.3 min (Table 2.2), matching with the period range of Pc5 waves. Resonance with the higher-frequency Pc4 oscillations is possible for larger azimuthal mode numbers. For a few tens to a few hundred keV

Toroidal-mode electric field ($m = 2$)

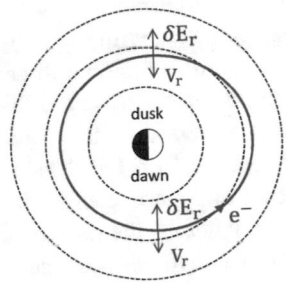

Poloidal-mode electric field ($m = 2$)

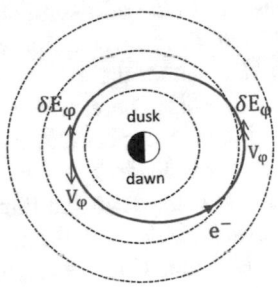

Fig. 6.3 Illustration of drift resonance of an electron with $m = 2$ toroidal (left) and poloidal (right) mode wave electric field. Noon is to the right

electrons (drifting around the Earth at $L = 6$ in timescales of an hour to about ten hours) to be in drift resonance with Pc4–Pc5 waves the azimuthal mode numbers have to be very large (~ 10–100). In summary, global large-amplitude ULF waves in the Pc4–Pc5 frequency band are natural agents of radial diffusion.

According to (2.68) the adiabatic (μ conserving) acceleration of equatorial electrons is given by

$$\frac{dW}{dt} = q\mathbf{E} \cdot \mathbf{v}_d + \mu \frac{\partial B}{\partial t} , \tag{6.37}$$

where $\mathbf{v}_d$ is the electron drift velocity around the Earth. The magnetic perturbation of the toroidal mode δB_ϕ and the dominant magnetic field component of the poloidal mode δB_r both have a node at the equator. In both cases the electric field of the ULF wave has a component along the direction of the of the electron's GC drift velocity in certain parts of drift path around the Earth resulting in drift-betatron acceleration. The compressional component $\delta B_\parallel$ of the poloidal mode is assumed to be so small, that the gyro-betatron acceleration $\mu\,\partial B/\partial t$ can be neglected and the energization is due to the drift-betatron effect only. In the toroidal mode $\delta B_\parallel = 0$ by definition. Figure 6.3 illustrates how a drift resonant ($\omega = \omega_d$) electron is accelerated by the toroidal (left) and poloidal (right) $m = 2$ ULF wave in a distorted dipole.

Let us first consider an electron interaction with the toroidal mode (Fig. 6.3, left). Start from the point in the dusk sector where the radial component of the electron velocity v_r reaches a maximum in inward direction. Here the electron encounters an outward electric field δE_r. Thus $dW/dt = q\delta E_r v_r > 0$, and the electron is accelerated. Half a drift period later the electron is in the dawn sector, where it has a maximal outward velocity component and encounters an inward electric field, and is accelerated again. The dawn and dusk sectors are in this case regions of maximal energy gain. The electron actually gains energy throughout of the orbit around the Earth, except at noon and midnight where $v_r = 0$. The asymmetric compression of the magnetic field is an important factor in the process. Increasing distortion increases v_r in the dawn and dusk sectors and thus increases the energy gain.

Efficient drift acceleration can also occur from the resonance with the poloidal mode δE_ϕ, where the electric field perturbation δE_ϕ is in the azimuthal direction (Fig. 6.3, right). The electron on the nightside encounters an electric field that is in the direction opposite to its velocity and is accelerated. On the other hand, if the electron is in drift resonance with the wave, it encounters on the dayside an electric field that is in the direction of the drift motion and loses energy. In the compressed dipole configuration $|\delta E_\phi v_\phi|$ is, however, smaller on the dayside than on the nightside. Thus the electron gains net energy over the drift period around the Earth. Adding a static convection electric field $\mathbf{E}_0$ weakens the net acceleration by the poloidal mode of this particular electron, because $\mathbf{E}_0$ is in the same direction as δE_ϕ when the electron moves on the nightside and on the nightside dayside.

It is important to realize that the two examples in Fig. 6.3 describe only one electron in drift resonance with a discrete single-frequency wave. Considering electrons in different drift phases with respect to the phase of the wave, some electrons gain, some others lose energy. Some of them are pushed closer to the Earth, others further away from the Earth. The net result is both radial and energy diffusion (see also the discussion by Lejosne and Kollmann 2020).

If the poloidal modes are distributed over a range of frequencies, or a non-static convection field is acting on the electron, the dominant component of the electron's drift velocity in the azimuthal direction may permit even more efficient acceleration than the interaction with purely toroidal modes of the same amplitude. Based on numerical calculations with a continuum of frequencies Elkington et al. (2003) concluded that the resonant mechanism can lead to very efficient inward diffusive radial transport of electrons and their acceleration from 100 keV to MeV energies.

It is evident that the radial transport and acceleration by ULF Pc4–Pc5 waves are closely related to each other. While the diffusive transport is generally considered as a relatively slow process (of the order of days) the ULF waves may also lead to fast radial diffusion. Jaynes et al. (2018) studied radiation belt electron response during the magnetic storm on 17 March 2015. The electron fluxes from a few hundred keV to relativistic energies recovered soon after the peak of the storm, while ultra-relativistic electron fluxes stayed low for a few days. While the energization up to relativistic energies could have been related to enhanced chorus wave activity to be discussed in Sect. 6.4.5, the reappearance and inward transport of ultra-relativistic electrons (up to 8 MeV) occurred when the observed chorus activity had already subsided, whereas empirical estimates of radial diffusion coefficients suggested fast diffusion. Because empirical diffusion coefficients are event-specific, also the amounts of radial diffusion and acceleration are event-specific.

6.4.3 Diffusion Coefficients in the (α, p)-Space

The challenges in the determination of diffusion coefficients in the pitch-angle momentum space are different from those of D_{LL}. The diffusion tensor is given by Eq. (6.20) but the calculation of its elements requires the use of an appropriate approximation of the dispersion equation and knowledge of the amplitude and polarization of the waves interacting with particles. In practical computations one needs to use realistic models of the spatial distribution and properties of the waves.

We consider the relativistic formulation of the diffusion equation (6.22) without external sources and losses following Lyons and Williams (1984)

$$\frac{\partial f}{\partial t} = \frac{1}{p \sin \alpha} \frac{\partial}{\partial \alpha} \sin \alpha \left(D_{\alpha\alpha} \frac{1}{p} \frac{\partial f}{\partial \alpha} + D_{\alpha p} \frac{\partial f}{\partial p} \right) +$$

$$+ \frac{1}{p^2} \frac{\partial}{\partial p} p^2 \left(D_{p\alpha} \frac{1}{p} \frac{\partial f}{\partial \alpha} + D_{pp} \frac{\partial f}{\partial p} \right) . \tag{6.38}$$

Here $D_{\alpha\alpha}$, $D_{\alpha p} = D_{p\alpha}$ and D_{pp} are the drift- and bounce-averaged diffusion coefficients

$$D_{\alpha\alpha} = \frac{p^2}{2} \left\langle \frac{(\Delta\alpha)^2}{\Delta t} \right\rangle$$

$$D_{\alpha p} = \frac{p}{2} \left\langle \frac{\Delta\alpha \Delta p}{\Delta t} \right\rangle \tag{6.39}$$

$$D_{pp} = \frac{1}{2} \left\langle \frac{(\Delta p)^2}{\Delta t} \right\rangle .$$

The multiplicative factors $p^2/2$, $p/2$ and $1/2$ normalize the units of all coefficients to momentum^2 s^{-1}. These coefficients can be computed from coefficients for given harmonic number n and perpendicular wave number $k_\perp$ as integrals over all wave numbers and sums over all harmonics as

$$D_{\alpha\alpha} = \sum_{n=-\infty}^{\infty} \int_0^{\infty} k_\perp \, dk_\perp \, D_{\alpha\alpha}^{nk_\perp}$$

$$D_{\alpha p} = \sum_{n=-\infty}^{\infty} \int_0^{\infty} k_\perp \, dk_\perp \, D_{\alpha p}^{nk_\perp} \tag{6.40}$$

$$D_{pp} = \sum_{n=-\infty}^{\infty} \int_0^{\infty} k_\perp \, dk_\perp \, D_{pp}^{nk_\perp} .$$

The integrals are calculated over perpendicular wave vectors only, as the resonance condition yields Dirac's delta in the parallel direction.

The expressions (6.40) are the components of the diffusion tensor (6.20) origi-nally derived by Kennel and Engelmann (1966) in the non-relativistic approximation and generalized to relativistic particles by Lerche (1968). The diffusion coefficients for given n and $k_\perp$ are related to the pure pitch-angle diffusion coefficients, which after a lengthy calculation turn out to be

$$
D_{\alpha\alpha}^{nk_\perp} = \lim_{\mathscr{V}\to\infty} \frac{q_j^2}{4\pi\,\mathscr{V}} \left(\frac{-\sin^2\alpha + n\omega_{cj}/(\gamma\omega)}{\cos\alpha} \right)^2 \frac{\Theta_{nk}}{|v_\parallel - \partial\omega/\partial k_\parallel|} \tag{6.41}
$$

for a given particle species j. Here $\mathscr{V}$ is the plasma volume, and the derivative $\partial\omega/\partial k_\parallel$ is to be evaluated at the resonant parallel wave number

$$
k_{\parallel,res} = (\omega - n\omega_{cj}/\gamma)/v_\parallel . \tag{6.42}
$$

The function Θ_{nk} contains the information of the amplitude and polarization of the wave electric field

$$
\Theta_{nk} = \left| \frac{E_{\mathbf{k},L}\,J_{n+s_j} + E_{\mathbf{k},R}\,J_{n-s_j}}{\sqrt{2}} + s_j \frac{v_\parallel}{v_\perp} E_{\mathbf{k},\parallel}\,J_n \right|^2 . \tag{6.43}
$$

Here L, R, and $\parallel$ refer to the left-hand, right-hand and parallel polarized components of the wave electric field for a given wave vector. The argument of the Bessel functions J_n is $(k_\perp v_\perp \gamma/\omega_{cj})$ and s_j is the sign of the particle species j. Finally, $D_{\alpha p}^{nk_\perp}$ and $D_{pp}^{nk_\perp}$ are

$$
D_{\alpha p}^{nk_\perp} = D_{\alpha\alpha}^{nk_\perp} \left(\frac{\sin\alpha\cos\alpha}{-\sin^2\alpha + n\omega_{cj}/(\gamma\omega)} \right) \tag{6.44}
$$

$$
D_{pp}^{nk_\perp} = D_{\alpha\alpha}^{nk_\perp} \left(\frac{\sin\alpha\cos\alpha}{-\sin^2\alpha + n\omega_{cj}/(\gamma\omega)} \right)^2 . \tag{6.45}
$$

The terms multiplying $D_{\alpha\alpha}^{nk_\perp}$ in these equations are smaller than 1, which is consistent with the conclusion of Kennel and Engelmann (1966) that pitch-angle diffusion dominates over diffusion in energy (or in the absolute value of momentum) as noted in Sect. 6.2.2.

In practical computations the distribution of the wave power as a function of frequency is often approximated by a Gaussian as

$$
B^2(\omega) = A^2 \exp\left(-\left(\frac{\omega - \omega_m}{\delta\omega} \right)^2 \right) . \tag{6.46}
$$

Here A is a normalization constant and ω_m and $\delta\omega$ are the frequency and band-width of the maximum wave power. However, the determination of the diffusion coefficients for waves fulfilling the electromagnetic dispersion equation is still a formidable technical task requiring heavy numerical computations (e.g., Glauert and Horne 2005). Restricting the analysis to parallel propagating ($k_\perp = 0$) whistler-mode and EMIC waves, the integrals assuming Gaussian distribution in frequency can be expressed in closed form (Summers 2005). This speeds up the computations significantly but means a neglection of effects of obliquely propagating waves, which are critical to the dynamics of radiation belts.

As will be discussed in Sect. 6.5, wave–particle resonances with the electron bounce motion also result in pitch-angle diffusion, which is found to be important for nearly equatorially mirroring particles. In that case the diffusion coefficients must be calculated without bounce-averaging. Detailed calculations of such diffusion coefficients have been presented by Tao and Li (2016) for equatorial magnetosonic waves, by Cao et al. (2017a) for EMIC waves and by Cao et al. (2017b) for the low-frequency plasmaspheric hiss.

6.4.4 Diffusion due to Large-Amplitude Whistler-Mode and EMIC Waves

When whistler-mode or EMIC waves grow to large amplitudes, the quasi-linear approach to calculate diffusion coefficients becomes invalid. Different schemes to estimate the diffusion by nonlinear wave–particle interactions have been introduced in the literature, e.g., the formation of electron phase-space holes discussed in Sect. 5.2.4 (Omura et al. 2013) and the dynamical systems approach (Osmane et al. 2016, and references therein). Here we present a straightforward approach to numerically integrate the equation of the electron motion in a wave field $(\mathbf{E}_w(\mathbf{r}, t), \mathbf{B}_w(\mathbf{r}, t))$ determined from observations or theoretical arguments. The relativistic equation of motion in a latitude-dependent background magnetic field $\mathbf{B}_0(\lambda)$ is

$$\frac{d\mathbf{p}}{dt} = -e\left(\mathbf{E}_w + \frac{1}{\gamma m_e}\mathbf{p} \times \mathbf{B}_w\right) - \frac{e}{\gamma m_e}\mathbf{p} \times \mathbf{B}_0(\lambda). \tag{6.47}$$

By launching a large number of electrons with different initial conditions representing the original $f(\alpha, p)$, it is possible to estimate the diffusion coefficients (6.39) from $\triangle\alpha$ and $\triangle p$ averaged over a time period $\triangle t$.

In practical computations (6.47) is convenient to transform to coupled differential equations for momentum parallel ($p_\parallel$) and perpendicular ($p_\perp$) to $\mathbf{B}_0$ and for the phase angle η between the perpendicular velocity of the electron $\mathbf{v}_\perp$ and the perpendicular component of the wave magnetic field $\mathbf{B}_{w\perp}$. After gyro-averaging, neglecting second order terms, and assuming parallel propagating ($k = k_\parallel$) waves the relativistic equations are (Albert and Bortnik 2009)

$$\frac{dp_\parallel}{dt} = \left(\frac{eB_w}{\gamma m_e}\right) p_\perp \sin\eta - \frac{p_\perp^2}{2\gamma m_e B_0} \frac{\partial B_0}{\partial s}$$

$$\frac{dp_\perp}{dt} = -\left(\frac{eB_w}{\gamma m_e}\right)\left(p_\parallel - \frac{\gamma m_e \omega}{k_\parallel}\right)\sin\eta + \frac{p_\perp p_\parallel}{2\gamma m_e B_0}\frac{\partial B_0}{\partial s} \qquad (6.48)$$

$$\frac{d\eta}{dt} = \left(\frac{k_\parallel p_\parallel}{\gamma m_e} - \omega + \frac{n\omega_{ce}}{\gamma}\right) - \left(\frac{eB_w}{\gamma m_e}\right)\left(p_\parallel - \frac{\gamma m_e \omega}{k_\parallel}\right)\frac{\cos\eta}{p_\perp}.$$

Here s is the coordinate along the magnetic field, the gradient ($\partial B_0/\partial s$) represents the mirror force, ω_{ce} is the electron gyro frequency in the background field $\mathbf{B}_0$ and n the harmonic number. The velocity along the magnetic field is given by $ds/dt = p_\parallel/(\gamma m_e)$ (For corresponding non-relativistic equations, see Dysthe 1971; Bell 1984). This formulation is applicable to both whistler-mode waves ($n \geq 1$) (e.g., Bortnik et al. 2008a) and EMIC waves, in which case it is sufficient to consider the first order resonance ($n = -1$) only (e.g., Albert and Bortnik 2009).

Because the electron gyro frequency is higher than the wave frequency, in case of EMIC waves much higher, the terms relative to $\sin\eta$ and $\cos\eta$ average to zero for small-amplitude waves after a few gyro periods and $d\eta/dt$ can be approximated as

$$\frac{d\eta}{dt} = \frac{k_\parallel p_\parallel}{\gamma m_e} - \omega + \frac{n\omega_{ce}}{\gamma}, \qquad (6.49)$$

In resonant interaction η is practically constant over a short period Δt and (6.49) reduces to the familiar resonance condition. As long as $\Delta v_\parallel$ remains small compared to the adiabatic motion, the interaction can be described using resonant ellipses (Sect. 5.1.3).

However, if B_w grows large enough, the nonlinear terms ($\propto \sin\eta$) are no more negligible and the scattering can be quite different. To understand the nonlinear resonant interaction in case of large-amplitude wave field, take the second time derivative of the last equation of (6.48) and insert $dv_\parallel/dt$ from the first. This gives an equation for a nonlinear driven oscillator, assuming here, for simplicity, non-relativistic motion ($\gamma = 1$)

$$\frac{d^2\eta}{dt^2} + k\left(\frac{eB_w}{m_e}\right)v_\perp \sin\eta = \left(\frac{3}{2} + \frac{\omega_{ce} - \omega}{2\omega_{ce}}\tan^2\alpha\right)v_\parallel \frac{\partial\omega_{ce}}{\partial s}, \qquad (6.50)$$

where the terms of smaller orders have been neglected (Bortnik et al. 2016).

The type of interaction depends on the relative effect of the nonlinear term on the LHS and the driving term on the RHS of (6.50). In addition to the amplitude of the wave the result depends on the particle's pitch angle (α) and on the latitude where the interaction takes place through the latitude-dependence of the mirror force

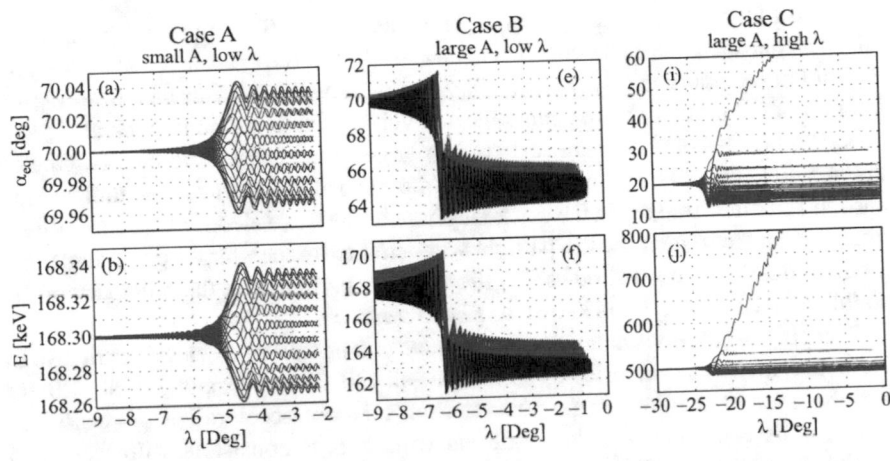

Fig. 6.4 Test-particle simulations of electron interaction with a whistler-mode wave packet. Small/large A refers to the wave amplitude and λ is the magnetic latitude. The spreading in the equatorial pitch angle (α_{eq}) and energy (E) is caused by the different phases of the particles when they interact with the wave packet. The oscillatory behavior of pitch angle and energy is due to their η-dependence close to the resonance $d\eta/dt \approx 0$ and decays when the particle moves further away from the site of the resonance (from Bortnik et al. 2008a, reprinted by permission from American Geophysical Union)

$\propto \partial\omega_{ce}(\lambda)/\partial s$. If the driving term dominates, the interaction remains linear. Note that at equator $\partial B_0/\partial s \to 0$ and the nonlinear interaction can become important also for small B_w (see also discussion at the end of Sect. 6.5.4).

Bortnik et al. (2008a) applied this method to whistler-mode chorus waves using both typical and very large amplitudes. They launched 24 test particles with initial phases η_0 distributed uniformly between 0 and 2π to move through a whistler-mode wave packet. The frequency of the wave packet representing the whistler-mode chorus elements was 2 kHz and the wave propagated away from the equator at $L = 5$ in the geomagnetic dipole field (Fig. 6.4).

In cases A and B of Fig. 6.4 the initial energy of the particles was 168 keV and the equatorial pitch angle $\alpha_{eq} = 70°$. The parameters of the particles were selected so that they started at the latitude $\lambda = -9°$ and were in resonance with the wave at $\lambda \approx -5°$ (case A) and $\lambda \approx -6.5°$ (case B). The interaction time Δt with a single wave packet was about 10–20 ms.

In case A the wave amplitude was 1.4 pT. The scattering from a single wave packet remained small as expected: in equatorial pitch angle about 0.03°–0.04° and in energy 30–40 eV. After several encounters with similar wave packages the result would be similar to quasi-linear diffusion.

In case B the interaction of the same particles with a large-amplitude wave $B_w = 1.4$ nT, corresponding to observations by Cattell et al. (2008), was quite different. The equatorial pitch angles of all particles dropped about 5° and the energies by 5 keV. In this case the initially uniform phases η became bunched by the wave at the time of interaction. Such a non-linear behavior is known as *phase bunching*.

In case C the initial energy was 500 keV and the initial equatorial pitch angle 20°. The particles were launched from $\lambda = -30°$ and they resonated at $\lambda \approx -23°$ with a large-amplitude ($B_w = 1.4$ nT) wave with WNA 50°. In this case a large fraction of the particles were again phase bunched with decrease in pitch angle and energy, whereas some of them scattered to larger pitch angles and energy. One of the particles (the red track in Fig. 6.4) became trapped in the wave potential and the particle remained trapped in the constant phase of the wave electric field for a longer time period. Consequently, the particle wandered to a much larger pitch-angle and the total energy gain was 300 keV at the time the particle met the boundary of the simulation. This behavior is known as *phase trapping*.

Trapping of electrons in the nonlinear wave potential is an essential result also in the above mentioned theories by Omura et al. (2013) and Osmane et al. (2016). As discussed at the end of the next section, efficient acceleration of relativistic electrons by nonlinear whistler-mode wave packets is consistent with wave and particle observations of *Van Allen Probes* (Foster et al. 2017).

6.4.5 Acceleration by Whistler-Mode Chorus Waves

Electron acceleration by chorus waves is different from the drift-resonant acceleration by ULF waves. It takes place through the gyro resonance between the waves and the electrons breaking the first adiabatic invariant. The right-hand polarized chorus waves interact with a fraction of energetic electrons through the Doppler-shifted gyro resonance $\omega - k_\parallel v_\parallel = n\omega_{ce}/\gamma$. Variables ω, $v_\parallel$ and, ω_{ce} can often be measured but $k_\parallel$ must, in practice, be determined by solving the dispersion equation, which in turn depends on plasma density and composition. As discussed in Sect. 5.1.3, the resonance with a wave of a particular ω and $k_\parallel$ defines a resonant ellipse in the $(v_\perp, v_\parallel)$-plane, which reduces to a resonant line in nonrelativistic case when the resonant condition depends only on electron's parallel velocity.

The chorus is a wide-band emission, so there is a continuum of resonant ellipses and, consequently, there is a finite volume in the velocity space intersected by single-wave characteristics (Eq. (5.9)). Thus a large number of electrons are affected as long as they fulfil the resonance condition. The resonant diffusion curves illustrate that gyro-resonant interactions with whistler-mode chorus waves can efficiently energize electrons from a few hundred keV to MeV energies (Summers et al. 1998). The energization takes place near the equator where chorus waves propagate almost parallel to the background magnetic field (small WNAs). This is important to the Doppler shift term $k_\parallel v_\parallel$ that must be large enough for the wave frequency and particle's (relativistic) gyro frequency to match.

An example of strong and rapid local acceleration of electrons in the heart of the outer radiation belt via interaction with whistler-mode waves was presented by Thorne et al. (2013b). They studied the geomagnetic storm on 9 October 2012 in the early phase of the *Van Allen Probes mission*. Intense chorus activity was observed from the dawn to dayside sector. The electron diffusion was calculated solving the Fokker–Planck equation (6.26) in the pitch-angle–momentum (α, p) space. The results of the diffusion calculations were consistent with *Van Allen Probes* data. The authors, however, noted that a definitive conclusion about the relative importance of chorus waves vs. other processes was not possible to give due to observational limitations.

Similarly to diffusion caused by ULF waves the critical issue is the determination of the diffusion coefficients, now in pitch angle and momentum. The procedure has to be based on empirical or modelled plasma and wave properties. The drift averaging is needed in practice although the wave distribution in local time is inhomogeneous and varies from one event to another. Thus the estimation of net acceleration is a challenge.

An example of determining the diffusion coefficients from several sources as the input to the analysis of the electron acceleration during the so-called Halloween storm in autumn 2003 was published by Horne et al. (2005). The interaction with chorus waves is most efficient when ω_{pe}/ω_{ce} is relatively small ($\lesssim 4$), which is the case outside the plasmapause. On 31 October 2003, this condition was met, as the high-density plasmasphere was confined inside $L = 2$ and remained inside $L = 2.5$ in the pre-noon sector (06–12 MLT) until November 4. In their numerical calculations the authors used relativistic electron data from the SAMPEX satellite, Kp and Dst indices, ground-based ULF observations and kHz-range wave observations from the *Cluster* spacecraft. They argued that the radial diffusion due to the ULF waves could not explain the strong increase of 2–6 MeV electron fluxes between L shells from 2 to 3 in the late phase of the storm after 1 November 2003. Instead, the Fokker–Planck calculations, based on diffusion rates calculated for chorus wave amplitudes measured by *Cluster* at somewhat higher drift shell ($L = 4.3$), suggested that the gyro-resonant interaction was sufficient to explain the establishment of very high electron fluxes in the slot region during this exceptionally strong storm period. We will discuss the slot region and the Halloween storm more thoroughly in Sect. 7.4.

While quasi-linear diffusion computations assuming linear whistler-mode waves seem to be able to produce observed acceleration of MeV electrons, the role of the large-amplitude rising-tone whistler elements (Sect. 5.2.4) raises interesting questions. For example, how well does a quasi-linear diffusion model represent the collective effect of nonlinear wave–particle interactions?

Foster et al. (2017) investigated the recovery of 1–5-MeV electrons after they had become depleted during the main phase of the storm on 17–18 March 2013. The recovery of MeV electrons was preceded by the occurrence of electrons in the energy range from a few tens to hundreds of keV that were injected by substorms in the storm recovery phase. Foster et al. (2017) used *Van Allen Probes* observations of the rising-tone whistler-mode wave packets to compute electron energization in the theory of Omura et al. (2015, and references therein, see Sect. 5.2.4). Figure 6.5 summarizes their results, according to which the nonlinear interaction turned out to be very efficient. For example, resonant 1-MeV electrons were found to be able to gain 100 keV in a single interaction with a wave packet of 10–20 ms duration. The observed wave packets had oblique WNAs from 5° to 20°, which was taken into

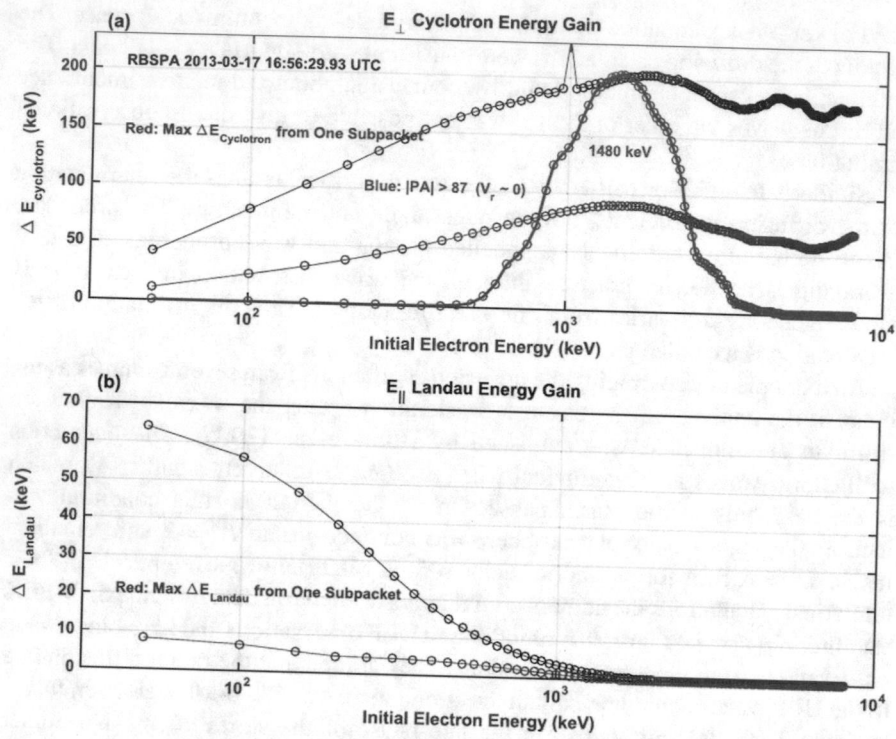

Fig. 6.5 The energy gain at different initial energies due to observed rising-tone whistler-mode wave packets. The upper panel shows the acceleration by gyro resonance and the lower panel by Landau resonance for different initial energies. The black lines indicate acceleration summed over all wave packets, the red lines the maximum energization by a single wave packet. The blue curve shows the most probable acceleration at pitch angles >87° (From Foster et al. 2017, reprinted by permission from American Geophysical Union)

account in the analysis, and both gyro and Landau resonances were accounted for. The Landau resonance was effective at energies below 1 MeV and comparable to the gyro resonance below 100 keV.

6.5   Electron Losses

In this section we discuss the basic features of electron loss processes from the outer radiation belt. The complexity of the dynamics of the belt is further discussed in Chap. 7 in the light of recent observations.

The main loss mechanisms of radiation belt electrons are magnetopause shadowing and pitch-angle scattering to the atmospheric loss cone via wave–particle interactions. Observations indicate that the electron flux in the belts can be strongly depleted in timescales of days or hours, sometimes even minutes. Also Coulomb collisions cause pitch-angle scattering, but they are much less efficient. For example, the lifetimes against Coulomb collisions of 100-keV electrons exceed one year beyond $L = 1.8$ and is about 30 years at $L = 5$ (Abel and Thorne 1998).

Practically all wave modes discussed in Chap. 5 can contribute to the losses of the outer belt electrons. What is the dominant scattering mechanism, depends on the electron energy and equatorial pitch angle. To remove an electron away from the belt it needs to be close to the loss cone, which is at the equator only a few degrees. Furthermore, gyro-resonant interactions are inefficient at equatorial pitch angles close to $90°$, where Landau- and bounce-resonant processes turn out to be important in scattering the electrons to smaller pitch angles where the gyro-resonant interactions can take over. Because a single interaction in the quasi-linear domain changes the pitch angle only by a very small amount (Sect. 6.4.4), a large number of interactions are needed to change the electron's pitch angle so much that it moves to the loss cone. Nonlinear interactions with large-amplitude waves can, however, lead to significant changes of pitch angle even in one interaction. If the electron interacts with a wave at higher latitude, where the loss cone width is much wider, it is easier to nudge it out of the belt.

6.5.1 Magnetopause Shadowing

Losses through magnetopause shadowing occur when the drift paths of electrons touch the magnetopause. Due to their large gyro radii the high-energy radiation belt electrons can cross the magnetopause even if the background plasma remains frozen-in the magnetospheric magnetic field. Figure 6.6 illustrates different factors contributing to the shadowing.

The nominal distance to the subsolar magnetopause during magnetospheric quiescence is about $10\,R_E$, which is well beyond the typical radiation belt electron drift shells (Fig. 6.6, left). Local inward ripples and excursions in the magnetopause can, however, allow electrons' drift paths to cross the magnetopause although the nominal distance would be beyond the drift shell. For example, during relatively quiet conditions Kelvin–Helmholtz vortices and/or flux transfer events at the magnetopause can cause such local inward excursions leading to losses.

During periods of large solar wind dynamic pressure the subsolar magnetopause can be compressed inside the geostationary distance ($6.6\,R_E$) as illustrated in the middle of Fig. 6.6. The compression enhances the shadowing losses as does the erosion of the magnetic field due to dayside reconnection. During the main phase of geomagnetic storms the ring current is enhanced, which leads to decrease of the equatorial magnetic field earthward of the current and on the surface of the Earth. Outside of the peak current on the equatorial plane the magnetic field is inflated. To conserve the third adiabatic invariant electrons move outward so that their drift shells enclose the same flux inside their drift paths illustrated on the right of Fig. 6.6.

Local Bow Shock Ripples Magnetopause Compression Drift Shell Expansion

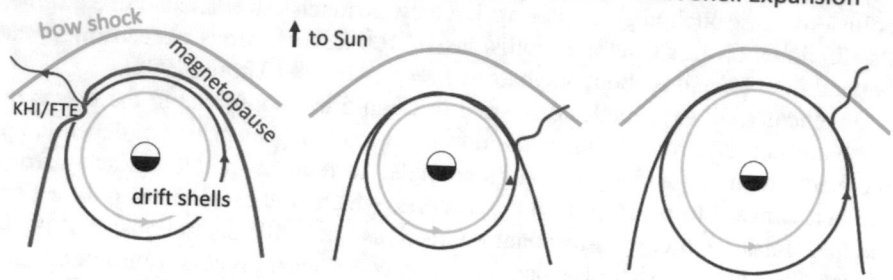

Fig. 6.6 Schematics of magnetopause shadowing during nominal (left) and strongly compressed (middle) magnetosphere as well as the inflated drift shells during the main phase of magnetospheric storms (right). The picture on the left reminds that also local perturbations of the magnetopause, such as Kelvin–Helmholtz instabilities and flux transfer events, can let radiation belt electrons to escape from the magnetosphere. The blue trace indicates the drift shell of a particle that crosses the magnetopause. The figure is a simplification of a similar picture in Turner and Ukhorskiy (2020)

As discussed in Sect. 2.6.2, the drift shell splitting due to the dayside compression of the magnetosphere shifts the electrons with large pitch angles furthest out. Consequently, such particles are lost most efficiently, which leads to butterfly-type of electron distribution function at large L. The electron energy also affects how effective shadowing losses are MeV electrons drift around the Earth in minutes (Table 2.2) and can get lost even in the case of a short time inward magnetopause excursion. The drift periods of lower-energy electrons can be hours and if the disturbance is short-lived, it can remove only a small fraction of the population.

6.5.2 Losses Caused by Whistler-Mode Waves in Plasmasphere

Interaction of electrons with whistler-mode waves scatters electrons both in energy and pitch angle. Whether this leads to acceleration or loss of radiation belt electrons depends on the shape of the particle distribution function close to the resonant velocity. It is important to keep in mind that in numerical and theoretical studies the detailed results depend on the chosen models of frequencies and WNA distributions of the wave amplitudes and on the properties of background plasma and magnetic field. For example, the frequency of the plasmaspheric hiss is less than 0.1 times the local electron gyro frequency, whereas outside the plasmapause, where the gyrofrequency is smaller, the frequencies of chorus waves are in the range 0.1 ω_{ce} < ω < 1.0 ω_{ce}. Thus different approximations of the dispersion equation need to be used in wave–particle interaction calculations if the complete dispersion equation is numerically too demanding to apply.

Plasmaspheric hiss plays a central role in the loss of electrons from the inner parts of the outer radiation belt and in the formation of the slot region between the inner and outer belts. Lyons et al. (1972) calculated diffusion coefficients based on, at time still relatively limited, observations. They were able to demonstrate that the core of the inner belt is not affected much by hiss-induced diffusion but its outer edge, i.e., the inner edge of the slot, is energy dependent, being closest to the Earth at highest energies, which is consistent with modern observations to be discussed more thoroughly in in Sect. 7.2 (see, Fig. 7.2).

The diffusion coefficients are proportional to the wave power, i.e., the square of the wave amplitude (6.41), and the estimated lifetimes depend on the distribution of wave power along the orbits of the particles. Lyons et al. (1972) used the amplitude of $B_w = 35$ pT and found electron lifetimes within the slot to be 1–10 days, increasing with increasing energy up to 2 MeV. Using a smaller hiss amplitude of $B_w = 10$ pT Abel and Thorne (1998) found electron lifetimes in the energy range 100 keV–1.5 MeV to be of the order 100 days, which they found to be consistent with several satellite observations in the outer electron belt but yet inside the plasmasphere. As shown in Fig. 5.11 the hiss amplitudes vary from a few to a few to a few tens of pT during quiet magnetospheric conditions to 100–300 pT during storms, resulting in large variations of radiation belt electron lifetimes in the plasmasphere.

Lyons et al. (1972) pointed out that to obtain correct electron lifetimes, in addition to the sum over a sufficient number of harmonic gyro-resonant ($n \neq 0$) terms, it is necessary to include the Landau resonance ($n = 0$) in the calculation of the diffusion coefficients. This is because for large WNA, where the whistler mode turns to the magnetosonic/X-mode (Fig. 4.3), the minimum gyro resonant velocity $v_{\parallel,res} = n\omega_{ce}/k_{\parallel}$ becomes larger than the velocity of the particles and the Landau resonance starts to dominate the scattering process. These conclusions have been confirmed and refined in several later investigations taking advantage of much more detailed and extensive modern observations (e.g., Ni et al. 2013; Thorne et al. 2013b, and references therein).

Ni et al. (2013) investigated the effects of gyro- and Landau-resonant terms on electron lifetimes up to ultra-relativistic energies for plasmaspheric hiss and oblique magnetosonic/X-mode waves. Figure 6.7 illustrates the diffusion coefficients. The green curves were calculated assuming quasi-parallel propagating whistler mode, the red curves with a model including a latitude-dependent WNA, to represent the observations that the WNA of the whistler mode is more oblique at higher latitudes (Chap. 5), and the blue curves indicate the diffusion caused by the magnetosonic/X-mode waves.

A striking feature in Fig. 6.7 is the so-called "bottleneck" of very small pitch-angle diffusion coefficients between nearly-perpendicular and smaller equatorial pitch angles. It is due to that gyro-resonant scattering, which dominates at small and intermediate pitch angles, is not efficient at $\alpha_{eq} \approx 90°$, where the Landau resonance takes over. Thus, the bottleneck slows down their transport from very large pitch angles toward the atmospheric loss cone. This problem was already recognized by Lyons et al. (1972) who pointed out that the combined gyro- and Landau-resonant interactions are not efficient enough to scatter electrons from large to the intermediate pitch angles. This results in larger non-relativistic electron fluxes at nearly-equatorial pitch angles than observed.

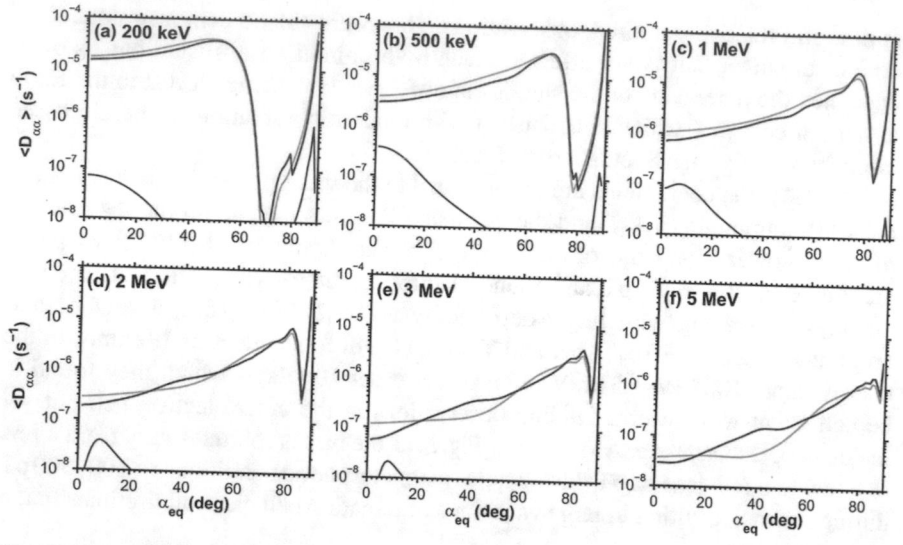

Fig. 6.7 Examples of pitch-angle diffusion coefficients at $L = 3.2$. The colors correspond to different propagation directions of plasmaspheric hiss: The green curves represent a model of quasi-parallel propagation of the whistler mode, the blue highly oblique propagation in the magnetosonic/X mode. The red curves are calculated using a model where the WNA of the whistler mode increases with increasing latitude. The "bottleneck" discussed in the text is the drop in the diffusion rates between gyro-resonant interaction at smaller equatorial pitch angles and Landau interaction close to 90° (From Ni et al. 2013, reprinted by permission from American Geophysical Union)

According Ni et al. (2013) the drop between the gyro-resonant and Landau-resonant diffusion is energy-dependent and extends also to relativistic energies. They found that below 2 MeV (top three panels of Fig. 6.7) the inclusion of the first-order gyro and Landau resonances and quasi-parallel propagation is an equally good approximation as calculations including higher-order terms. At ultra-relativistic energies realistic latitude-dependent WNAs and higher harmonics need to be taken into account. Ni et al. (2013) noted that above 3 MeV the higher harmonics even become dominant at intermediate pitch angles. The diffusion due to the nearly-perpendicular propagating magnetosonic/X-mode waves (blue curves in Fig. 6.7) was found to be weaker at all pitch angles, being most notable for lower-energy (<1 MeV) electrons at both small ($\lesssim 40°$) and large ($\gtrsim 80°$) pitch angles.

Ni et al. (2013) estimated the lifetimes of equatorially mirroring electrons to be days at 500 keV, a few tens of days at 2 MeV and more than 100 days at 5 MeV. Thus, in those rare cases where ultra-relativistic electrons get access to the slot region as a result of strong magnetospheric perturbations, they can remain trapped for weeks or months as will be discussed in Sect. 7.4.

A possible way to overcome the bottleneck is the pitch-angle scattering due to resonance with the bounce periods of the electrons (Sect. 6.1). Because the electron bounce frequencies in the plasmasphere are of the order of a few Hz (Table 2.2), which are smaller than the lowest observed hiss frequencies of a few tens of Hz, the interaction can only take place at high multiples of the (angular) bounce frequency $\omega = l\omega_b$. Cao et al. (2017b) calculated the diffusion coefficients including bounce terms up to $l = 50$ in the L-range 4–5. They found the diffusion rates to be comparable with the Landau resonance at energies below 0.5 MeV. At energies above 1 MeV the bounce resonance exceeds the Landau resonance in particular at intermediate pitch-angles $\gtrsim 50°$. They concluded that the Landau and bounce resonances are critical to move electrons from $\alpha_{eq} \approx 90°$ to smaller pitch angles.

The different electron lifetimes due to scattering by plasmaspheric hiss are demonstrated in high-resolution observations by *Van Allen Probes*. Zhao et al. (2019b) studied the high-energy electron spectra beyond $L \approx 2.6$ and found that instead of decreasing monotonically as a function of energy they tend to peak around 2 MeV. They called these *reverse* or *bump-on-tail* spectra similar to the familiar gentle-bump of the elementary Vlasov theory (Fig. 5.1). In the plasmasphere the energy density of relativistic particles is, however, much smaller than that of the dense and massive background plasma and the bump is too gentle to drive an instability, being a consequence rather than a driver of plasma waves. The reverse high-energy spectra form during a few days after the storm main phase when the plasmaspheric hiss scatters electrons of a few hundred keV to 1 MeV to the atmospheric loss cone followed by slow inward transport of >1-MeV electrons. Hiss scatters also >1 MeV electrons, but very slowly. Numerical simulation of the spectral evolution solving the diffusion equation (6.39) following the big ($Dst_{min} = -222\,nT$) Saint Patrick's day storm on 17 March 2015 was found to reproduce the observations very well (Fig. 6.8).

Other whistler-mode waves potentially leading to pitch-angle diffusion in the plasmasphere are lightning-generated whistlers and emissions from ground-based VLF transmitters. The lightning-generated whistlers have maximum amplitudes at frequencies 3–5 kHz, which are higher than the typical plasmaspheric hiss frequencies. Meredith et al. (2009) added a model of lightning-generated whistler spectra to their diffusion calculations and found that they introduce a possible way to overcome the bottleneck by resonating with high pitch-angle ultra-relativistic electrons (2–6 MeV) and scattering them to lower pitch angles.

The strongest artificial signals propagating in the whistler mode to the inner magnetosphere arise from U.S. Naval communication transmitters at frequencies around 25 kHz. As to be discussed in Sect. 7.4.2, the emissions form a radio bubble around the Earth, which has been proposed as an explanation why ultra-relativistic electrons only seldom penetrate to the L-shells below 2.8.

6.5.3   Losses due to Chorus Waves and Electron Microbursts

Outside the plasmapause the whistler mode appears as chorus emissions. The basic wave–particle interactions with radiation belt electrons are similar to those in the plasmasphere but here the background plasma is both hotter and much more

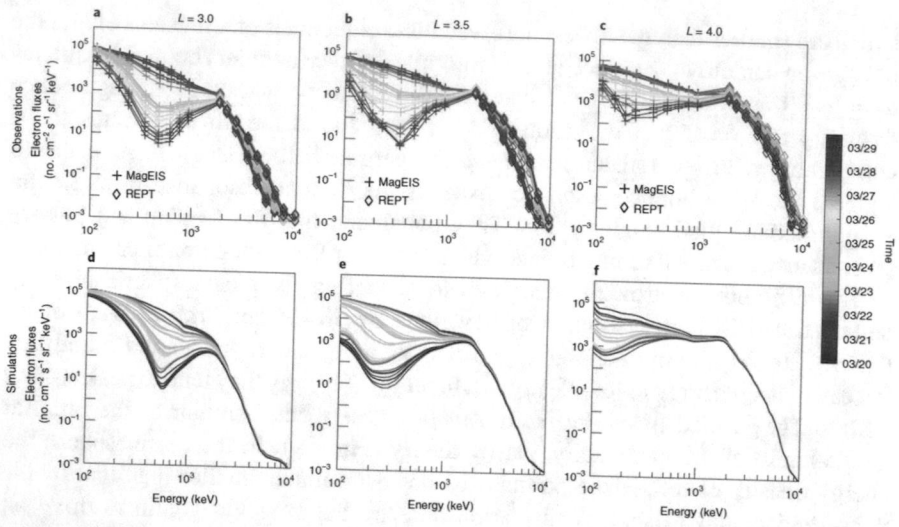

Fig. 6.8 Comparison of the evolution of observed and simulated spectra during 20–29 March 2015. The upper panels show the *Van Allen Probe* A MagEIS and REPT spectra and the lower panels the Fokker-Planck simulation of equatorially trapped electrons using a time-varying plasmaspheric hiss model. The colors indicate the time from 20 March (blue) to 29 March (red) (From Zhao et al. 2019b, reprinted by permission from Springer Nature)

tenuous, and the frequencies of chorus waves are closer to the local electron gyro frequency than in the plasmasphere. All these factors affect the propagation characteristics of the waves and how they interact with electrons of different energies and pitch angles. When suprathermal electrons advect from the magnetospheric tail, their distribution function becomes anisotropic in the velocity space, which leads to instability and energy transfer from the electrons to the whistler-mode chorus waves (Sect. 5.2). At higher energies chorus waves are considered as efficient accelerators of electrons up to relativistic energies (Sect. 6.4.5).

The chorus emissions are also important agents of pitch-angle diffusion toward the atmospheric loss cone, because the pitch-angle diffusion coefficients are larger than the energy diffusion coefficients (Eq. (6.40)). While the scattering of $\lesssim 100\,\text{keV}$ electrons is efficient near the equator, the chorus-wave losses of MeV electrons is most efficient at higher latitudes ($\lambda \gtrsim 15°$), where the waves propagate increasingly obliquely (e.g., Thorne et al. 2005, and Fig. 5.5).

A specific feature of chorus waves, in particular of the lower-band chorus below 0.5 f_{ce}, is that they are composed of short nonlinear rising-tone emissions in frequency (Sect. 5.2.4). These large-amplitude wave packets may lead to the brief *electron microbursts*. The microbursts were originally identified in balloon-borne observations as X-ray bremsstrahlung of $\gtrsim 200$-keV electrons precipitating into the atmosphere (Anderson and Milton 1964). The microbursts have later been observed with instruments on several high-altitude balloons and sounding rockets and using fast-sampling electron detectors looking upward into the atmospheric loss cone onboard spacecraft traversing the high-latitude ionosphere at low altitudes where the loss cone is wide.

The microbursts occur in timescales of milliseconds and they are observed in the energy range from a few tens of keV to several MeV. At lowest energies they are related to the generation of chorus waves, while at higher energies microbursts have been shown to be able to empty the outer radiation belt in a timescale of one day. Both quasi-linear gyro-resonant interaction with small-amplitude chorus waves and nonlinear interaction with large-amplitude chorus wave packets have been suggested to cause the very rapid pitch-angle scattering of electrons, even from a brief interaction with a single chorus wave packet (Bortnik et al. 2008a). In addition, the nonlinear Landau trapping by large-amplitude oblique whistler-mode chorus at high geomagnetic latitudes has been suggested to play a significant role in losses as they increase efficiently the parallel energy of electrons in a region where the loss cone is relatively wide (e.g., Osmane et al. 2016, and references therein).

The number of events with simultaneous high-resolution observations of large-amplitude chorus emissions and microburst precipitation in close magnetic field conjunction is limited. Mozer et al. (2018) investigated an event on 11 December 2016 when high-resolution wave data from *Van Allen Probes* B was available. The observed wave amplitude exceeded occasionally 1 nT, being in the nonlinear regime. The cross-correlation between 1-s averaged precipitating electron flux observed with the low-altitude *AeroCube* 6B microsatellite and the *Van Allen Probes* wave magnetic field was close to 0.9, which is an exceptionally high correlation in this context.

Mozer et al. (2018) calculated the standard bounce-averaged quasi-linear pitch-angle diffusion coefficient for an average amplitude of 100 pT. They found that the observed precipitating electron flux corresponded remarkably well to the estimated flux from quasi-linear diffusion once the data was averaged over 1 s, which extends over several periods of both chorus elements and microbursts. This result suggests that, on the average, the Fokker–Planck approach may describe quite well the pitch-angle scattering although the underlying scattering process may be nonlinear interaction with high-amplitude elements of whistler-mode chorus waves.

Another interesting conjugate event occurred on 20 January 2016. The CubeSat FIREBIRD II observed microbursts of 200-keV to 1-MeV electrons, and *Van Allen Probe* A detected lower band chorus of similar cadence and duration (Breneman et al. 2017). As microbursts were dispersionless, the scattering was considered to be a nonlinear first-order gyro resonance. *AeroCube* and FIREBIRD observations illustrate that even CubeSat-class satellites can have great scientific value.

The Japanese *Arase* satellite, launched in December 2016, made it possible to conduct together with *Van Allen Probes*, high-resolution magnetically conjugate wave and particle observations simultaneously close to the equator and at higher magnetic latitudes. An example of conjugate observations between *Arase* and *Van Allen Probes* A on 21 August 2017 was published by Colpitts et al. (2020) (Fig. 6.9). This was the first time when the propagation of individual whistler-mode wave packets from the lower (12°) to higher (21°) magnetic latitude was directly observed.

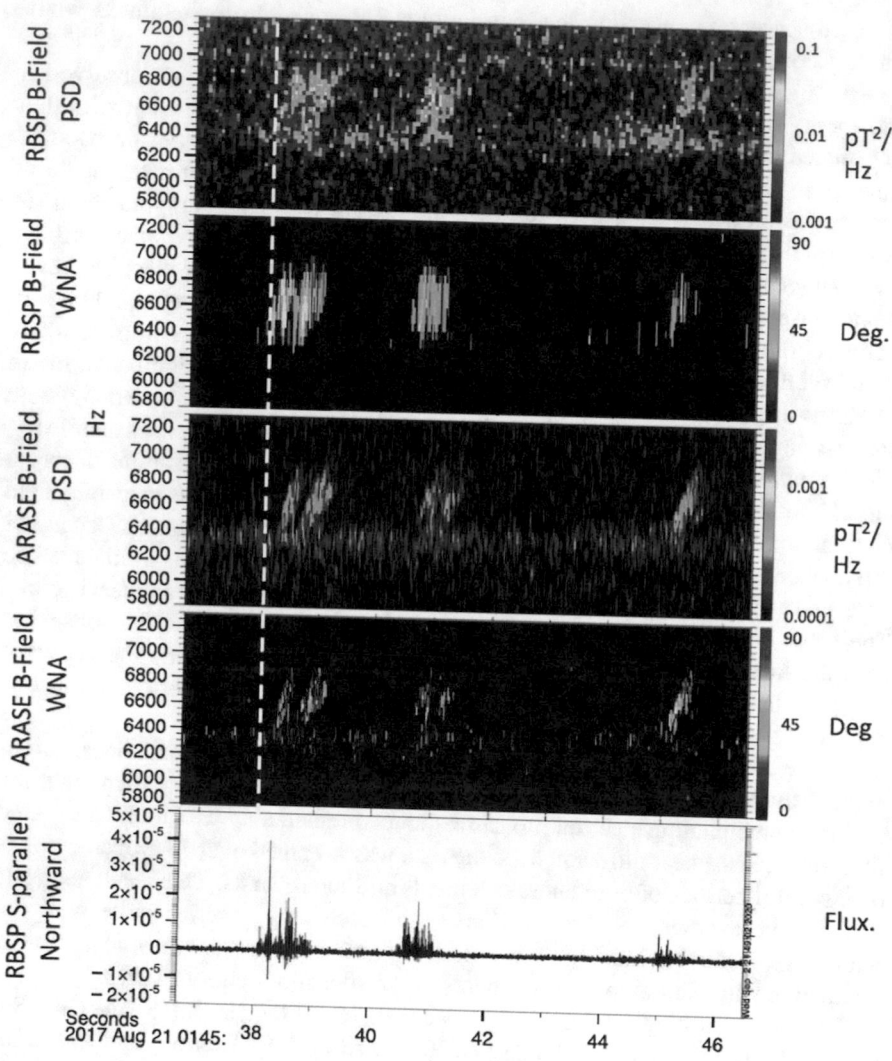

Fig. 6.9 Observations of the propagation of whistler-mode wave packets from magnetic latitude $\lambda = 12°$ (*Van Allen Probes* A) to $\lambda = 21°$ (*Arase*) over a period of 10 s. The top four panels show the *Van Allen Probes* A power spectral density (here PSD does not mean phase space density!) and wave normal angle (WNA), *Arase* PSD and WNA, all these in the frequency range 5.8–7.3 kHz, i.e., in the lower-band whistler mode ($f_{ce}/2 \approx 7.9$ kHz). The lowest panel shows the magnetic field-aligned component of the Poynting vector calculated from *Van Allen Probes* A electric and magnetic field data (From Colpitts et al. 2020, reprinted by permission from American Geophysical Union)

Figure 6.9 shows a few chorus elements of the lower-band whistler mode observed by both satellites. The Poynting vector in the bottom panel indicates that wave energy was propagating at the location of *Van Allen Probes* A ($\lambda = 12°$) toward the higher latitude. The dashed vertical line indicates the time when the first chorus element arrived at *Van Allen Probes* A. The same element arrived 0.2 s later at *Arase* at $\lambda = 21°$, consistent with a ray-tracing study presented by Colpitts et al. (2020). The wave normal angle became increasingly oblique while the wave propagated to the higher latitude, thus making the wave more efficient to scatter the relativistic electrons toward the loss cone.

Due to observational limitations it is difficult to answer the question how large fraction of the total electron precipitation losses beyond the plasmapause are in the microbursts. Greeley et al. (2019) investigated their role during storm recovery phases using SAMPEX observations from 1996 to 2007. They found that the microburst losses had a high correlation with the global loss of 1–2 MeV electrons, in particular during storms driven by interplanetary coronal mass ejections (ICME), when the microbursts may even be the main loss process. The correlation was weaker for stream interaction region (SIR) driven storms. (For further discussion of the different storm drivers, see Sect. 7.3.)

6.5.4   Losses Caused by EMIC Waves

The quasi-linear pitch-angle diffusion of relativistic electrons due to whistler-mode chorus waves is a relatively slow process. On the other hand, electromagnetic ion cyclotron waves have been found to lead to enhanced electron losses at L-shells close to the plasmapause where the waves are frequently observed, in particular during storm-time conditions. Summers et al. (1998) demonstrated that EMIC waves lead to almost pure pitch-angle diffusion. Contrary to chorus, the EMIC waves are not efficient electron accelerators.

According to the resonance condition $\omega - k_\parallel v_\parallel = n\omega_{ce}/\gamma$, in which n can be both a positive and negative integer (or zero), both right-hand and left-hand polarized waves can be in resonance with right-hand gyrating electrons. The whistler-mode resonances correspond to $n \geq 1$. For the resonance with left-hand polarized EMIC waves it is, in practice, sufficient to consider the lowest order term ($n = -1$) only, due to much smaller wave frequency compared to the electron gyro frequency. The relative direction of the wave propagation and the parallel electron velocity must, of course, be such that in the electron's guiding center frame the wave rotates in the same sense as the electron. Furthermore, the resonance condition requires that the energy of the electron is sufficiently high. Not only the Lorentz factor γ has to be large but also the parallel velocity must be large enough to Doppler shift the wave frequency close to ω_{ce}/γ. The minimum resonant energies are of the order of 1 MeV or larger, assuming that the plasma frequency is considerably higher than the electron gyrofrequency ($\omega_{pe}/\omega_{ce} \geq 10$) (Summers and Thorne 2003). This condition is met close to the plasmapause in the afternoon sector where the EMIC waves are frequently observed.

In numerical diffusion studies it is essential to apply an appropriate background plasma model. An example is the investigation by Jordanova et al. (2008) of the intense storm on 21 October 2001. The model included all major loss processes and was coupled with a dynamic plasmasphere model with 77% H^+, 20% He^+ and 3% O^+. The EMIC wave amplitudes were calculated self-consistently with evolving plasma populations, resulting to He^+ band amplitudes $B_w \approx 5$ nT at $L = 4.5$ and $B_w \approx 10$ nT at $L = 6.25$. The analysis was performed considering separately EMIC scattering alone, all processes except EMIC waves, and all scattering processes including EMIC waves. The highest pitch-angle diffusion coefficients for relativistic electrons were found to be in the range 0.1–$5\,s^{-1}$ and limited to equatorial pitch angles $\lesssim 60°$. Considering that the applied He^+ band frequencies were below 1 Hz, so strong diffusion is at the limit of the quasi-linear approach. Jordanova et al. (2008) concluded that scattering by EMIC waves enhances the loss of >1-MeV electrons and can cause significant electron precipitation during the storm main phase. This conclusion has been verified observationally during the *Van Allen Probes* era when phase space densities have been possible to calculate with better accuracy than before (e.g., Shprits et al. 2017, and references therein).

In theoretical calculation of the resonant energy for interaction of EMIC waves with *electrons* both electron and ion terms in (4.63) must be retained and the derivation is a bit more complicated than the derivation of the *ion* resonant energy (5.19) (e.g., Summers and Thorne 2003; Meredith et al. 2003). Considering nearly parallel propagating hydrogen band EMIC waves and assuming small He^+ ($<10\%$) and O^+ ($<20\%$) concentrations Mourenas et al. (2016) derived a simplified equation for the minimum resonant energy of the electrons

$$W_{res,min} \approx \frac{\sqrt{1+K}-1}{2}, \qquad (6.51)$$

where

$$K = \frac{1}{\cos^2 \alpha_{eq}} \frac{\omega_{ce,eq}^2}{\omega_{pe,eq}^2} \frac{\omega_{cp,eq}^2 (1 - \omega/\omega_{cp,eq})(m_p/m_e)}{\omega^2 (1 - \omega_{cp,eq}(1 - \eta_p)/\omega)}. \qquad (6.52)$$

Here the electron pitch angle and electron and proton gyro and plasma frequencies are given at equator and η_p is the proton concentration (typically $\eta_p > 0.7$). The main message of these equations is that the minimum resonant energy is approximately proportional to $B/\sqrt{n_e}$ and inversely proportional to $\cos \alpha_{eq}$ and to the wave frequency ω.

Based on CRRES observations Meredith et al. (2003) concluded that minimum energy conditions of 1–2 MeV electrons was met during about 1% of the electron drift motion around the Earth. They noted that while 1% may sound small, it actually is enough to keep diffusion significant and, at the same time, loss timescales in the range of hours to one day. If the interaction would take place within a much wider part of the drift motion, the electrons would disappear too fast compared to the observations.

During favorable conditions the electron loss may also be faster. Kurita et al. (2018) analyzed *Van Allen Probes* and *Arase* observations following each other during moderate substorm activity on 21 March 2017. They concluded that the relativistic electrons where lost in the L-shell range 4–5 in a timescale of 10 min or even faster, which corresponds to only a couple of drift periods. From the satellite and ground-based observations the EMIC wave activity was estimated to occur within a few-hour period around the magnetic midnight. It is possible that in this particular case the interaction took place during a much longer fraction of the drift path than estimated by Meredith et al. (2003).

Similar to the whistler-mode waves, the gyro-resonant scattering due to EMIC waves is limited to small and intermediate pitch angles because the minimum resonant energy increases beyond the electron energies when $\alpha_{eq} \rightarrow 90°$. The WNAs of the waves are $\lesssim 30°$, but if the wave amplitude is large enough, the parallel component of the wave electric field may be sufficient to lead to pitch-angle scattering of the electrons through the bounce resonance. The H^+ band waves can fulfil the resonance condition $\omega = l\omega_{be}$ at low resonant numbers in the outer radiation belt up to $L \lesssim 6$. Cao et al. (2017a) calculated pitch-angle diffusion coefficients at energies >100 keV using oblique EMIC waves with the amplitude of 1 nT. They found that at equatorial pitch angles $>80°$, where the gyro-resonant diffusion became weak, the bounce-resonant diffusion took over and exceeded 10^{-3} s^{-1} close to $90°$. At $L = 3$ the dominant bounce harmonic number was $l = 2$, whereas $l = 1$ dominated at $L = 4 - 5$.

Using typical plasma, wave and particle observations from the *Van Allen Probes* Blum et al. (2019) demonstrated that 50–100 keV electrons can be scattered efficiently by bounce-resonant interaction with both He^+ and H^+ band EMIC waves. They found pitch-angle diffusion coefficients exceeding 10^{-3} s^{-1} for electrons with equatorial pitch angles approaching $90°$.

The nonlinear interaction between EMIC waves and electrons can result in resonant pitch-angle scattering of $\alpha_{eq} = 90°$ electrons (with $v_{\parallel} = 0$) at the equator even in case of an exactly parallel propagating wave. To see this, write the electron's equation of motion in the wave field $(\mathbf{E}_w, \mathbf{B}_w)$ in the form (6.47). The Lorentz force due to the wave magnetic field $\propto \mathbf{p} \times \mathbf{B}_w$ has a component parallel to $\mathbf{B}_0$. According to Eq. (6.48) the acceleration due to the nonlinear term is proportional to $B_w \sin \eta$, where η is the phase angle between the perpendicular velocity of the electron $\mathbf{v}_\perp$ and $\mathbf{B}_{w\perp}$. Because the gyro frequency is much higher than the frequency of the wave, η is highly oscillatory. For a small-amplitude wave the nonlinear term averages out rapidly and its effect is negligible causing just a small oscillation of α around $90°$. But for larger amplitudes the small oscillation may hit the bounce resonance and scatter the electron to off-equatorial motion.

Lee et al. (2020) performed test-particle simulations of 5-MeV electrons with an initial $\alpha_{eq} = 90°$ and different phase angles η integrating Eq. (6.47) in a dipolar $\mathbf{B}_0(\lambda)$. In case of $B_w/B_0 = 0.05$ and wave normal angle $0°$ the diffusive effect on the pitch angle remained small (about $5°$) and constant during the length of the simulation (1600 gyro periods). Increasing the wave amplitude to $B_w/B_0 = 0.1$ increased the diffusive $\triangle\alpha$ to about $20°$. Increasing the wave normal angle, when also the wave electric field causes a parallel force, led to rapid growth of the pitch-angle scattering. Consequently, the interaction with large-amplitude EMIC waves can contribute to the pitch-angle diffusion of equatorial and nearly equatorial ultra-relativistic electrons.

6.6   Different Acceleration and Loss Processes Displayed in Phase Space Density

It is likely that both local acceleration by whistler-mode chorus waves and ULF wave driven inward radial transport contribute to electron energization. Both wave modes can be significantly enhanced during geomagnetically active periods, but there is no one-to-one correlation between them making individual events different from each other. An important aspect is also the energy-dependence of acceleration to the highest energies. A plausible scenario is that electrons are first accelerated to MeV energies by chorus waves and then further to ultra-relativistic energies through inward transport by ULF waves (e.g., Jaynes et al. 2018; Zhao et al. 2019a).

Which one of these mechanisms is more important, and under which conditions, has, however, remained a highly controversial subject where new observations and refined computer simulations have been found to be in favor of one or the other. In cases when the phase space density (PSD, Sect. 3.5) as a function of adiabatic integrals $f(\mu, K, L^*)$ can be determined from multisatellite observations with sufficient accuracy and wide enough coverage, its temporal evolution can be used to investigate the relative roles of the mechanisms that are fully adiabatic (e.g., the *Dst* effect, Sect. 2.7) and processes that break one or more adiabatic invariants.

The method is illustrated schematically in Fig. 6.10, in which the temporal evolution of PSD as a result of different processes is sketched as a function of L^* for a given μ. In a fully adiabatic process conserving all adiabatic invariants the PSD does not change. Radial transport and local acceleration/losses show different time evolution of the PSD.

In the case of inward radial transport alone the source is typically at large radial distances, and a wide range of energies over a wide domain in L^* is affected (Fig. 6.10, top left). The PSD increases with time at all drift shells and maintains its monotonous gradient $\partial f/\partial L^* > 0$ during the transport toward the Earth. Since the inward transport brings new electrons into the radiation belt region from larger distances, the PSD increases in absolute sense, as indicated by the upward arrow.

The local acceleration through wave–particle interactions, in turn, enhances the PSD within a limited radial distance leading to a temporally growing peak as a function of L^* (Fig. 6.10, top middle). The subsequent radial diffusion spreads the peak to both directions. Note, however, that a local peak (Fig. 6.10, top right) may also appear after the radial transport has first enhanced the PSD (time t_0 to t_1) and

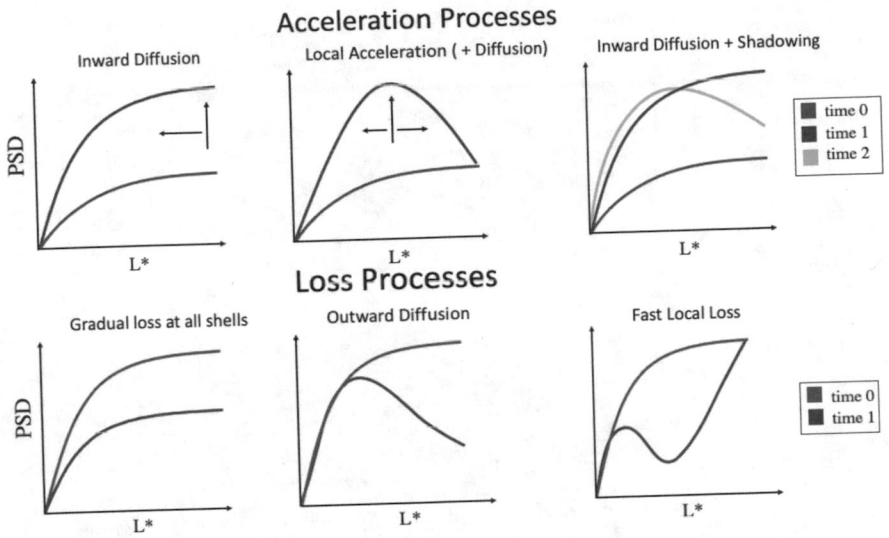

Fig. 6.10 Illustration of how temporal evolution of the phase space density can be used to distinguish between acceleration by radial diffusion or an internal mechanism. See the text for explanation (The figure is drafted following similar pictures of Chen et al. 2007; Shprits et al. 2017)

then magnetopause shadowing (Sect. 6.5.1) removes electrons from the outer parts of the belts (time t_1 to t_2).

Similarly, it is possible to distinguish between different loss mechanisms in the the temporal evolution of the PSD. The sketch in the bottom left of Fig. 6.10 illustrates the gradual loss due to pitch-angle scattering to the atmospheric loss cone through interaction with plasmaspheric hiss and whistler-mode chorus waves at a wide range of L-shells. The picture in the bottom middle describes outward radial diffusion and subsequent loss to the magnetopause. The time evolution of the PSD on the bottom right illustrates fast local loss by the EMIC waves.

It is important to understand that the PSD is not a magic wand. The method is constrained by the resolution and spatial coverage of the observations and, in particular, by the accuracy of the applied magnetic field model in the process to convert particle fluxes to PSD (Sect. 3.5).

The importance of wide enough L^* coverage was emphasized by Boyd et al. (2018) who combined *Van Allen Probes* and THEMIS observations. Of the 80 events they investigated only 24 featured a clear peak in the PSD as a function of L^* when the PSD was calculated from *Van Allen Probes* data alone. However, when THEMIS data from larger distances were included in the analysis, 70 of the 80 events indicated local acceleration.

Figure 6.11 shows two examples of the Boyd et al. (2018) study. The event on 13–14 January 2013 was one where the gradient was clearly positive as a function of L^* in the *Van Allen Probes* data but turned negative at larger distances and there

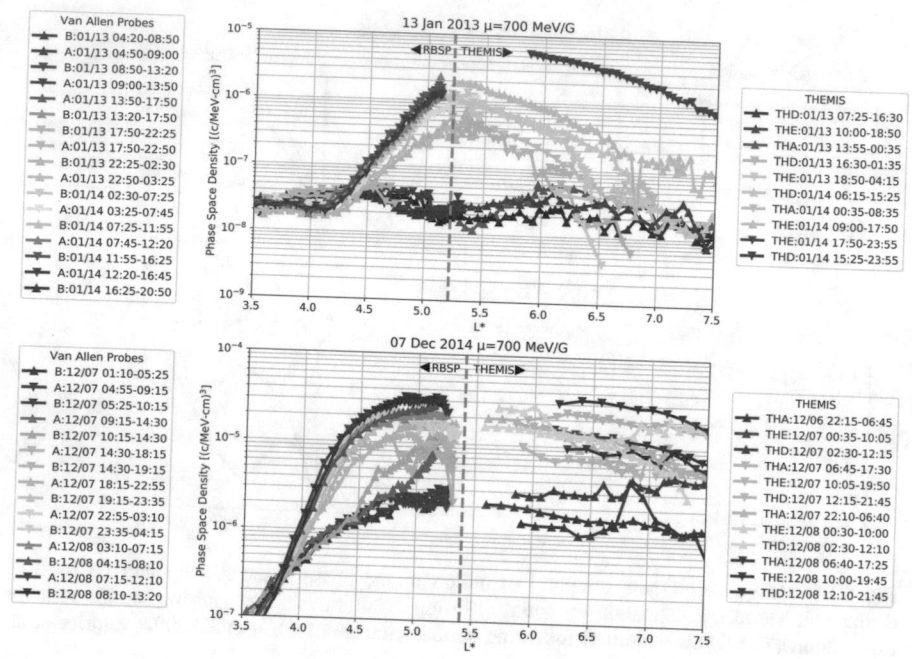

Fig. 6.11 Evolution of the phase space densities combined from *Van Allen Probes* and THEMIS observations on 13–14 January 2013 (top) and on 6–8 December 2014 (bottom). The different times are given with different colors increasing from blue to red (From Boyd et al. 2018, Creative Commons Attribution-NonCommercial-NoDerivs License)

was a clear growing peak from $L^* \approx 4.3$ to at least $L^* \approx 7.5$, when THEMIS data were included. The PSD from *Van Allen Probes* observations alone thus suggested radial transport but the wider radial coverage supported the interpretation that local acceleration was the dominant one.

The event on 6–8 December 2014 was different. In that case the *Van Allen Probes* data hint a local peak slightly earthward of the apogee of the spacecraft. However, there was no clear negative gradient in the PSD calculated from THEMIS observations.

6.7 Synergistic Effects of Different Wave Modes

In the previous sections we have mostly considered the source and loss effects of various wave modes one at a time. In reality the picture is more complicated. During its drift motion an individual electron encounters different wave environments at different MLT sectors. For example, whistler-mode chorus can accelerate the electron in the dawn side and the same electron may be scattered toward the loss cone by EMIC waves in the afternoon sector. Note also that these emissions are not strictly limited to these sectors and, as discussed in Chap. 5, may occasionally be observed at all local times, also simultaneously in the same location.

As discussed earlier, the combined effect of different wave modes on charged particles can be *additive*. Examples of this are processes in which a bounce resonance may first move electrons from large equatorial pitch angles to smaller pitch angles, where gyro resonance can take over, or in which the electrons are first accelerated to MeV energies by chorus waves, whereafter ULF waves may take care of the acceleration to ultra-relativistic energies.

The effect of wave modes can also be *synergistic*, where nonlinear interaction between the waves modifies the properties of the wave that scatters the particles either in energy or pitch angle. In particular, large-amplitude ULF waves have been found to modulate the key parameters of particle interactions with EMIC, chorus and plasmaspheric hiss emissions.

An early suggestion that ULF oscillations might modulate the electron scattering by whistler-mode waves was presented by Coroniti and Kennel (1970). They found that such modulations ought to be found in a wide range of periods 3–300 s. This corresponds to observed precipitation pulsations with periods 5–300 s in X-ray emissions and riometer absorptions caused by >30-keV electrons.

Modern observations of poloidal mode ULF oscillations with mirror-like magnetic and density oscillations were illustrated in Fig. 5.19. Xia et al. (2016) found that the ULF modulation strengthened both upper- and lower-band chorus emissions in the troughs of the magnetic fluctuation, whereas chorus waves weakened at the crests of the fluctuation. Careful analysis of electron and proton pitch-angle distributions suggested that the chorus emissions below $0.3\ f_{ce}$ were consistent with linear growth by enhanced low-energy electrons, whereas some, likely nonlinear, mechanism may be required to excite the chorus at higher frequencies, perhaps similar to the formation of the chirping emissions discussed in Sect. 5.2.4.

The mirror-type appearance can also affect the minimum resonant energy of electrons with EMIC waves. According to (6.51) the minimum resonant energy depends on the background magnetic field and plasma density as

$$W_{res,min} \propto \frac{\omega_{ce}}{\omega_{pe}} \propto \frac{B}{\sqrt{n_e}}. \tag{6.53}$$

Consequently, during the half period of the ULF wave, when the oscillation reduces the magnetic field and enhances the plasma density, $W_{res,min}$ is reduced from the constant background level. In order this to be effective requires, of course, that the amplitude of the modulating wave is large enough.

Using THEMIS (in 2007–2011) and *Van Allen Probes* (in 2012–2015) observations Zhang et al. (2019) investigated in total 167 large-amplitude ULF wave events colocated with hydrogen-band EMIC waves in the L-shell range 4–7. The events were found at all MLTs, mostly in the evening sector. The magnetic fluctuations were in the range $0.01 < \Delta B/B < 2$ and the density fluctuation was in most cases even larger. Under the background conditions $5 < \omega_{pe0}/\omega_{ce0} < 25$ the average fluctuation ratio $(\Delta n_e/n_e)/(\Delta B^2/B^2)$ was in the range 1–3.

Theoretically these levels of ULF fluctuations could reduce $W_{res,min}$ up to about 30% from the constant background level. Thus assuming that the background plasma conditions without the ULF fluctuation would suggest minimum resonant energy of 1 MeV, the presence of such fluctuations would reduce it to 0.7 MeV. Zhang et al. (2019) also noted that their event selection criteria limited the frequency range of the ULF waves to higher than 5 mHz in THEMIS observations and higher than 10 mHz in *Van Allen Probes* observations, thus missing the lowest-frequency end of Pc5 waves where $\triangle B/B$ can be expected to be larger facilitating scattering of electrons of even smaller energy.

How important this mechanism is to the loss of sub-MeV electrons remains unclear. The occurrence rate of the events with simultaneous EMIC and ULF oscillations studied by Zhang et al. (2019) was not very large. It peaked at L-shells 5.5–6, where it was $(3 \pm 1) \times 10^{-3}$.

Also plasmaspheric hiss has been found to be modulated by ULF fluctuations. Breneman et al. (2015) investigated global-scale coherent modulation of the electron loss from the plasmasphere using *Van Allen Probes* hiss observations, balloon-borne X-ray counts due to precipitating electrons in the energy range 10–200 keV, and ground-based ULF observations. The intensity of the hiss emission was modulated by the ULF oscillation and there was an excellent correspondence between hiss intensity and electron precipitation during the two in detail analyzed events on 3 and 6 January 2014. Thus the global-scale forcing of the plasmasphere by ULF waves can lead to enhanced hiss emission and consequent scattering of plasmaspheric electrons to the atmospheric loss cone.

Simms et al. (2018) performed an extensive statistical analysis of the effects of ULF Pc5, lower-band VLF chorus, and EMIC waves on relativistic and ultra-relativistic electron fluxes in four energy bands (0.7–1.8 MeV, 1.8–4.5 MeV, 3.5–6.0 MeV and 6.0–7.8 MeV) observed at geostationary orbit during 2005–2009. They used autoregressive models where the daily averaged fluxes were correlated with the fluxes and wave proxies observed in the previous day. The models were constructed separately for different pairs of the wave modes including both linear and quadratic terms of each wave, and a cross-term of the waves to represent the synergistic effects.

The regression coefficients contain a lot of information about linear and nonlinear influences of the individual modes and of their mutual interactions, thorough discussion of which is beyond the scope of this book. Here we focus on the synergetic effects suggested by the analysis.

The influence of Pc5 waves was found to be largest at midrange power and decreased due to the negative effect of the nonlinear term, and this was more pronounced when Pc5 waves were paired with VLF chorus. The synergistic interaction of Pc5 and chorus emissions was found to mutually increase their effects being statistically significant at higher energies. This is consistent with the idea that both Pc5 and chorus waves contribute to the electron acceleration to relativistic and ultra-relativistic energies and that the combined effect is not only additive but also synergistic. Simms et al. (2018) suggested that the nonlinearity of the Pc5 influence may be responsible for different conclusions found in different studies of its effectiveness relative to VLF chorus.

The EMIC waves had a negative effect on electron fluxes, in particular, at the highest energy ranges. This is consistent with the fact that the electron energy must exceed the minimum resonant energy. The negative effect was enhanced when both the EMIC waves and either Pc5 or chorus waves were at high levels, and again there was clear indication of synergistic interaction.

6.8 Summary of Wave-Driven Sources and Losses

Table 6.3 summarizes the sources and primary regions of occurrence (MLT, L-shell and latitude) of the wave modes presented in the previous sections as well as the resonances of radiation belt electrons with different wave modes.

Because radiation belt electron gyro frequencies are higher than the wave frequencies, the electron energies and the Doppler-shift of the wave frequency $k_\parallel v_\parallel$ must be large enough to fulfil the resonance condition. Near the equator most of the relevant wave modes (hiss, chorus, EMIC) propagate indeed quasi-parallel to the background magnetic field with small wave normal angles (Table 4.2). The requirement for high $v_\parallel$ implies that resonance occurs mostly from small to intermediate pitch angles (up to 60–70°).

Chorus waves can resonate with a wide range of equatorial pitch angles of the electrons (up to nearly 90°) and over wide range of energies due to their wide range of frequencies. The gyro resonances with lower-frequency hiss and EMIC waves are, in turn, limited to lower/intermediate pitch angles and to the highest energies. Landau and bounce resonances may work from intermediate to 90° pitch angles.

Additionally, nonlinear interactions with large-amplitude waves can lead to rapid acceleration and scattering losses. The waves often propagate obliquely, typically at high-latitudes where they can most efficiently scatter electrons close to the loss cone.

Recall that the relation of wave frequency and parallel wave number depends on the dispersion equation. For example, Eq. (5.11) indicates that the resonant energy of whistler-mode waves is inversely proportional to the wave frequency.

It is, in general, a highly important but complex question on what timescales electrons are lost from the belts due to wave–particle interactions. But on the other hand, observed loss timescales can give insight to the scattering wave mode. Particularly interesting are the cases where the whole high-energy belt population is lost as fast as in ten minutes at low L-shells where the magnetopause shadowing is unlikely to occur. Hiss, chorus and magnetosonic waves all scatter relativistic electrons, but in timescales from a day to months. Even in case of nonlinear and strong interaction with large-amplitude waves the effect to the whole population is expected to take time. The most plausible cause for the fast wave–particle scattering are EMIC waves. Another possible cause for fast depletions is magnetopause shadowing and drift shell splitting.

Table 6.3 A summary of sources, dominant regions of occurrence, and possible resonances, including the approximate ranges of energies and equatorial pitch angles α_{eq} of electrons interacting with different wave modes in the inner magnetosphere

Wave	Source	Region[a]	Possible resonances
Chorus	Anisotropic velocity distributions of 1–100 keV electrons	Midnight to early dusk (strongest dawn), outside the plasmasphere up to magnetopause	*Gyro*: 30 keV to MeVs, wide α_{eq} range. Near the equator scatter $\lesssim 100$ keV electrons and accelerate higher energy electrons, at high latitudes scatter MeV electrons
			Landau: 30 keV to MeVs.
			At high latitudes microburst losses by non-linear trapping by large-amplitude oblique chorus.
EMIC	Anisotropic velocity distributions of 1–100 keV protons	Noon to dusk H$^+$ band, also dawn outside the plasmasphere up to magnetopause	*Gyro*: $\gtrsim 1$–2 MeV, intermediate[b] and small α_{eq} (non-linear trapping can affect also sub-MeV electrons)
			Landau: 30 keV to MeVs, large $\alpha_{eq}(\gtrsim 85°)$
			Bounce: 50–100 keV, large α_{eq}
Hiss	Local generation & non-linear growth, penetration of chorus	Dawn to post-noon, inside plasmasphere	*Gyro*: ~ 100 keV to MeVs, intermediate and small α_{eq}
			Landau: 100s keV, intermediate and large α_{eq} ,
			Bounce : $\gtrsim 1$ MeV, intermediate and large α_{eq}
Magneto-sonic/X-mode	Ion ring distributions	Noon to dusk, both inside and outside the plasmasphere, confined close to equator ($\lambda \lesssim a$ few degrees)	*Landau*: ~ 30 keV–1 MeV, intermediate and large α_{eq}
			Bounce: ~ 100 keV to MeVs, wide α_{eq} range
Pc5	Solar wind drivers (e.g., KHI, FTE, pressure pulses), RC ions and ion injections from tail	Global, most frequent dawn and dusk sectors	*Drift*: small m: MeV electrons large m: 10s to 100s keV electrons (m is the azimuthal wave number)

[a] Information on typical geomagnetic latitudes λ is limited due to the lack of high latitude observations. Most wave modes are found close to the equator, but extend to at least 20°–30°
[b] Intermediate pitch angle defined here $30° \lesssim \alpha_{eq} \lesssim 70°$

Open Access This chapter is licensed under the terms of the Creative Commons Attribution 4.0 International License (http://creativecommons.org/licenses/by/4.0/), which permits use, sharing, adaptation, distribution and reproduction in any medium or format, as long as you give appropriate credit to the original author(s) and the source, provide a link to the Creative Commons license and indicate if changes were made.

The images or other third party material in this chapter are included in the chapter's Creative Commons license, unless indicated otherwise in a credit line to the material. If material is not included in the chapter's Creative Commons license and your intended use is not permitted by statutory regulation or exceeds the permitted use, you will need to obtain permission directly from the copyright holder.

Chapter 7
Dynamics of the Electron Belts

In this chapter we discuss the overall structure and dynamics of the electron belts and some of their peculiar features. We also consider the large-scale solar wind structures that drive geomagnetic storms and detail the specific responses of radiation belts on them. Numerous satellite observations have highlighted the strong variability of the outer electron belt and the slot region during the storms, and the energy and L-shell dependence of these variations. The belts can also experience great variations when interplanetary shocks or pressure pulses impact the Earth, even without a following storm sequence.

We start by describing the main electron populations in the belts and then cover the nominal quiet- and storm-time structure of the belts. Then, we move on to describe the penetration of the electrons at different energies into the slot region and the inner belt, including the seemingly impenetrable barrier for ultra-relativistic electrons, storage ring and three-part radiation belt configuration. We conclude the chapter with a brief discussion of consequences of energetic electron precipitation to the upper atmosphere.

7.1 Radiation Belt Electron Populations

Radiation belt electron populations can be divided into the inner belt electrons and to four different energy ranges of outer belt electrons (Table 7.1) that reflect different sources and different responses to magnetospheric perturbations and plasma waves. Also, the technical hazards of these populations differ. Although the number density of highest-energy electrons is small, even a single electron can have harmful effects to sensitive satellite electronics. The highest-energy electrons are sometimes called "killer electrons" because their increased fluxes have been associated with some of the serious satellite failures. Also lower-energy electrons can have adverse effects since they contribute to charging of spacecraft surfaces and solar panels.

© The Author(s) 2022
H. E. J. Koskinen, E. K. J. Kilpua, *Physics of Earth's Radiation Belts*,
Astronomy and Astrophysics Library, https://doi.org/10.1007/978-3-030-82167-8_7

Table 7.1 Main electron populations in the radiation belts and their key sources. The energy ranges are indicative and vary among different studies. All populations can be lost due to magnetopause shadowing and through wave-particle interactions with chorus, EMIC, hiss, and ULF waves. For details of the source loss processes, see Chap. 6

Population	Energy range	Sources
Source	30 keV–200 keV	Substorm injections, global convection
Seed	200 keV–500 keV	Substorm injections, global convection, acceleration by chorus
Relativistic, core	500 keV–2 MeV	Acceleration by chorus, inward transport by ULF waves
Ultra-relativistic	>2 MeV	Acceleration by chorus, inward transport by ULF waves

The suprathermal electrons with energies from a few tens of keV to about 200 keV are called the *source population*. They can equally well be identified as ring current electrons although they contribute to the net electric current much less than ring current ions due to their much smaller energy density. The source electrons originate mostly from substorm injections and large-scale magnetospheric convection transporting electrons from the tail closer to the Earth and energizing them through adiabatic heating. The paramount importance of the 30–200 keV suprathermal electrons to the radiation belt dynamics is that they provide *source of free energy* to whistler-mode chorus waves (Sect. 5.2), which motivates them to be characterized as a source population. When they give rise to the chorus waves primarily through gyro resonance, they scatter toward the atmospheric loss cone. They are also scattered by Landau and bounce resonances with EMIC and magnetosonic/X-mode waves.

The *seed population* is in the medium energy range up to several hundred keVs. They also originate primarily from the substorm injections and inward transport by global convection. Seed electrons are further accelerated to higher energies by the chorus waves (Sect. 6.4.5) generated by the source population and by large-scale ULF Pc4–Pc5 waves (Sect. 6.4.1). Because substorms or global convection are not able to directly inject MeV electrons into the inner magnetosphere, the few hundred keV population acts as the crucial *seed* to the highest energy population. They scatter toward the atmospheric loss cone primarily due to interactions with hiss waves in the plasmasphere and by gyro resonance with chorus waves outside the plasmasphere.

The highest-energy electrons are relativistic. The energies of the *core population* reach from about 500 keV to 2 MeV with Lorentz factors 2–5. We call electrons, whose kinetic energy exceeds 2 MeV, *ultra-relativistic*. Their $\gamma > 5$, which implies that their velocities are larger than 98% of the speed of light.

The acceleration of the core and ultra-relativistic electrons is among the most important questions in radiation belt physics. The acceleration proceeds gradually first from seed to relativistic energies, and then further to ultra-relativistic energies. Throughout the process both local acceleration by chorus and energization related to the inward radial transport via Pc4–Pc5 ULF waves are viable mechanisms. Chorus waves can be in resonance with electrons of large range of energies. The drift resonance of ULF waves with seed electrons requires that the wave azimuthal mode number m has to be large, while for the highest energies resonance occurs with lower m. However, the relative roles of these processes remain unclear at the time of writing this book.

7.2 Nominal Electron Belt Structure and Dynamics

The structure and dynamics of the electron belts are continuously driven by variations in the solar wind and the consequent changes in the magnetic field, plasma and plasma wave conditions in the inner magnetosphere. While the observations indicate widely different properties of the belts under different conditions, certain nominal features and typical storm-time responses can be identified.

The location of the plasmapause plays an important role in the structure of the outer electron belt. During quiet times the plasmasphere extends up to $L = 4$–5 (Fig. 1.4), and thus, a significant part of the outer belt is embedded in high-density low-energy plasma being under the influence of plasmaspheric hiss waves. When the geomagnetic activity increases, the plasmasphere shrinks and most of the outer belt remains outside the plasmasphere, where chorus and EMIC waves have the main influence on the electron dynamics. As a consequence of the plasmaspheric plume (Sect. 1.3.2) in the afternoon sector the drift shells of the outer belt electrons can cross regions both inside and outside of the plasmasphere.

During geomagnetic activity a large number of fresh particles are brought from the magnetotail to the outer belt region and energized. In the storm main phase, during the rapid enhancement of the ring current (decrease of the Dst index), sustained convection, substorm injections and inward radial diffusion driven by ULF Pc5 waves can all contribute to the enhancement of radiation belt populations. In the recovery phase the adiabatic convection weakens, but substorm injections and inward radial transport can still replenish the high-energy particle population. Furthermore, even in the absence of strong large-scale convection and related ring current enhancement, relatively strong *auroral storms*, displayed in the AE index, can take place in association with intense substorms.

At the same time when fresh particles are injected and energized during the storm main phase, different loss processes are in action. The dayside magnetopause often gets eroded and/or compressed earthward. Simultaneously, electron drift shells expand outward because the geomagnetic field is weakened by the enhanced ring current, leading to enhanced magnetopause shadowing (Sects. 2.6.2 and 6.5.1). Simultaneously EMIC and chorus waves scatter electrons into the atmospheric loss cone. Indeed, the main enhancements of relativistic electrons are usually not observed before the storm recovery phase, when the loss processes weaken but efficient acceleration via chorus waves and inward radial transport by Pc5 waves continues, often over extended periods.

The expansion of drift shells due to enhanced ring current in the storm main phase often appears to cause a disappearance of electrons in satellite data. However, the electrons are not necessarily lost from the inner magnetosphere but only move outward, gaining energy, and return back during the recovery phase when the ring current weakens, losing the same amount of energy. This essentially adiabatic process is the *Dst effect* introduced in Sect. 2.7.

Figure 7.1 illustrates the statistical response of radiation belt electrons to 110 geomagnetic storms observed by *Van Allen Probes*. Let us walk through the Figure starting from the typical quiet-time structure of the belts about 12 h before the zero epoch time at the peak storm activity in the center of each panel. Following clear

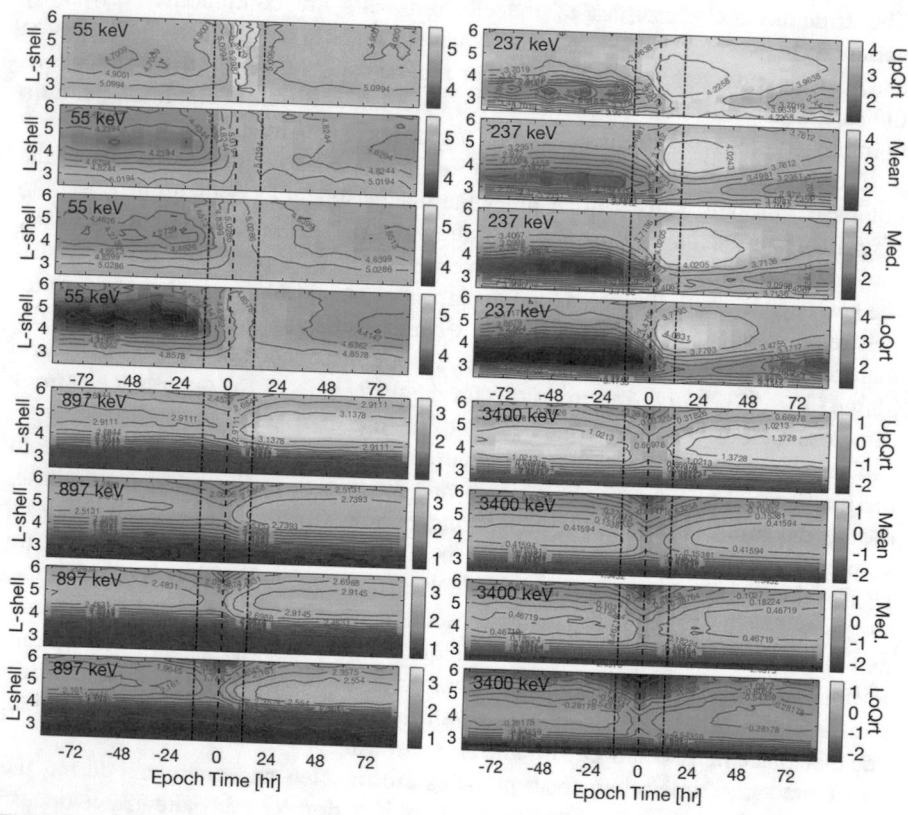

Fig. 7.1 Statistical response of the outer radiation belt for 110 geomagnetic storms observed with *Van Allen Probes*. Zero epoch time indicated by the dashed vertical line corresponds to the peak of the storm determined from the minimum of the *SYM-H* index. The dash–dotted vertical lines are 12 h before and after the epoch time. The fluxes are shown for four energies representing source (55 keV), seed (237 keV), core (897 keV) and ultra-relativistic (3400 keV) populations. The four panels at each energy show the upper, mean, median, and the lower quartile of the fluxes. Note that because the lowest *L*-shell in the Figure is 2.5, the outer boundary of the inner belt is visible only at source and seed energies (From Turner et al. 2019, Creative Commons Attribution-NonCommercial-NoDerivs License)

trends can be discerned: At source (top four panels on the left) and seed (top four panels on the right) energies the inner belt extends to $L \approx 3$. Their fluxes are also enhanced at the highest L-shells shown, in particular this trend is clear for the seed population (top right). The core electrons (897 keV; bottom left panels) are found to peak at $L \approx 5$, whereas ultra-relativistic electrons (3400 keV; bottom right) peak at slightly lower L-shells ($L \approx 4.5$). The L-shell range where ultra-relativistic electrons are enhanced is also wider and extends to lower L-shells. While the relativistic electrons may disappear completely from the outer belt, some amount of lower-energy electrons remain there even during quiet times. The *slot* between the inner and outer belts is, however, clear from seed to ultra-relativistic energies, but its location and range in L are energy-dependent.

During the storm main phase, source electrons flood rapidly to the outer belt and the slot region due to enhanced magnetospheric convection and substorm injections. Thereafter the source electron fluxes decay relatively quickly. This is attributed to the weakening convection and the consequent expansion of the plasmapause. The hiss waves in the plasmasphere now rapidly scatter trapped low-energy electrons toward the atmospheric loss cone. The post-storm fluxes, however, stay elevated compared to the pre-storm fluxes due to injections related to substorms, which can be frequent during the storm recovery phase.

Seed electrons enhance in the heart of the outer belt after the peak of the storm. Substorm injections, inward ULF wave driven transport and acceleration by chorus waves may all contribute to this enhancement. As the recovery phase progresses, the band of enhanced seed electron fluxes weakens and narrows, and its peak moves to higher L-shells ($L > 5$). This behavior is consistent with observations that substorm injections and associated chorus waves move to higher L-shells in the storm recovery phase (Turner et al. 2013). The reformation of the slot at seed energies is also clear as the plasmaspheric hiss-related scattering gradually becomes effective. The peak flux of seed electrons is, however, larger and the slot narrower than before the storm, but the peak narrows and the slot widens as the time progresses.

The core and ultra-relativistic electrons, in turn, feature a clear depletion in the storm main phase followed by an enhancement in the recovery phase. Losses in the main phase can occur due to magnetopause shadowing and wave–particle interactions, the former of which is strengthened due to the inflation of drift shells and outward radial transport. In some cases the apparent depletion can also be mostly due to the adiabatic Dst effect, i.e., no true losses to the magnetopause or to the atmospheric loss cone occur and the particles return to their original domain after the inflation of the drift shells is over. The reason, why source and seed electrons do not show strong depletion in the main phase, is that a large amount of fresh electrons are injected to the inner magnetosphere already during the main phase. In addition, they drift more slowly around the Earth (Fig. 2.7), and therefore, any short-time inward magnetopause incursion removes more efficiently higher-energy electrons. The plasmasphere is also typically confined close to the Earth in the main phase and thus electrons are not affected by hiss that could rapidly scatter lower-energy electrons. Neither do EMIC waves resonate with lower-energy electrons.

Figure 7.1 shows that the time when the recovery phase enhancement of particle fluxes begins depends on electron energy; seed electrons enhance first, followed by core and finally by ultra-relativistic electrons. The peak of the core electron flux also widens and moves to slightly lower L-shells when compared to the pre-storm flux. These features are in agreement with the progressive acceleration scheme (e.g., Jaynes et al. 2018), according to which the chorus-wave acceleration of seed electrons is responsible for the inward widening of the relativistic electron fluxes. The peak flux of ultra-relativistic electrons, in turn, moves to slightly higher L-shells when compared to the pre-storm flux, but the inner boundary of enhancement stays at approximately the same L. Statistically both core and ultra-relativistic electron fluxes stay stable during the 4-day period after the peak of the storm according to Fig. 7.1. This reflects the slower timescales of hiss waves in scattering high-energy electrons and the lack of electron loss processes in the outer parts of the outer belt during quiet periods.

Further insight into the differences in the structure of the electron belts during quiet and active magnetospheric conditions is illustrated in Fig. 7.2, where electron fluxes from two orbits of *Van Allen Belt Probes* are presented. During the geomagnetically quiet-time orbit (Fig. 7.2, left) the inner and outer belts are separated by a clear slot region, featured by distinct decreases in electron fluxes over a wide range of energies. The inner boundary of the outer belt (i.e., the outer boundary of the slot) is at roughly similar L-shells ($L \approx 4.5$) for all energies. The outer boundary of the inner belt (i.e., the inner boundary of the slot) exhibits, in turn, clear energy dependence, moving to higher L with decreasing energy. For instance, at the lowest energies (37- and 57-keV channels; yellow curves on the top) the outer edge of the inner belt reaches to $L \approx 4$, while at 400–500 keV (blue curves) the boundary is much closer to the Earth, at $L \approx 2$–2.5. In other words, the slot region widens with increasing electron energy, and finally, the inner belt is almost completely void of core and relativistic electrons.

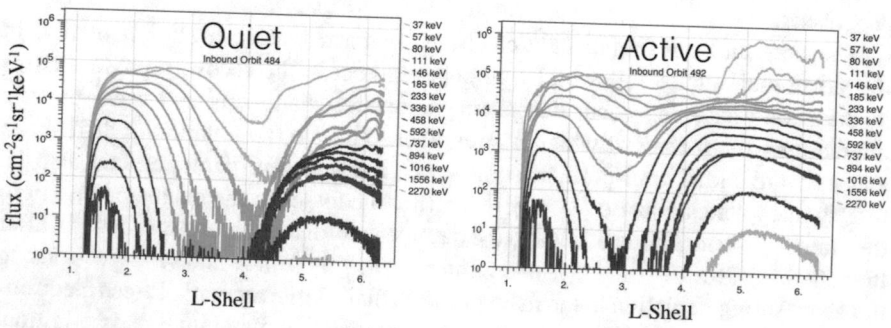

Fig. 7.2 Electron fluxes at different energies during two *Van Allen Probes* orbits featuring geomagnetically quiet (left) and active (right) times as a function of L-shell and energy. HOPE and MAGEIS data are combined from 1 keV to 4 MeV between $L \simeq 1.5$–6 R_E (From Reeves et al. 2016, Creative Commons Attribution-NonCommercial-NoDerivs License)

During the active-time orbit (Fig. 7.2, right) the slot region is flooded by low-energy (source) electrons. At energies above a few hundred keV the slot still exists but is confined to a considerably narrower L-range when seed- and core-energy electrons penetrate closer to the Earth. Now also the inner boundary of the outer belt shows a clear energy-dependence; the electrons penetrate to lower the L-shells with decreasing energy.

The *overall* flux variations as response to storms are featured in Fig. 7.3 based on the same set of 110 storms as Fig. 7.1. The response is determined by comparing the maximum fluxes from the 3-day intervals before and after the storm *SYM-H* peak, excluding a 1-day interval around the peak of the storm. "Enhancing" events are characterised by at least a factor of two increase from pre- to post-storm maximum fluxes, while in "depleting" events there is at least a factor of two decrease. The strategy is the same as that of Reeves et al. (2003) who, using geostationary measurements, found that the likelihoods of electron fluxes to enhance, experience no-change or deplete are approximately equal (33%, 30% and 37%, respectively). Now, however, the *Van Allen Probes* data allow investigating how the overall response depends on energy and L-shell.

Figure 7.3 shows that depleting events are rare at source and seed energies. In the outer belt, seed and source populations typically show enhancement or no-change, consistent with Fig. 7.1. Substorm injections refill the belts in the recovery phase despite losses by wave scattering into the atmospheric loss conc. Seed electrons feature enhancements in the majority of cases (75%) in the heart of the belt, also in agreement with the previous discussion that their enhancement in the storm recovery phase can be attributed to chorus acceleration. Enhancing events are also observed in the slot and the inner belt at source and seed energies. Figure 7.3 showcases the overall stability of the inner belt and the slot region as no-change events clearly dominate at $L \lesssim 3$ at most of the shown energies. At core energies at the lower L-shells ($L \lesssim 3.5$) no-change events clearly dominate. At higher L-shells core electrons, in turn, typically either enhance (approximately half of the cases) or deplete (approximately one third of the cases). The highest-energy electrons (>5 MeV), however, most often experience no change at all.

One should, however, keep in mind that the overall statistical response described here ignores strong variations in electron fluxes that may occur during the storm period itself.

7.3 Solar Wind Drivers of Radiation Belt Dynamics

The large-scale heliospheric structures that drive magnetospheric storms were briefly introduced in Sect. 1.4. They are the interplanetary coronal mass ejections (ICMEs), including their shocks and sheath regions, slow–fast stream interaction regions (SIRs), and fast solar wind streams with Alfvénic fluctuations. Examples of typical ICME and SIR events are sketched in Fig. 7.4. Table 7.2 summarizes solar wind drivers, their typical durations, solar wind conditions and radiation belt

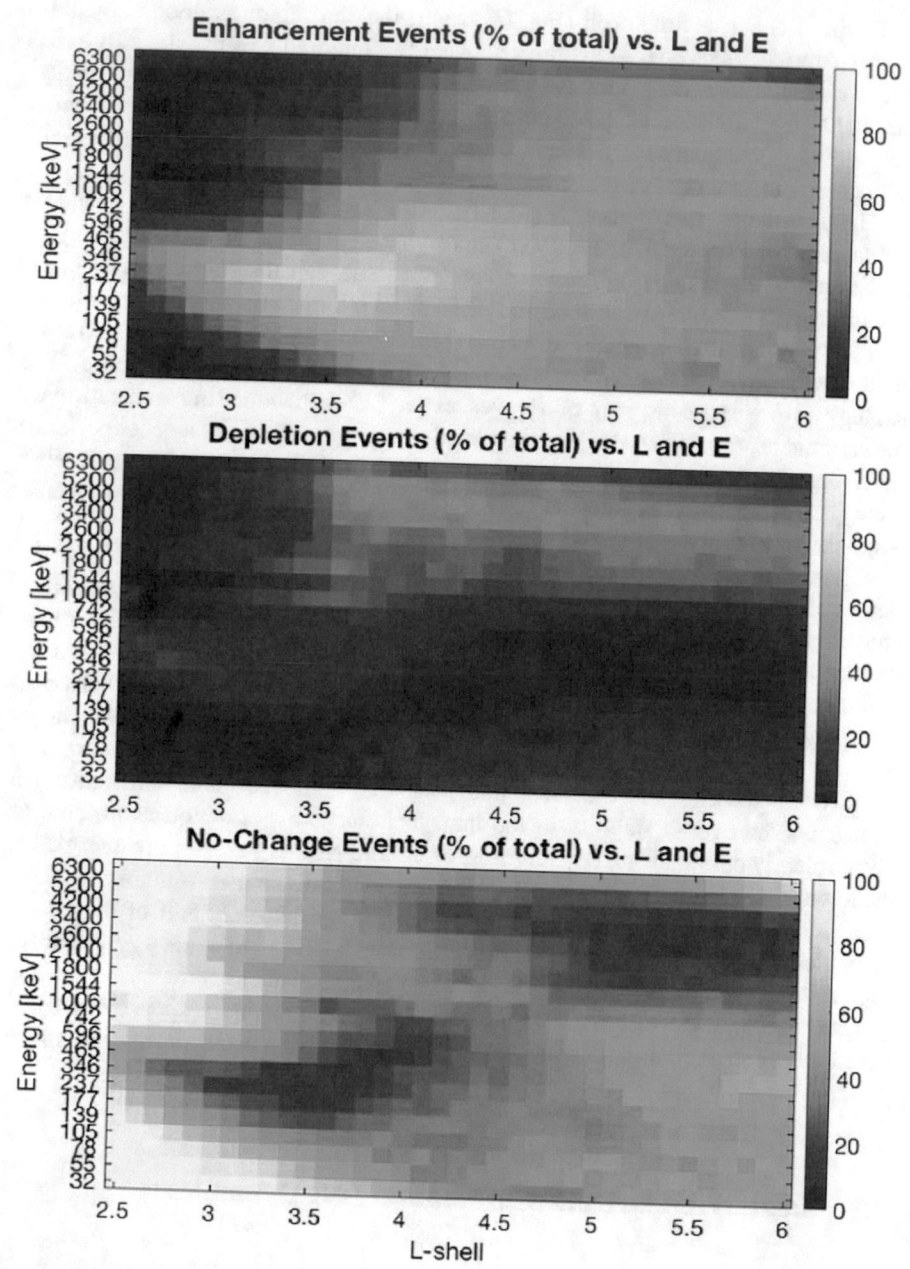

Fig. 7.3 The statistical overall response of the electron fluxes to storms. The panels show the events that enhanced (top), depleted (middle), or caused less than a factor of two change in the fluxes (bottom) (From Turner et al. 2019, Creative Commons Attribution-NonCommercial-NoDerivs License)

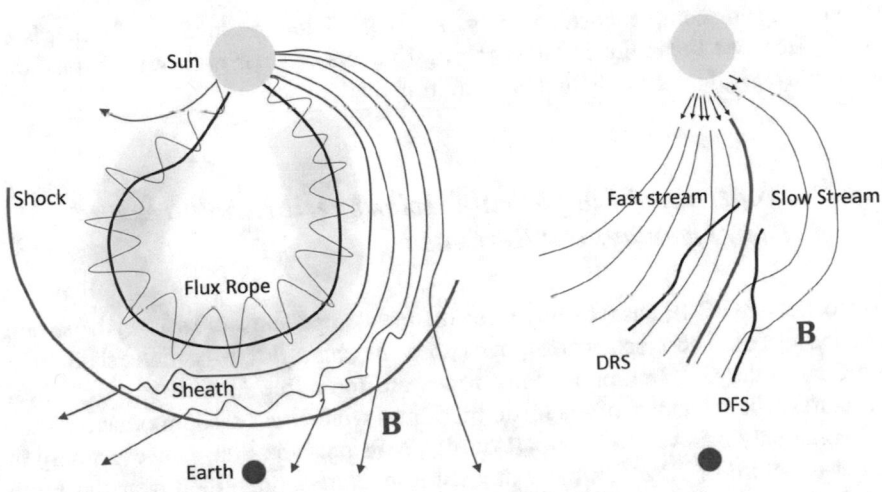

Fig. 7.4 Sketches of an ICME, shock and sheath (left) slow and fast solar wind streams with an SIR in between (right). DFS indicates the fast shock ahead the stream interface and DRS the reverse shock propagating backward in the solar wind frame (Modified from a figure published earlier by Kilpua et al. 2017)

Table 7.2 Large-scale solar wind drivers affecting the belts, their typical durations, solar wind conditions and outer radiation belt electron responses. These structures can occur isolated (except the combination of shock and sheath), but they often come in sequences, e.g., ejecta following shock and sheath, and a considerable fraction of ejecta is followed by a SIR/fast stream. Note also that not all SIRs are followed by a fast stream

Driver	Duration	Solar wind conditions	Typical outer belt relativistic electron response
Shock	Instantaneous	Jump in plasma and field parameters	Rapid acceleration in the heart of the belt ($L \approx 4$)
Sheath	~8–9 h	High dynamic pressure, large-amplitude IMF variations, high variability (compressive)	Sustained and deep depletions at wide range of energies and L
Ejecta (flux rope)	~1 day	Low dynamic pressure, smooth field rotation, low variability	Both deplete (at high L-shells) and enhance (in particular in the heart of the outer belt, $L \approx 4$)
SIR	~1 day	High dynamic pressure, intermediate-amplitude IMF fluctuations, high variability, gradually increasing speed (compressive)	Depletions
Fast stream	Days	Low dynamic pressure, Alfvénic fluctuations (relatively lower amplitude and faster), high speed	Enhancements (in particular at high L-shells)

responses. (For comprehensive reviews see, e.g., Kilpua et al. 2017; Richardson 2018.) Here we focus on the most relevant factors of their ability to disturb the geospace and affect the radiation belt environment.

7.3.1 Properties of Large-Scale Heliospheric Structures and Their Geomagnetic Response

A geoeffective ICME has typically three distinct components: the shock, the sheath and the ejecta. The ejecta corresponds to a *magnetic flux rope* unleashed from the Sun in the CME eruption. Flux ropes are force-free ($\mathbf{J} \times \mathbf{B} = 0$) helical structures where bundles of magnetic field lines wind about a common axis. They are commonly believed to be a part of all CMEs near the Sun. However, only in about one-third of ICMEs a clear flux rope structure is identified near the Earth orbit. This is due to evolution, deformations and interactions that the CME/ICMEs experience while they travel from the Sun to the Earth. Furthermore, the observing spacecraft may sample only the outskirts of the ICME missing a clear signature of the flux rope. Those ICMEs that show flux rope signatures are commonly called *magnetic clouds*. Their key defining characteristics in *in situ* observations are a smooth and coherent rotation of the magnetic field direction over a large angle during an average passage of the ICME lasting about 1 day, magnetic field magnitude considerably larger than nominally in the solar wind and a low plasma beta. The smooth rotation of the field and enhanced field magnitude are clear in the sample ICME in Fig. 7.5.

A leading shock forms when the ICME propagates so fast that its speed difference with respect to the ambient solar wind exceeds the fast magnetosonic speed, whereas the MHD perturbations cannot propagate upstream faster than the ICME. ICME-driven shocks are relatively weak in astrophysical context, their average Mach numbers being 1.89 ± 0.98 (fast forward shocks observed between 1995 and 2017 in the Heliospheric Shock Database[1] maintained at the University of Helsinki).

The sheath is a turbulent region that forms between the shock and the leading edge of the ICME ejecta. In some cases the only observed signatures of a CME in the solar wind are the shock and the disturbed IMF trailing the shock. This is because the shock has a considerably wider extent than the driving ejecta. ICMEs that are not fast enough to develop a shock but propagate faster than the ambient solar wind also perturb the upstream plasma flow and the IMF. However, the perturbed regions are less turbulent than the sheaths behind the shocks but can nevertheless cause magnetospheric disturbances.

A stream interaction region (SIR) is a compression region that forms when a high-speed stream from the Sun catches a slower stream ahead, as depicted in the

---

[1] http://ipshocks.fi.

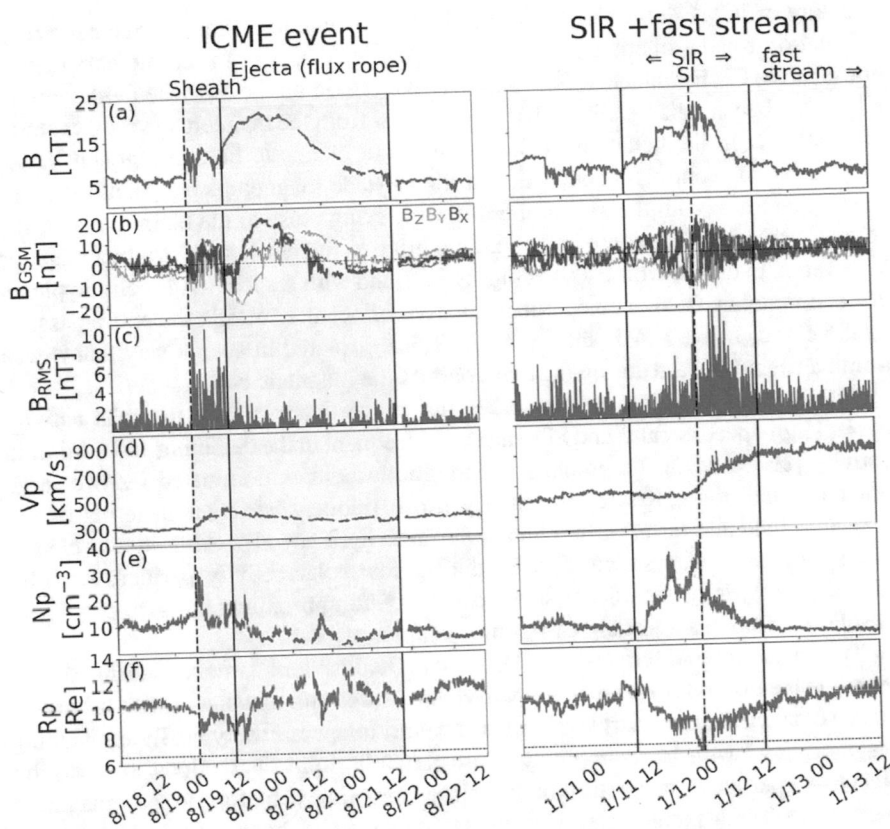

Fig. 7.5 Solar wind plasma conditions during an ICME (left) and SIR with a fast stream (right). The panels give the (**a**) magnetic field magnitude, (**b**) magnetic field components in GSM coordinates, (**c**) 1-min root-mean-square field variations, (**d**) solar wind speed, (**e**) solar wind density and (**f**) subsolar magnetopause position determined using the Shue et al. (1998) model (Data source: CDAWeb, https://cdaweb.gsfc.nasa.gov/index.html/)

right-hand panels of Fig. 7.5. Unlike fast ICMEs, most SIRs are not associated with shocks at the Earth orbit. SIRs can develop fast forward–fast reverse (forward and reverse in the solar wind frame) shock pairs, but typically only when they have propagated at heliospheric distances larger than 1 AU. Figure 7.5 shows that the solar wind speed increases gradually within an SIR and reaches about $750 \, \mathrm{km \, s^{-1}}$ in the fast stream. Although there is a positive velocity gradient across an SIR, not all of them are trailed by a particularly fast stream. The stream interface separates the denser and slower solar wind from the faster and more tenuous solar wind.

The average properties of ICMEs and SIRs near the Earth orbit vary considerably from event to event. The magnetic field in ICMEs ranges from only a few nanoteslas to about $100 \, \mathrm{nT}$ and their speeds from that of the slowest solar wind of about $300 \, \mathrm{km \, s^{-1}}$ to 2000–$3000 \, \mathrm{km \, s^{-1}}$. In SIRs the magnetic field can reach up to about 30 to $40 \, \mathrm{nT}$ and the peak speed to about $800 \, \mathrm{km \, s^{-1}}$. The internal magnetic field structure of ICMEs also varies significantly. In magnetic clouds the magnetic field orientation rotates smoothly, while in complex ejecta, e.g., in the mergers of two consecutive CMEs launched from same active region on the Sun, the field configuration can be very disorganized. The key interest from the magnetospheric dynamics viewpoint is in the behavior of the north-south magnetic field component (B_Z), as its direction controls reconnection at the dayside magnetopause (Sect. 1.4.1). In magnetic clouds with the axis of the flux rope lying close to the ecliptic plane, B_Z rotates either from the south at the leading edge to the north at the trailing edge, or vice versa. In clouds whose axis is highly inclined with respect to the ecliptic plane, B_Z can maintain its sign being either positive or negative during the entire passage of the cloud. In the ICME shown in Fig. 7.5 B_Z rotated in such a way that it was dominantly northward during the main part of the magnetic cloud.

The SIR and ICME occurrence rates and properties vary with the solar activity cycle. High-speed streams and SIRs are most frequent in the declining and minimum activity phases when the global solar magnetic field is dominated by two large polar coronal holes. ICMEs are clearly more frequent, stronger in terms of the magnetic field and faster during high solar activity. They also drive shocks clearly more frequently than slower ICMEs during low solar activity periods. At solar minimum CMEs of significant size leave the Sun approximately once per week, at solar maximum several big CMEs may erupt every day.

Typical solar wind properties of large-scale heliospheric structures differ significantly; sheaths and SIRs are compressive structures and associated with relatively large solar wind density, dynamic pressure and temperature, typically embedding magnetic field with large-amplitude and relatively rapid fluctuations. In turn, the ejecta—in particular magnetic clouds—show smoother variations in their magnetic field and plasma parameters. The ejecta also tend to have clearly lower solar wind density and dynamic pressure than sheaths and SIRs because CMEs expand significantly after being released from the Sun. The fluctuations in fast streams are Alfvénic, but the associated magnetic field amplitudes are typically smaller than those in sheaths and SIRs, and their temporal variations are faster. Fast streams have low density and generally also low dynamic pressure. Due to high dynamic pressure, the magnetopause is typically strongly compressed during the sheaths and SIRs, but is closer to its nominal position during the ICME ejecta and fast streams. During ICME ejecta, however, the magnetopause can erode significantly due to enhanced magnetic reconnection if interplanetary magnetic field is southward for extended periods, even though dynamic pressure would not be particularly large.

Nearly all intense and big geomagnetic storms ($Dst_{min} < -100\,\text{nT}$) are caused by ICMEs, and can be driven by both sheaths and ejecta, or a combination of them. The most geoeffective ICMEs are those that drive shocks and contain a flux rope. Shock-driving ICMEs are typically fast with a strong magnetic field, and both the sheath and the ejecta can contribute to a geomagnetic storm. The enhanced magnetic field and smoothly rotating flux ropes can, in turn, provide sustained periods of strongly southward IMF. The geoeffectiveness of an ICME is further strengthened when it is followed by a fast solar wind stream. Fast trailing wind increases the speed of the ICME and compresses its rear part leading to increased field magnitude and plasma density. This can, in particular, enhance geoeffectiveness of a magnetic cloud, whose magnetic field rotates from north to south. Furthermore, the trailing fast stream typically prolongs the storm recovery phase. Slow ICME ejecta that have mainly northward field may, in turn, pass the Earth almost unnoticed.

ICME-driven storms are often preceded by a clear storm sudden commencement (Sect.1.4.2) due to the shock wave compression of the magnetosphere. The initial phase before the main phase may be prolonged if the high dynamic pressure sheath has mainly northward magnetic field.

An SIR followed by a fast stream leads usually to a weak or moderate storm followed by a long recovery phase, sometimes lasting up to a week. The long recovery phase can be attributed to Alfvénic fluctuations in fast streams, which lead to sustained substorm activity.

Interacting ICMEs are particularly challenging objects of study, as interactions between the ICMEs can result in widely different structures (Manchester et al. 2017; Lugaz et al. 2017). The outcome depends on the relative speed, on the internal magnetic field direction and orientation, and on the structure of the CMEs involved. If the magnetic fields of the consecutive CMEs have opposite orientations at their interface, a magnetic reconnection can lead to merging of the CMEs, resulting in a complex structure at the Earth orbit where individual characteristics of the original eruptions are no more discernible. If the fields on the interface are in the same direction, the leading ICME is compressed by the following ICME, forming two separate but closely spaced each other following ejecta where the first ejecta can maintain strong magnetic field and high speed. Such cases can lead to particularly strong southward fields and dynamic pressure, and consequently to most extreme geomagnetic storms. At angles between parallel and opposite directions of the magnetic fields, the efficiency of merging varies analogous to the solar wind–magnetosphere interaction at different clock-angles of the interplanetary magnetic field.

7.3.2 Typical Radiation Belt Responses to Large-Scale Heliospheric Transients

As described above, different large-scale heliospheric structures have widely different properties. They put the magnetosphere under variable forcing that affects the resulting magnetospheric activity and conditions in the inner magnetosphere, and consequently, also the electron flux response in the radiation belts. On average, storms driven by SIRs are considered more effective in generating MeV electrons at the geostationary orbit than storms driven by ICME-related southward B_Z, a concept introduced by Paulikas and Blake (1976) based on geostationary observations. After ICME-associated storms it takes a longer time (about 2 days) for the fluxes to recover to pre-storm levels than after SIR-associated storms (about 1 day) (Kataoka and Miyoshi 2006). The picture is, however, more complicated. A *Van*

Allen Probes study by Shen et al. (2017) found that within $L^* = 3.5\text{--}5.5$ fluxes of >1-MeV electrons enhance more during ICME-driven than SIR-driven storms. At lower energies (<1 MeV) ICME-driven storms resulted in larger flux enhancements at lower L-shells ($L^* = 2.5\text{--}3.5$), while SIR-driven storms generated more flux enhancements in the outer parts of the outer belt.

The impact of an interplanetary shock causes an abrupt and global compression of the dayside magnetosphere and launches a compressional magnetosonic impulse that propagates radially inward and azimuthally around the Earth. This compressional disturbance has an azimuthal electric field (Sect. 4.4.2), which extends from the dayside to the nightside and can lead to a rapid (1 min timescale) acceleration and inward transport through breaking the third adiabatic invariant of drift-resonant electrons (Foster et al. 2015; Kanekal et al. 2016). Primarily MeV electrons are accelerated in the heart of the belt at $L \approx 4$, as their drift times around the Earth match the frequency of the electric field impulse and they can thus experience acceleration over a significant part of their orbit. Low-energy electrons cannot be in resonance, because they drift slowly compared to the timescale of the accelerating electric field. Relativistic electrons, in turn, drift so fast that they can encounter shock-induced azimuthal electric field during multiple orbits that leads to effective acceleration. *Drift echos* of electrons bunched in the same drift phase can be observed several times before the phase mixing redistributes the particles over different drift phases.

An example of an abrupt enhancement of ultra-relativistic electrons measured by the *Van Allen Probe* A as a response to an interplanetary shock impact on the Earth on 8 October 2013 is shown in Fig. 7.6. Since the drift period depends on particle energy, drift echoes appear with different time lags for particles of different energies. The observed spectra also change with the magnetic field magnitude because the spacecraft moves to a larger radial distance from the Earth during the event. The amplitudes of the electric field impulse in the inner magnetosphere are typically a few mV m^{-1} (Zhang et al. 2018), but during particularly strong shocks much larger amplitudes have been observed. For example, the shock on 24 March 1991, discussed in Sect. 7.4.2 below, caused an electric field impulse of an order of magnitude stronger, approximately 40–80 mV m^{-1} (Blake et al. 1992; Wygant et al. 1994).

The passages of sheath regions cause deep and sustained depletions of relativistic and ultra-relativistic electrons from the radiation belts over a wide range of L-shells (Hietala et al. 2014; Kilpua et al. 2015). Therefore, if high-energy electrons are accelerated due to the shock impact as discussed above, they get quickly lost when the sheath arrives. Strong depletions during sheath regions can be explained through various loss processes acting in concert. First, the sheath strongly compresses the dayside magnetosphere due to its high dynamic pressure. Second, the wave activity in the inner magnetosphere is enhanced during the sheath (Kalliokoski et al. 2020). Resulting wave–particle interactions with EMIC, hiss and chorus waves can scatter the electrons to the loss cone over a wide range of energies, whereas Pc5 ULF waves can transport them radially outward. As the dayside magnetopause is compressed, diffusion to higher L-shells leads to effective magnetopause shadowing losses. The

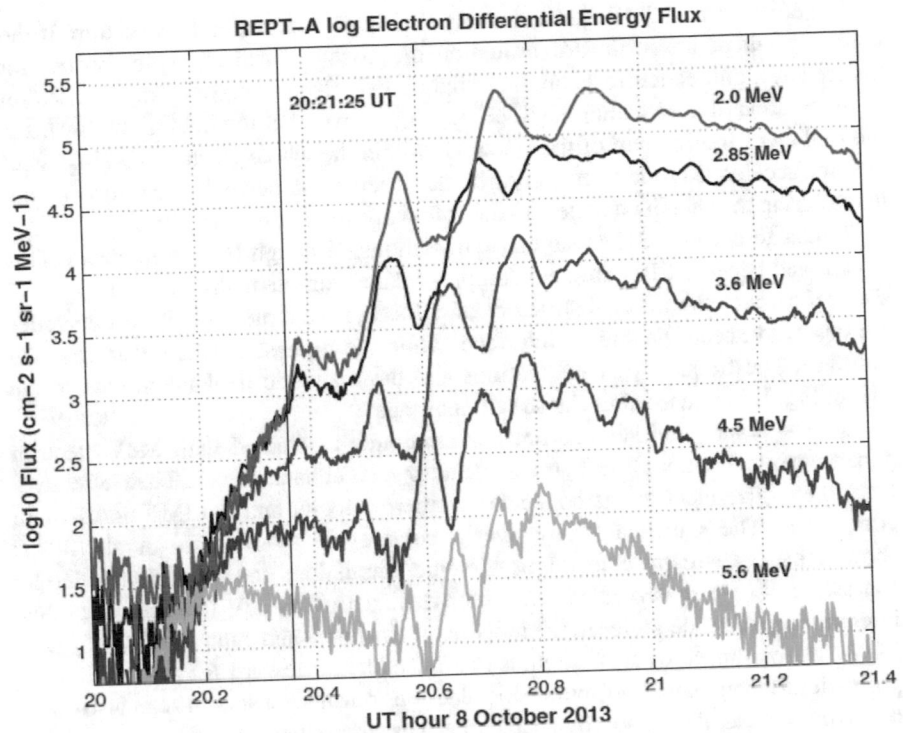

Fig. 7.6 Response of ultra-relativistic electrons to an impact of an interplanetary shock on 8 October, 2013 as detected by *Van Allen Probe* A. The solid vertical line shows the time of the shock impact (From Foster et al. 2015, reprinted by permission from American Geophysical Union)

inflation of drift shells can further enhance the shadowing losses in case the sheath causes a geomagnetic storm but also non-geoeffective sheaths can cause drastic responses. Large-amplitude magnetic field and dynamic pressure variations typical to sheaths trigger substorms which, in turn, lead to the enhancement of source and seed electrons. As the sheath passages are relatively short (on average 8–9 h) and during them loss processes of relativistic electrons dominate, progressive acceleration of seed electrons to MeV energies does not typically occur during the sheath passages.

Also the ICME ejecta often deplete the relativistic electron fluxes, but not as efficiently as sheaths, and the depletions are expected to be confined to higher L-shells. While sustained southward field in the ejecta can erode the dayside magnetopause inward, the ejecta have significantly lower dynamic pressure than the sheaths and, consequently, the magnetopause stays further away from the Earth. The inner magnetospheric wave activity, in particular ULF Pc5 and EMIC waves, is present during the ejecta, but is, on average, weaker than during sheaths. These properties imply that both magnetopause shadowing and precipitation losses are, on average, less effective during ICME ejecta than sheath passages.

The effect of an ICME ejecta strongly depends on its magnetic structure. If the magnetic field points northward throughout the passage, the ejecta typically does not induce any significant effects on the radiation belt electron fluxes. The distribution of the southward field within the CME ejecta, i.e., whether the field is southward at the leading or trailing part of the ejecta, or within the whole ejecta, may also affect electron acceleration and losses, in particular when combined with the effects taking place during the sheath passage and the following solar wind structures.

Similar to sheaths, SIRs also cause depletions, although less pronounced. This is expected because SIRs are also compressed and turbulent structures. Figure 7.5 above illustrates the internal structure of SIRs. For example, the highest densities are observed ahead the stream interface, while the highest speeds occur after the interface. The frequency of fluctuations and the magnetic field magnitude in the SIR of Fig. 7.5 increase after the stream interface.

The fast solar wind streams are key structures associated with MeV electron enhancements, but not all fast streams lead to such enhancements. Their effectiveness to energize electrons is connected to their velocity and the IMF north-south component. The statistical study over a solar cycle by Miyoshi et al. (2013) showed that fast streams embedding Alfvénic fluctuations and being dominated by southward B_Z were associated with intense fluxes of 30-keV (source) electrons, intense whistler-mode waves in the inner magnetosphere and enhancement of $\gtrsim 2.5$-MeV electron fluxes, whereas streams dominated by northward B_Z lacked all these properties. Enhancements of relativistic electrons during fast streams can also occur due to inward radial transport by ULF Pc5 waves strengthened by increased Kelvin–Helmholz instability at the magnetopause.

It is noteworthy that fast streams lack properties that would cause effective losses (Kilpua et al. 2015). Due to their low dynamic pressure and lack of sustained and strong southward magnetic field, the magnetopause is close to its nominal position. The lack of strong ring current enhancement means that the Dst effect is not significant. The EMIC wave activity and associated scattering losses are also expected to be relatively weak due to weak ring current and lack of dynamic pressure variations exciting the waves. However, the plasmapause is expected to reach higher L-shells due to weakening global convection. This means that outer belt electrons at wider L-range can be subject to scattering by hiss waves. For relativistic and ultra-relativistic energies the scattering timescales by hiss are, however, long.

ICMEs interacting with SIRs and fast streams can lead to a variety of responses that depend on the characteristics of individual structures composing these complex events. The interacting events have regularly sheath-like structures that consist of both disturbed solar wind and ICME-related plasma. Such events can cause drastic changes in radiation belts over intervals of a few days. Another interesting scenario is the case where the shock of an ICME propagates within a previous ICME and enhances the magnetic field within. Such cases have been found to effectively deplete the outer belts from MeV electrons (Lugaz et al. 2015; Kilpua et al. 2019).

7.4   The Slot Between the Electron Belts

The slot between the inner and outer electron belts is an intriguing region. While it is clear that the loss of electrons from the slot must be due to wave–particle scattering to the atmospheric loss cone, its temporal and spatial evolution pose several unanswered questions.

7.4.1   Injections of Source and Seed Electrons into the Slot

As discussed Sect. 7.2, the location, width and particle content of the slot strongly depends on the electron energy and geomagnetic activity. This implies that also the dominant mechanisms that inject electrons into and remove them from the slot depend on electron energy. During geomagnetic storms the slot is commonly filled with electrons up to a few hundred keV and it gradually reforms when the storm subsides.

   One possible way to energize electrons and bring them closer to Earth is the ULF Pc4–Pc5 driven inward radial transport. For seed electrons the drift resonance is met only in the case of a wave with high azimuthal mode number m. This is because their drift periods at the distance of the slot are of the order of hours (Table 2.2), i.e., much longer than the oscillation periods of ULF Pc4–Pc5 waves, which are shorter than 10 min.

   Substorms inject source and seed electrons into the inner magnetosphere, but typically at distances much larger than the slot region. The injections are related to earthward propagating dipolarization fronts launched by the near-Earth reconnection process (Sect. 1.4.3). The dipolarization front is associated with sharp and large-amplitude electric field variations up to several $mV\,m^{-1}$, which is capable of accelerating particles to high energies and transporting them simultaneously closer to the Earth. Dipolarization fronts can propagate from the tail to geostationary distance, or somewhat closer to the Earth, and the related acceleration/transport results in *dispersionless injections* where all energies arrive to the observation point at the same time. The injections are relatively localized azimuthally, spanning about 1–3 h in MLT, beyond which the energy-dependent gradient and curvature drifts introduce dispersion to the observed spectra.

   Turner et al. (2015) studied 47 events during the *Van Allen Probes* era when the source and seed electron (<250 keV) injections were observed inside $L = 4$, all the way down to $L = 2.5$, a phenomenon known as *Sudden Particle Enhancement at Low L Shells* (SPELLS). The injections were preceded by significant substorm activity and the authors suggested that injections to the slot region resulted from interaction of electrons with compressional magnetosonic waves launched by the braking of the dipolarization front. As the observed wave periods were of the order of 100 s (Pc4 range), the waves cannot be in drift resonance with the electrons, unless the azimuthal mode number is very large ($m \sim 30$). However, as originally suggested by Southwood et al. (1969), a fraction of electrons in this energy range can be in bounce–drift resonance with the compressional waves (Eq. (6.2)) depending on their equatorial pitch angles.

Another mechanism commonly invoked to allow source and seed electron penetration into the slot is the inward radial transport driven by the global convective electric field. During enhanced convection the Alfvén layers (Sect. 2.3) shrink allowing electrons to access lower L-shells. The stronger the geomagnetic storm is, the closer to the Earth charged particles from the tail can enter. This convective transport is efficient only for lower energy electrons because the motion of higher energy electrons ($\gtrsim$100 keV) is controlled by gradient and curvature drifts (Eq. (2.30)). The lower energy electrons are also energized by the drift-betatron acceleration during convective transport toward stronger geomagnetic field (Eq. (2.69)). Based on *Van Allen Probes* data and modeling of the intense magnetic storm on 14 February 2014 Califf et al. (2017) showed that the inward radial transport by the observed global convection electric field of the order of 1–$2\,\mathrm{mV\,m^{-1}}$ was sufficient to explain the enhancements of 100–500-keV electrons in the slot region at $L < 3$.

Seed electron fluxes in the slot region may also enhance if the lower-energy electrons in the slot were accelerated locally by chorus waves. It is, however, not clear if chorus waves can accelerate about 100-keV electrons or do the waves mainly cause electron scattering toward the loss cone.

7.4.2 Impenetrable Barrier

The access of MeV electrons to low L-shells is more limited than the access of seed and source populations. In particular, ultra-relativistic electrons are only very rarely observed in the slot region. Baker et al. (2014) noted that ultra-relativistic electrons (>2 MeV) were not observed inside $L = 2.8$ during the first 20 months of *Van Allen Probe* measurements. Although this period coincides with Solar Cycle 24 maximum, it was a geomagnetically relatively quiet period and void of big storms. The existence of this (almost) *impenetrable barrier* has been confirmed in subsequent studies (e.g., Ozeke et al. 2018).

Figure 7.7 from Baker et al. (2014) illustrates the barrier against 7.2-MeV (6.5–7.5 MeV) electrons seen in observations of the *Van Allen Probes* REPT instrument. The inner boundary of the ultra-relativistic electrons stayed strikingly clear and stable near $L = 3$. The most notable features in the figure are a depletion of fluxes on 1 October 2012 and another associated with a slight earthward movement of the barrier to $L = 2.8$ on 1 October 2013.

Due to the relative geomagnetic quiescence the plasmapause was most of the time beyond $L = 4$ and never inside $L = 3$. Thus, the plasmapause and the impenetrable barrier usually were not co-located and the barrier resided well within the plasmasphere. The existence of a steep and stable inner boundary not co-located with the plasmapause is an outstanding question. Doppler-shifted cyclotron resonances of trapped radiation belt electrons with VLF electromagnetic waves from

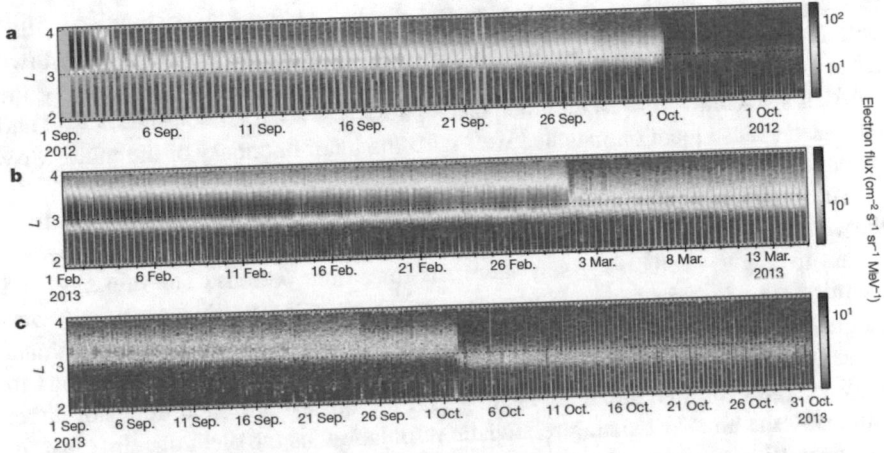

Fig. 7.7 Electron fluxes at 7.2 MeV (6.5–7.5 MeV) during three selected periods from 1 September 2012 to 31 October 2013. The plot combines REPT instrument data from both *Van Allen Probes* (From Baker et al. 2014, reprinted by permission from Springer Nature)

ground-based radio transmitters could cause precipitation to the atmosphere, but this process has been expected to occur only at lower energies (<500 keV) (e.g., Koons et al. 1981; Abel and Thorne 1998).

Foster et al. (2016) argued, however, that VLF waves could play a crucial role in shaping the impenetrable barrier during strong geomagnetic storms when the VLF bubble extends beyond the contracted plasmapause. The high-energy electrons become thus exposed to VLF waves outside the plasmasphere where significantly lower densities increase the resonant energies. As a consequence, relativistic and ultra-relativistic electrons could be in resonance with the VLF waves and be lost to the atmosphere. Foster et al. (2016) also noted that during the 17 March 2015 big geomagnetic storm the outer edge of the VLF wave bubble matched very closely with the inner edge of ultra-relativistic radiation belt electrons. Since energetic electrons can access low L-shells during geomagnetically very disturbed times only, the barrier stays there when geomagnetic storm subsides and plasmapause extends back beyond the VLF bubble.

Another suggested mechanism for the sharp barrier is the balance between slow radial diffusion by the ULF waves and faster scattering by hiss waves (Ozeke et al. 2018). This would, however, imply that a true barrier would not really exist.

Because the slot region is not void of source and seed electrons, the barrier is not impenetrable to non-relativistic electrons. It is not completely impenetrable to relativistic electrons either, and there have been a small number of events when even ultra-relativistic electrons have been found inside $L = 2.8$. Such events have been associated with very strong magnetospheric storms. One example is the so-called Halloween storm in October–November 2003 that was caused by a series of successive interacting ICMEs driving strong magnetospheric activity (Baker et al. 2004). During the Halloween storm electrons in the range 2–6 MeV filled the whole slot region and remained there for several weeks. Also the inner belt was filled with high-energy electrons. This event was unique during the period of almost two solar cycles of SAMPEX observations. Extreme ultraviolet images taken by the IMAGE spacecraft showed that the plasmapause was contracted inside $L = 2$ and the plasmapause location matched well with the inner boundary of the multi-MeV outer belt.

The Halloween storm was stronger ($Dst_{min} = -383\,\text{nT}$) than any storm during the *Van Allen Probes* era. It also illustrates that once ultra-relativistic electrons get access to low L-shells, they stay there for long time periods. The presence and energization of ultra-relativistic electrons so close to the Earth remain, however, enigmatic. The global convective electric field, which is one of the dominant causes for the source and seed electron injections, cannot energize electrons to MeV energies through adiabatic heating and the quasi-static field does not affect efficiently the already existing relativistic population, as the electrons drift fast, in less than 10 min, around the Earth. The mechanisms that have been suggested to be the source of relativistic and ultra-relativistic electrons close to the Earth are local acceleration by chorus waves, when the plasmasphere is highly compressed (e.g., Baker et al. 2014), and radial inward transport by ULF Pc5 waves and related energization (e.g., Ozeke et al. 2018), or a combination of these (e.g., Zhao et al. 2019a).

None of these mechanisms may not be efficient enough to inject relativistic electrons to so low L-shells. The impact of a strong interplanetary shock to the magnetosphere (Sect. 7.3.2) can also create particularly large electric field impulses that reach low L-shells and may rapidly energize electrons to ultra-relativistic energies. Li et al. (1993) suggested this to have been the source of ultra-relativistic electrons inside $L = 2.8$ during the famous storm on 24 March 1991, when a strong interplanetary shock driven by an ICME impacted the Earth.

The left panel in Fig. 7.8 shows how the CRRES satellite, traversing through the slot region at the time of the shock arrival, observed multiple peaks of ultra-relativistic electrons. This feature would suggest an initial injection/acceleration of these extremely energetic electrons, followed by their drift echos. The figure shows that the counting rates of all three energy channels exhibit very similar behavior. Further investigation of the data after conversion from the count rates to electron fluxes (Blake 2013) revealed that curves lied on top of each other (right panel of Fig. 7.8), implying that all channels measured ultra-relativistic electrons >15 MeV, whereas there were no electrons in the range 6–15 MeV. The drift echoes indicate a drift period of 2 min, which is the drift period of 17-MeV electrons at $L = 2$. The ultra-relativistic electrons remained trapped in the nominal slot region over several months. To explain the acceleration of electrons to so high energies is evidently a major challenge.

The access of 1-MeV electrons to low L-shells is more common than the penetration of ultra-relativistic electrons. The strongest storms during Solar Cycle 24 when *Van Allen Probes* were in operation occurred on 17 March 2015 ($Dst_{min} = -222\,\text{nT}$) and 23 June 2015 ($Dst_{min} = -204\,\text{nT}$). During these events no multi-MeV electrons were observed inside the impenetrable barrier, but 1-MeV electrons

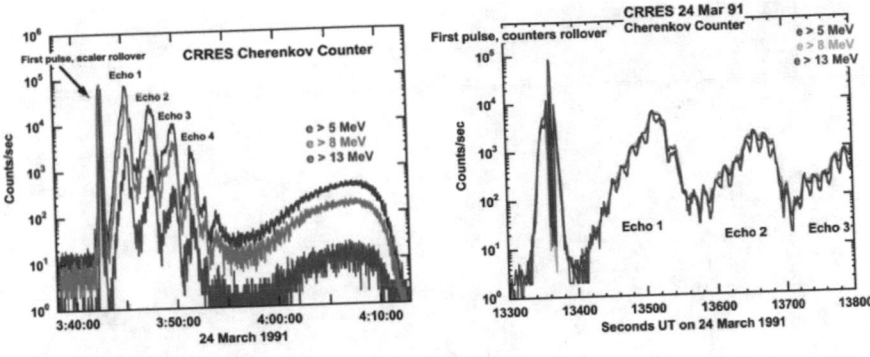

Fig. 7.8 CRRES spacecraft observations of ultra-relativistic electrons in three energy channels when the spacecraft was traversing through the slot region (roughly from $L = 2.5$ to $L = 2$). Left: Observed data in counts. Right: Counts converted to fluxes (note erroneous text on the vertical axis as in the original figure) (From Blake 2013, reprinted by permission from American Geophysical Union)

entered the slot and the inner belt (Claudepierre et al. 2017; Hudson et al. 2017). The energization in these cases has again been attributed to the impact of strong interplanetary shocks. A similar example was the so-called Bastille Day storm ($Dst_{min} = -300\,\mathrm{nT}$) on 15–16 July 2000 when 1-MeV electrons were first injected to $L = 2.5$ from where they slowly diffused to $L = 2$.

7.5 Storage Ring and Multiple Electron Belts

Soon after the *Van Allen Probes* were launched on 30 August 2012, an interesting discovery was made (Baker et al. 2013). The striking feature was a *three-part energetic electron belt* structure that persisted for about a month. The inner electron belt ($L^* \lesssim 2.5$) remained stable, but the outer belt was divided into two distinct parts with a newly formed gap in between. Figure 7.9 reproduces the ultra-relativistic (3.4 MeV) electron fluxes from the period 3–15 September 2012 using data from the *Van Allen Probes* REPT instrument, taken from CDAWeb.[2]

The event commenced with an abrupt removal of electrons from the outer belt as a consequence of a passage of an interplanetary shock wave. In particular, the energetic electrons at the outer L-shells of the outer belt were almost completely removed. Only a relatively thin band of energetic electrons remained in the outer belt around $3 < L^* < 3.5$. This period coincided with the main phase of a magnetic storm. About 2 days later energetic electrons reappeared at $L^* \gtrsim 4$, but the zone $3.5 < L^* < 4$ remained void of ultra-relativistic electrons. While the electron

---

[2] https://cdaweb.gsfc.nasa.gov/index.html/.

234

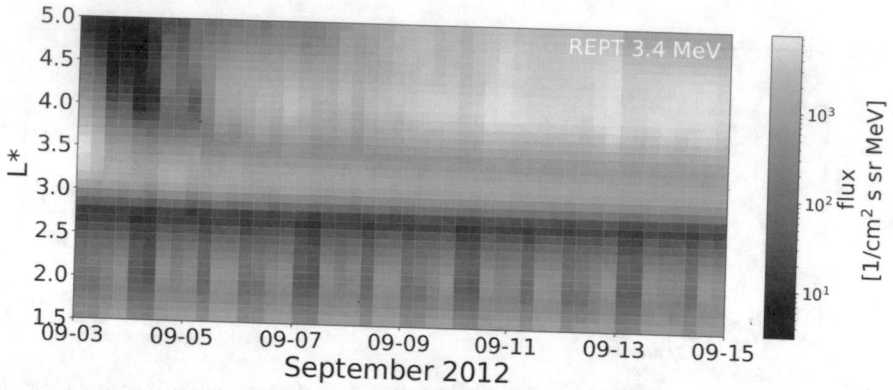

Fig. 7.9 Three-part radiation belt structure during the first half of September 2012. The color scale of electron fluxes (cm^{-2} s^{-1} sr^{-1} MeV^{-1}) is logarithmic. The "intermittency" of the inner belt flux is due to the 9-h orbits of the *Van Allen Probes*, of which only a fraction was spent earthward of $L = 2.5$ (Data source: CDAWeb)

population at $L^* \gtrsim 4$ experienced some variations over the following weeks, the gap between $3.5 < L^* < 4$ and the narrow belt at $3 < L^* < 3.5$, called the *storage ring* or *remnant belt*, experienced very little variations. During the period when the three-part belt structure was observed, the plasmasphere extended beyond $L^* = 4$.

The three-belt structure illustrates the importance of making high-resolution observations in the right place. Namely, afterwards the three-part belt structure has actually been found to be a relatively common phenomenon, and even four-belt structure, where the outermost of the three belts exhibits two distinct bands, has been identified (A talk by A. Jaynes et al. at the AGU Fall Meeting, 2019).

A statistical study by Pinto et al. (2018) covering 5 years of *Van Allen Probe* data found in total 30 three-belt events. This configuration was observed most frequently at electron energies 3.4–5.2 MeV, but occurred over a wider range of relativistic and ultra-relativistic energies as well. The three-part belt structure emerged during both geomagnetic storms and quieter conditions. About 18% of all geomagnetic storms featured this configuration. The largest fraction of three-belt events was caused exclusively by SIRs and fast streams (76%), while the rest occurred during pure ICMEs (17%) or during a combination of an ICME and SIR/fast stream (7%). Due to the small number of events these percentages are just indicative.

The essential feature of the formation of the three-belt structure is the removal of most of the outer belt. Suggested mechanisms include pitch-angle scattering by EMIC waves (Shprits et al. 2018) and the magnetopause shadowing due to inward magnetopause excursion with simultaneous outward radial transport by ULF waves (Mann et al. 2016). Both mechanisms can cause permanent losses of electrons from the radiation belts, after which new high-energy electrons are needed to appear into the outer parts of the outer belt. These new electrons cannot, however, penetrate too deep, otherwise the gap would not remain between the storage belt and outer parts. This may be the reason why most three-belt events are associated to SIRs.

The ICME related storms are known to result enhancements of MeV electrons deeper in the outer belt than SIR related storms (Sect. 7.3.2). Mann et al. (2016) suggested that radial inward transport with associated energization could account for the reappearance of high-energy electrons and final formation of the three-belt structure. As the three-belt configuration is not uncommon, it is possible that several mechanisms can lead to its formation and may act in concert.

The stability of the storage ring over long time periods has been associated with the expanded plasmasphere and slow losses due to hiss-induced scattering at relativistic energies. This, however, requires that the storage ring stays inside the plasmasphere and is not exposed to the faster loss processes outside the plasmasphere. Considering the dependence of the plasmapause on the geomagnetic activity, the storage ring is expected to prevail through weak and moderate geomagnetic conditions, but be erased relatively rapidly during intense geomagnetic storms when the plasmapause is pushed closer to the Earth. Pinto et al. (2018) calculated the empirical lifetimes of the storage rings in the above mentioned three-belt events. The lifetimes increase with increasing energy, being on average a few days at energies of 1.8 MeV and several months at energies of 6.3 MeV. This is consistent with the decrease of pitch-angle diffusion rates due to plasmaspheric hiss with increasing energies (Thorne et al. 2013a).

In summary, the three-belt structure is formed by first removing the outer belt electrons leaving only a narrow storage ring at low L-shells ($L^* \lesssim 3.5$) followed by a recovery of electrons at the outer parts of the belt ($L^* \gtrsim 4$). The processes can occur both temporally very close to each other or have longer time periods in between. The remnant belt may also be pre-existing or created during the event due to inward movement of the peak of the flux (Pinto et al. 2018).

7.6 Energetic Electron Precipitation to Atmosphere

Scattering of outer radiation belt electrons to the atmospheric loss cone leads to observable effects in the middle and upper atmosphere. The middle atmosphere consists of the *stratosphere* and the *mesosphere*. In polar regions the stratosphere extends approximately from 8 to 50 km. Above the stratosphere the mesosphere reaches to the *mesopause* at about 80–100 km. Above the mesopause begins the upper atmosphere consisting of the *thermosphere* up to about 600 km, beyond which the collisionless neutral gas is called the *exosphere*. The ionosphere is the partly ionized lower part of the upper atmosphere.

Radiation belt electrons with energies $\gtrsim 30$ keV can penetrate down to altitudes of 50–90 km. This *energetic electron precipitation* (EEP; often referred to as energetic particle precipitation, EPP, when also proton precipitation is included) can lead to significant ionization of the neutral atmosphere and, consequently, affect atmospheric chemical composition and dynamics, leading ultimately to regional climate forcing. Most importantly, the energetic particle precipitation adds to the solar EUV and soft X-ray generation of odd nitrogen and odd hydrogen (NOx and

HOx) molecules in the polar atmosphere. These molecules have a crucial role in the ozone balance in the middle atmosphere via catalytic ozone destruction (e.g. Verronen et al. 2011; Seppälä et al. 2014). NOx molecules are produced by EEP in the upper polar atmosphere, from where they are transported to lower stratospheric altitudes within the winter-time polar vortex. The effects of EEP generated HOx molecules are, in turn, direct in the mesosphere.[3]

Short-term ozone depletions related to EEP can be quite dramatic. Almost all ozone can be locally wiped away at altitudes 70–80 km. Figure 7.10 shows the effects on mesospheric ozone of three strong EEP events in which depletions reached as far down as 60 km. The individual EEP events last typically only a few days, but during times of high solar activity they can occur frequently and cause longer lasting effects. In particular, direct ozone effects of EEP generated HOx are observable on solar cycle timescales, accounting approximately to a few tens of percent of the variations.

Precipitating electrons can be measured either directly by using spacecraft orbiting the Earth on polar orbits at altitudes of about 700–800 km, or indirectly by X-ray detectors on stratospheric balloons. Another method is to observe the ionization due to the precipitation in the D-layer (60–90 km) of the atmosphere using ground-based riometers and incoherent scatter radars. Incoherent scatter radars measure the electron density in the ionosphere, whereas riometers record cosmic radio noise that reaches the ground. When ionization in the D-layer is enhanced during an EEP event, the absorption of cosmic noise increases. Subtracting the quiet day curve from the riometer recordings gives the *cosmic noise absorption* (CNA), which is proportional to the enhanced D-layer electron column density. Ground-based information of EEP energies and fluxes remains, however, limited compared to direct satellite observations of precipitating electrons.

Energetic electron precipitation originates both from scattering of stably-trapped radiation belt electrons and from freshly injected quasi-trapped electrons before they have completed a full drift around the Earth. The efficiency of EEP is strongly connected to the presence and intensity of plasma waves in the inner magnetosphere that can scatter electrons via wave–particle interactions. All wave modes discussed in the context of electron scattering in Chap. 6 may contribute to the precipitation, but their relative importance in different situations remains currently an open issue.

Precipitation of electrons from a few tens to a few hundred keV, which in the ionosphere cause diffuse aurora, is considered to be predominantly due to resonant interactions with upper-band whistler-mode waves driven by low-energy electrons (Sect. 5.2.3). Precipitation in this energy range may also result when the electrons drive the lower-band chorus losing their perpendicular momentum to the waves. Relativistic electrons can, in turn, experience abrupt (<1 s) scattering into the loss cone as electron microbursts due to their nonlinear interactions with large-amplitude

---

[3] For more comprehensive information on the chemistry of the upper atmosphere we refer to the web-site of the Chemical Aeronomy in the Mesosphere and Ozone in the Stratosphere (CHAMOS) collaboration: http://chamos.fmi.fi.

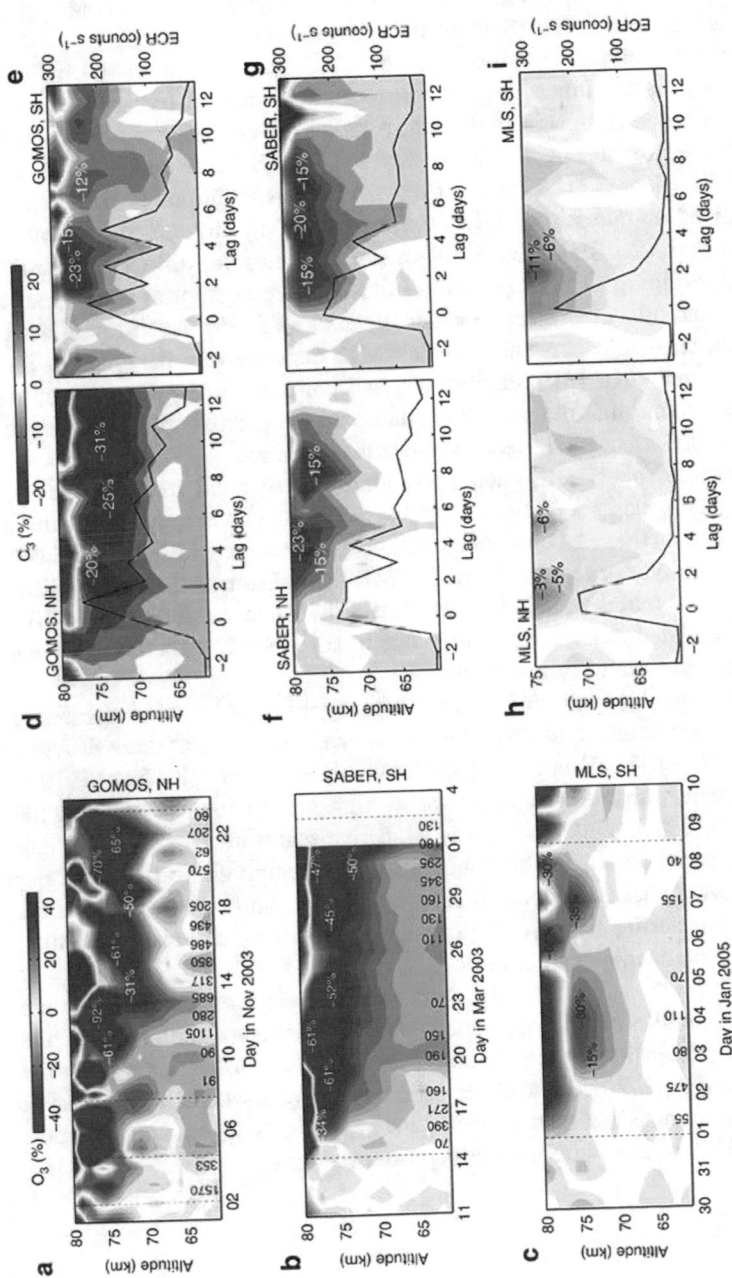

Fig. 7.10 Observations of the ozone depletion during three EEP events in March 2003, November 2003 and January 2005 in the northern (NH) and southern (SH) hemispheres using the Global Ozone Monitoring by Occultation of Stars (GOMOS) instrument onboard Envisat, the Sounding of the Atmosphere using Broadband Emission Radiometry (SABER) onboard TIMED and the Microwave Limb Sounder (MLS) onboard EOS-Aura. The color coding shows the O_3 anomalies (%) and the black solid line the daily mean electron precipitation counting rates (counts s^{-1}) estimated from the Medium-Energy Proton and Electron Detector (MEPED) onboard POES (From Andersson et al. 2014, Creative Commons Attribution 4.0 International License)

elements of the lower-band chorus (Sect. 6.5.3). Relativistic electron microbursts have been related to strong enhancements of HOx and NOx in the atmosphere and with significant short- and long-term effects on mesospheric ozone accounting to about 10–20% of the total losses (Seppälä et al. 2018).

The effectiveness of precipitation also correlates positively with ULF Pc4–Pc5 wave activity in the inner magnetosphere and on ground (e.g., Spanswick et al. 2005). This connection is traditionally linked to an indirect effect: ULF waves enhance the growth rate of chorus waves (Coroniti and Kennel 1970) and consequently EEP (Sect. 6.7). A direct effect to EEP has additionally been established by localized compressional ULF waves modulating the equatorial bounce loss cone (Rae et al. 2018). Global ULF waves, furthermore, displace electrons radially inward, bringing them closer to the Earth where equatorial geomagnetic field is stronger and thus the loss cone wider (Brito et al. 2015). As different wave modes in the inner magnetosphere have distinct MLT dependences, the precipitation signatures also show clear MLT dependence as demonstrated by Grandin et al. (2017), who also found that the substorm-related events peaked strongly close to midnight, while ULF-associated events peaked close to noon.

The inner magnetospheric wave activity is controlled by solar wind driving and consequent magnetospheric activity conditions, and the occurrence and magnitude of EEP also depend on these factors. Precipitation as estimated using riometer CNA recordings is generally more intense during ICME-related storms than during SIR-related storms, but remains elevated for considerably longer periods during SIR-driven storms (Longden et al. 2008), likely due to sustained chorus activity often related to high-speed streams following the SIR.

Figure 7.11 shows the contribution of different types of solar wind streams to POES satellite observations of electron fluxes and average fluxes at three different energies (>30 keV, >100 keV, and >300 keV) over three solar cycles from 1979 to 2013. The solar wind streams are categorized as high-speed streams (including the SIR contribution), ICMEs, and slow solar wind (including unclear cases). High-speed streams clearly dominate the contribution to precipitation at all energies during all solar cycle phases except near solar maximum, while the largest average fluxes are observed during ICMEs. The frequency and amplitude of precipitation are largest during the declining solar cycle.

Finally, we note that while the effects of energetic electron precipitation are important in the middle and upper atmosphere, the actual significance of the precipitation to the intensity of the radiation belt electrons is not clear (e.g., Turner et al. 2013; Gokani et al. 2019). However, while at higher L-shells magnetopause shadowing can dominate the losses, at the lower L-shells ($\lesssim 4$), wave–particle scattering should be the most important loss process.

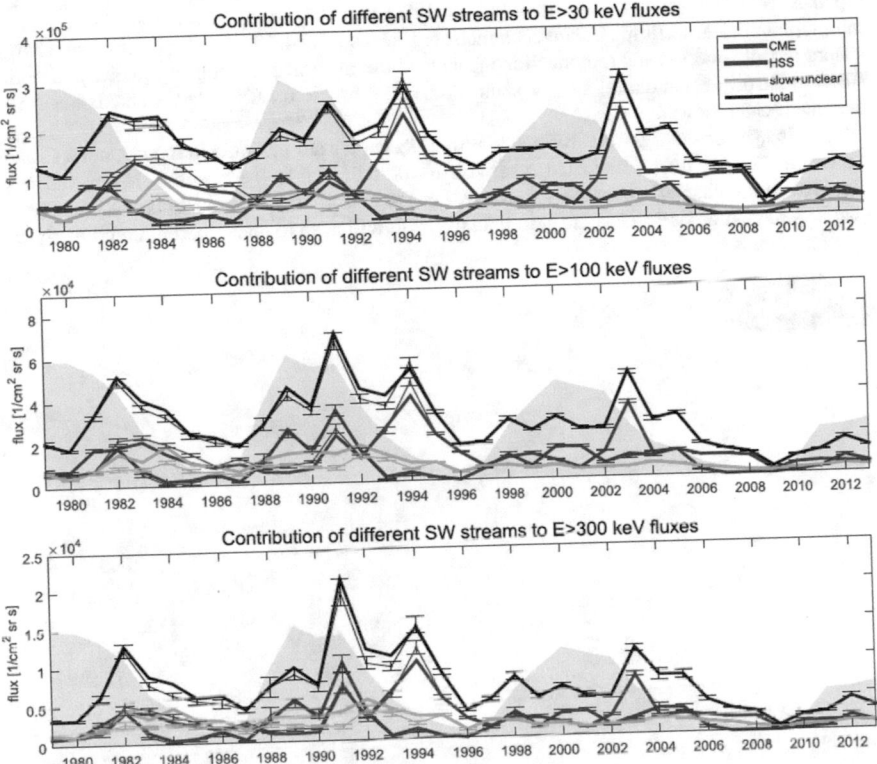

Fig. 7.11 Annual contribution to total electron fluxes (black) from high-speed streams/SIRs (blue), ICMEs (red) and slow/undefined events (green). The thick lines show fluxes computed including data points with missing solar wind data, whereas these are excluded from the thin lines. The error bars of the thin lines give the standard error of mean for the annual contributions. The sunspot number is shown by the grey shading (From Asikainen and Ruopsa 2016, reprinted by permission from American Geophysical Union)

Open Access This chapter is licensed under the terms of the Creative Commons Attribution 4.0 International License (http://creativecommons.org/licenses/by/4.0/), which permits use, sharing, adaptation, distribution and reproduction in any medium or format, as long as you give appropriate credit to the original author(s) and the source, provide a link to the Creative Commons license and indicate if changes were made.

The images or other third party material in this chapter are included in the chapter's Creative Commons license, unless indicated otherwise in a credit line to the material. If material is not included in the chapter's Creative Commons license and your intended use is not permitted by statutory regulation or exceeds the permitted use, you will need to obtain permission directly from the copyright holder.

Appendix A
Electromagnetic Fields and Waves

We summarize here some basic concepts of elementary electrodynamics. This also serves as an introduction to the notations and units used in the book.

A.1 Lorentz Force and Maxwell Equations

The motion of charged particles in the *electric* (**E**) and *magnetic fields* (**B**) is governed by Newton's second law where the force is the *Lorentz force*

$$\frac{d\mathbf{p}}{dt} = \mathbf{F} = q(\mathbf{E} + \mathbf{v} \times \mathbf{B}) \tag{A.1}$$

acting on a particle with charge q and velocity **v**. The electric and magnetic fields fulfil the *Maxwell's equations*, which we write in SI units as

$$\nabla \cdot \mathbf{E} = \rho/\epsilon_0 \tag{A.2}$$

$$\nabla \cdot \mathbf{B} = 0 \tag{A.3}$$

$$\nabla \times \mathbf{E} = -\frac{\partial \mathbf{B}}{\partial t} \tag{A.4}$$

$$\nabla \times \mathbf{B} = \mu_0 \mathbf{J} + \frac{1}{c^2}\frac{\partial \mathbf{E}}{\partial t}, \tag{A.5}$$

where the *charge* (ρ) and *current densities* (**J**) are determined by the particle distribution functions (Chap. 3). Their SI units are $[\rho] = \mathrm{A\,s\,m^{-3}} = \mathrm{C\,m^{-3}}$ and $[J] = \mathrm{A\,m^{-2}}$.

© The Author(s) 2022
H. E. J. Koskinen, E. K. J. Kilpua, *Physics of Earth's Radiation Belts*,
Astronomy and Astrophysics Library, https://doi.org/10.1007/978-3-030-82167-8

The SI unit of the electric field is $[E] = V\,m^{-1}$ and of the magnetic field $[B] = V\,s\,m^{-2} \equiv T$. $\mathbf{B}$ can be described as the density of *magnetic flux* through an area S

$$\Phi = \int_S \mathbf{B} \cdot d\mathbf{S}. \tag{A.6}$$

The natural constants in Maxwell's equations are in SI units

$\epsilon_0 \approx 8.854 \times 10^{-12}\,A\,s\,V^{-1}\,m^{-1}$,   vacuum permittivity
$\mu_0 = 4\pi \times 10^{-7}\,V\,s\,A^{-1}\,m^{-1}$,   vacuum permeability
$c = 1/\sqrt{\epsilon_0 \mu_0} = 299\,792\,458\,m\,s^{-1}$, definition of the speed of light.

The electric and magnetic fields are often convenient to express in terms of the *scalar* (φ) and *vector potentials* ($\mathbf{A}$) as

$$\mathbf{E} = -\frac{\partial \mathbf{A}}{\partial t} - \nabla \varphi \tag{A.7}$$

$$\mathbf{B} = \nabla \times \mathbf{A}. \tag{A.8}$$

The vector potential has a central role in the definition of the action integrals and adiabatic invariants in Chap. 2.

Maxwell's equations form a set of 8 partial differential equations. If we know the source terms, we have more than enough equations to solve the six unknown field components. If we, however, want to treat all 10 functions ($\mathbf{E}, \mathbf{B}, \mathbf{J}, \rho$) self-consistently, we need more information of the medium. In a conductive medium it is customary to use *Ohm's law*

$$\mathbf{J} = \sigma \cdot \mathbf{E}, \tag{A.9}$$

where the *conductivity* σ ($[\sigma] = A\,(V\,m)^{-1} = (\Omega\,m)^{-1}$) is, in general, a tensor and may in nonlinear media also depend on $\mathbf{E}$ and $\mathbf{B}$.

Ohm's law is not a fundamental law of nature in the same sense as Maxwell's equations. It is an empirical relationship to describe the conductivity of the medium similar to the constitutive relations for the *electric displacement* ($\mathbf{D}$) or the *magnetic field intensity* (electrical engineer's magnetic field, $\mathbf{H}$) given by $\mathbf{D} = \epsilon \cdot \mathbf{E}$ and $\mathbf{B} = \mu \cdot \mathbf{H}$ where the permittivity and the permeability of the medium, ϵ and μ, are, in general, tensors.

The medium is called *linear* if ϵ, μ, and σ are scalars and constant in space and time. Note that they usually are, also in linear media, functions of the wave number and frequency of electromagnetic waves propagating in the medium. Much of plasma physics deals with the properties of $\epsilon(\omega, \mathbf{k})$, for example, our discussion of magnetospheric plasma waves and their interaction with radiation belt particles in Chaps. 4–6.

A large fraction of energetic electrons in the Earth's radiation belts and the most energetic inner-belt protons are relativistic and the equation of motion must be written relativistically

$$\frac{d}{d\tau} p^\mu = K^\mu ,$$
(A.10)

where p^μ is the four-momentum, K^μ the electromagnetic four-force, τ the proper time and $\mu = \{0, 1, 2, 3\}$ are the coordinates of the Minkowski space. The space components of the four-momentum form the momentum vector $\mathbf{p} = \gamma m \mathbf{v}$, and the equation of motion is

$$\frac{d}{dt}(\gamma m \mathbf{v}) = q(\mathbf{E} + \mathbf{v} \times \mathbf{B}) ,$$
(A.11)

where $\gamma = (1 - \beta^2)^{-1/2}$ is the *Lorentz factor* and $\beta = v/c$.

The time component of (A.10) gives the rate of change of energy W performed by the electromagnetic field on the charged particles

$$\frac{dW}{dt} = \frac{d}{dt}(\gamma m c^2) = q\mathbf{E} \cdot \mathbf{v} .$$
(A.12)

Thus, in absence of external forces, only the electric field can change the energy of charged particles, as

$$\mathbf{v} \cdot \mathbf{F} = q(\mathbf{v} \cdot \mathbf{E} + \mathbf{v} \cdot \mathbf{v} \times \mathbf{B}) = q(\mathbf{v} \cdot \mathbf{E}) .$$
(A.13)

Note that we write $\mathbf{p} = \gamma m \mathbf{v}$, where m is the mass of the particle measured in its rest frame (e.g., electron mass $m_e = 511 \, \text{keV} \, c^{-2}$ and proton mass $m_p = 931 \, \text{MeV} \, c^{-2}$). We prefer to avoid using the concepts of "rest mass" or "relativistic mass" and simply replace m by γm in case of relativistic motion.

Here it is also useful to recall the expression for *relativistic kinetic energy*

$$W = mc^2(\gamma - 1)$$
(A.14)

i.e., the total energy minus the rest energy mc^2. This formula provides an easy tool to calculate the velocity of a relativistic particle. For example, for a 1-MeV electron $\gamma \approx 3$, from which $v^2 = (8/9)c^2 \Rightarrow v = 0.94 \, c$.

Sometimes the Lorentz factor is useful to express in terms of momentum

$$\gamma = \sqrt{1 + \left(\frac{p}{mc}\right)^2} .$$
(A.15)

From (A.14) and (A.15) we get the relationship of relativistic momentum and energy

$$c^2 p^2 = W^2 + 2mc^2 W ,$$
(A.16)

which is useful, e.g., in calculation of the charged particle's rigidity (Sect. 2.5) or the phase space density from observed particle flux (Sect. 3.5).

A.2   Electromagnetic Waves in Linear Media

To introduce the electromagnetic waves let us start with waves in vacuum where the charge and current densities ρ and $\mathbf{J}$ are zero. From Maxwell's equations we get

$$\nabla^2 \mathbf{H} - \frac{1}{c^2} \frac{\partial^2 \mathbf{H}}{\partial t^2} = 0 \tag{A.17}$$

$$\nabla^2 \mathbf{E} - \frac{1}{c^2} \frac{\partial^2 \mathbf{E}}{\partial t^2} = 0 \,, \tag{A.18}$$

where we write $\mathbf{H} = \mathbf{B}/\mu_0$ for notational convenience.

The solutions of these equations give the waves propagating at the speed of light. Most of our treatise deals with *plane waves*, for which there is a plane where the electric field of the wave is constant. An important exception are the large-scale ULF waves in the quasi-dipolar magnetic field, which cannot be described as plane waves. A plane wave propagating in the z-direction can be represented by a sinusoidal function

$$E_x(z, t) = E_0 \cos(kz - \omega t) \,, \tag{A.19}$$

where E_0 is the *amplitude*, $\omega = 2\pi f$ the *angular frequency*, and $k = 2\pi/\lambda$ the *wave number*. f denotes the *oscillation frequency* and λ the *wavelength*. The *phase velocity* of the wave is $\omega/k = c$.

Note that it is common to use just the word "frequency" to refer to both ω and f. The theoretical treatment is more logical to write in terms of the angular frequency, whereas the oscillation frequency is used to represent the observations. Thus one should be careful with the factor of 2π. We use the SI unit s^{-1} for the angular frequency and hertz (Hz) for the oscillation frequency.

In the vector form the wave electric field is

$$\mathbf{E}(\mathbf{r}, t) = \mathbf{E}_0 \cos(\mathbf{k} \cdot \mathbf{r} - \omega t) \,, \tag{A.20}$$

where $\mathbf{k}$ is the *wave vector* that points to the direction of wave propagation. Throughout this book we use the complex notation for plane waves with the following sign convention for the argument of the exponential function:

$$\mathbf{E} = \mathbf{E}_0 \, e^{i(\mathbf{k} \cdot \mathbf{r} - \omega t)} \;;\; \mathbf{B} = \mathbf{B}_0 \, e^{i(\mathbf{k} \cdot \mathbf{r} - \omega t)} \,. \tag{A.21}$$

When $\mathbf{E}_0$ and $\mathbf{B}_0$ are constant, the temporal and spatial dependencies are said to be *harmonic* and Maxwell's equations, including the source terms, can be transformed to an algebraic form

$$\begin{aligned}
i\mathbf{k} \cdot \mathbf{D} &= \rho \\
\mathbf{k} \cdot \mathbf{B} &= 0 \\
\mathbf{k} \times \mathbf{E} &= \omega\mathbf{B} \\
i\mathbf{k} \times \mathbf{H} &= \mathbf{J} - i\omega\mathbf{D} ,
\end{aligned} \tag{A.22}$$

where we have introduced the *electric displacement* $\mathbf{D} = \epsilon \cdot \mathbf{E}$ in order to have the equations applicable to dielectric media. Recall that in a general dielectric medium ϵ is a tensor. From (A.22) it is clear $\mathbf{k} \perp \mathbf{E}$, $\mathbf{k} \perp \mathbf{H}$, and $\mathbf{E} \perp \mathbf{H}$. Such a wave is called *transverse*. In plasmas also *longitudinal* ($\mathbf{k} \parallel \mathbf{E}$) electrostatic waves can propagate (Sect. 4.2).

Assume next that $\rho = 0$, $\mathbf{J} = 0$, $\sigma = 0$, and ϵ and μ are constant scalars, but not necessarily equal to ϵ_0 and μ_0. Now the phase velocity of the electromagnetic wave becomes $v_p = 1/\sqrt{\epsilon\mu}$ instead of the speed of light in vacuum. ω and $\mathbf{k}$ are related through a *dispersion equation* (or *dispersion relation*)

$$k = \frac{\omega}{v_p} = \sqrt{\epsilon\mu}\,\omega = \frac{n}{c}\omega , \tag{A.23}$$

where

$$n = \sqrt{\frac{\epsilon\mu}{\epsilon_0\mu_0}} \tag{A.24}$$

is the *refractive index* of the medium. In tenuous space plasmas $\mu \approx \mu_0$ is a very good approximation and we can write $n = \sqrt{\epsilon/\epsilon_0}$. As the dielectric function $\epsilon(\omega, \mathbf{k})$ describes the response of the medium to wave propagation, it is customary to speak of *index of refraction* of the wave in question.

The *group velocity* of the wave is defined by

$$v_g = \frac{\partial\omega}{\partial k} . \tag{A.25}$$

In this special case $v_g = v_p = c/n$ and independent of frequency and wave number. Thus the medium is not dispersive. Plasma, in turn, is a dispersive medium with more complicated dispersion equations (Chap. 4).

In three-dimensional space the phase and group velocities are vector quantities. We write the wave vector as $\mathbf{k} = k\,\mathbf{e}_n$, where $\mathbf{e}_n$ is the unit vector defining the direction of the *wave normal*, which is perpendicular to the surface of constant wave phase. The wave normal direction is the *direction of wave propagation*, which is the direction of the phase velocity as a vector

$$\mathbf{v}_p = \frac{\omega}{k}\mathbf{n} . \tag{A.26}$$

In isotropic media the direction of wave propagation is the same as the direction of energy flux expressed by the *Poynting vector*

$$S = \frac{1}{2} \mathbf{E} \times \mathbf{H}^*,$$

(A.27)

where * denotes the complex conjugate.

The background magnetic field makes magnetospheric plasma anisotropic. In anisotropic media the wave electric field may have a component $\parallel \mathbf{k}$, implying that $\mathbf{S} \nparallel \mathbf{k}$. *Ray-tracing* is a method of following the ray of the wave in order to find the direction of energy and information propagation. The term "ray" derives from the light ray in optics. The propagation velocity of the ray is the group velocity

$$\mathbf{v}_g = \frac{\partial \omega}{\partial \mathbf{k}},$$

(A.28)

i.e., the gradient of the frequency in the **k**-space. Various results of ray-tracing are discussed in Chap. 5.

A.3   Dispersion Equation in Cold Non-magnetized Plasma

From the full set of Maxwell's equations (A.22) together with Ohm's law we can derive the dispersion equation for plasma waves in *cold plasma approximation*. This approximation can be used when the phase velocity of the wave is much larger than the thermal velocity $\sqrt{2k_B T/m}$ in the plasma ($k_B = 1.38065 \times 10^{-23}\,\text{m}^2\,\text{kg}\,\text{s}^{-2}\,\text{K}^{-1}$ is the *Boltzmann constant*).

Consider the motion of an electron in the electric field of a plane wave. As the time dependence of plane waves electric field is harmonic, we can assume the same for the electron velocity and replace d/dt by $-i\omega$. Thus the equation of motion becomes

$$m_e \frac{d\mathbf{v}}{dt} = -i\omega m_e \mathbf{v} = -e\mathbf{E}.$$

(A.29)

The velocity is related to the electric current as

$$\mathbf{J} = -n_e e \mathbf{v} = \frac{\omega_{pe}^2}{\omega^2} i\omega \epsilon_0 \mathbf{E},$$

(A.30)

where n_e is the electron number density and we have introduced the *electron plasma frequency* (Sect. 3.1.2)

$$\omega_{pe}^2 = \frac{n_e e^2}{\epsilon_0 m_e}.$$

(A.31)

Interpreting (A.30) in terms of Ohm's law the expression for the conductivity is

$$\sigma = \frac{\omega_{pe}^2}{\omega^2} i\omega\epsilon_0 . \tag{A.32}$$

Assume that, except for conductivity, the medium has the electromagnetic properties of vacuum ($\epsilon = \epsilon_0$ and $\mu = \mu_0$). The Ampère–Maxwell law can now be written as

$$i\mathbf{k} \times \mathbf{H} = \frac{\omega_{pe}^2}{\omega^2} i\omega\epsilon_0\mathbf{E} - i\omega\epsilon_0\mathbf{E} = -i\omega \left(1 - \frac{\omega_{pe}^2}{\omega^2}\right)\epsilon_0\mathbf{E} . \tag{A.33}$$

Thus the medium *looks like* a dielectric with permittivity

$$\epsilon = \left(1 - \frac{\omega_{pe}^2}{\omega^2}\right)\epsilon_0 . \tag{A.34}$$

Note that opposite to the ordinary dielectrics, where electrons are bound to nuclei, in a plasma of free electrons $\epsilon < \epsilon_0$.

In plasma physics we often write $\omega_{pe}^2/\omega^2 = X$. The refractive index (A.24) is thus

$$n = \sqrt{1 - X} . \tag{A.35}$$

This is the dispersion equation relating ω and k

$$c = \frac{\omega}{k}\sqrt{1 - X} . \tag{A.36}$$

The phase and group velocities are

$$v_p = \frac{\omega}{k} = \frac{c}{\sqrt{1 - X}} \tag{A.37}$$

$$v_g = \frac{\partial\omega}{\partial k} = c\sqrt{1 - X} . \tag{A.38}$$

In this case the phase velocity is larger than the speed of light in vacuum. Because the group velocity is the velocity of energy propagation, it cannot exceed the speed of light.

248

When k increases (short wavelengths), the dispersion equation approaches that of an electromagnetic wave in free space $\omega = ck$, because the response of the electrons at high frequencies weakens due to their finite inertia.[1]

At the long wavelength limit $(k \to 0)$ we find the standing plasma oscillation $\omega = \omega_{pe}$. If the frequency becomes smaller than the local plasma frequency $(X > 1)$, the frequency becomes imaginary. The wave cannot propagate into such a domain and is reflected. The plasma frequency is said to be a *cut-off* for this wave, below which the wave becomes evanescent.[2] Cold plasma waves in a background magnetic field are discussed in Sect. 4.3 and used thereafter extensively in discussion of wave–particle interactions.

---

[1] Even at frequencies much higher than the electron plasma frequency the electromagnetic waves interact, although weakly, with electrons through Thomson scattering. Thomson scattering is the fundamental mechanism behind the widely used ionospheric incoherent scatter radars.

[2] Also the cut-off phenomenon is utilized in ionospheric physics. Ionosondes, which transmit signals with increasing frequency are used to determine the altitude profile of ionospheric plasma density by measuring the time delay of the signals of different frequencies reflected back from different altitudes.

Appendix B
Satellites and Data Sources

This appendix is a brief introduction to spacecraft that have been mentioned in the text. The list is not exhaustive and several other satellites have made important contributions to the physics of the radiation belts. While the links listed here will some day become inactive, information of different satellite missions, their instruments and data bases are easy to find in the internet.

When looking for data obtained with various satellites, a recommendable starting point is the Coordinated Data Analysis Web (CDAWeb)[1] maintained at the Goddard Space Flight Center of NASA (National Aeronautics and Space Administration). In addition to spacecraft data CDAWeb contains a wealth of ground-based data and various supplementary products, e.g., pre-generated data and orbit plots. However, it is always important to contact the original data sources before rushing into scientific conclusions.

While the second Soviet satellite *Sputnik* 2 had in November 1957 carried, in addition to the famous dog Laika, radiation detectors to a Low-Earth Orbit (LEO), the seminal observations of radiation belt particles were made in 1958 with—according to the present standards—simple radiation detectors based on Geiger–Müller tubes onboard *Explorer* I and III, and *Pioneer* III[2] by James A. Van Allen and his team in Iowa. The monograph *Origins of Magnetospheric Physics* by Van Allen (1983) contains a detailed description of the instrument development and data analysis leading to these observations and also includes a discussion of the early Soviet experiments. *Explorer* I and III were on LEO whereas *Pioneer* III was an unsuccessful Moon mission, whose main contribution came from crossing the radiation belts both when going up and falling down.

---

[1] https://cdaweb.gsfc.nasa.gov/index.html/.

[2] Whether the names of spacecraft are capitalized or not varies from one source to another, even different co-investigators of a joint mission do not always follow the same conventions. This applies also to roman or arabic numbering of satellites of the same family.

© The Author(s) 2022
H. E. J. Koskinen, E. K. J. Kilpua, *Physics of Earth's Radiation Belts*,
Astronomy and Astrophysics Library, https://doi.org/10.1007/978-3-030-82167-8

During the 1960s a large number of satellites contributed to the increasing observational basis of energetic particles, plasmas and plasma waves in the magnetosphere. Particularly important for radiation belt studies was the series of *Orbiting Geophysical Observatories*, of which OGO 1 and 3 are referred to in Chap. 5.

OGO 1 was launched in September 1964 and made observations until November 1969. Its initial orbit was highly elliptical (HEO) with the initial perigee of 281 km, the apogee of 149,385 km (about 24.5 R_E geocentric) and inclination of 31°. The satellite was equipped with 20 different instruments with the main focus on the radiation belts. It, however, suffered from serious technical problems and the data set remained limited.

Also OGO 3 was launched to HEO (295 × 122,291 km, with inclination of 31°) in June 1966. Its instruments and objectives were the same as those of OGO 1. It was much more successful and provided a wealth of high-quality data and its routine operations continued until December 1969. The mission was terminated in February 1972.

The geostationary orbit (GEO, 6.6 R_E) close to the geographic equator has been populated by numerous satellites—mostly commercial but also scientific—since the early space age. The satellites have made significant contributions to the knowledge of radiation belts. Particularly important has been the joint NASA and NOAA (National Oceanic and Atmospheric Administration) *Geostationary Operational Environmental Satellite* Program (GOES). The first GOES satellite was launched in October 1975. Currently there are always two GOES satellites in operation: one above the east coast and another above the west coast of the United States. Their main objective is atmospheric research and meteorology but most of them have carried scientific magnetometers and energetic particle detectors enabling also long-term studies of plasma waves, in particular in the ULF range, and variations in particle fluxes. Data products are available at the Space Weather Prediction Center[3] of NOAA.

The *Active Magnetospheric Particle Tracer Explorer* (AMPTE) was an International (Germany, UK, US) three-satellite mission launched in August 1984. Its goals were to study the access of solar wind ions to the magnetosphere and the transport and energization of magnetospheric particles. It consisted of a German *Ion Release Module* (AMPTE/IRM), an American *Charge Composition Explorer* (AMPTE/CCE) and a UK-provided sub-satellite called AMPTE/UKS following close to AMPTE/IRM. The perigee of AMPTE/CCE was about 1100 km and the (geocentric) apogee 8.8 R_E with a low (4.8°) inclination. AMPTE/CCE provided a wealth of radiation belt and ring current data until its failure in 1989 (examples are discussed in Chaps. 5 and 6).

The U.S. Air Force *Combined Release and Radiation Effects Satellite* (CRRES), although short-lived, was a major radiation belt satellite. It was launched in July 1990 to the geosynchronous transfer orbit (GTO) with perigee of 347 km and (geocentric) apogee 6.2 R_E. The inclination of the orbit was about 18° and it reached

---

[3] https://www.swpc.noaa.gov.

maximally to $L \approx 9$. In addition to radiation belt studies the spacecraft was used to measure radiation effects on state-of-the-art electronics devices. The mission ended prematurely after only 14 months of operations probably due to a failure of onboard battery. Despite its short lifetime the satellite has provided material for numerous important radiation belt studies. Particularly interesting was the strong storm on 24 March 1991 when the slot region became filled with ultra-relativistic electrons. Examples of CRRES contributions are in Chaps. 5, 6, and 7.

The longest (almost) continuous radiation belt data set is available from NASA's *Solar, Anomalous, and Magnetospheric Particle Explorer* (SAMPEX, Baker et al. 1993). It was launched to an almost circular LEO (520×670 km) polar (inclination 82°) orbit in July 1992. The advantages of polar orbits compared to near-equatorial orbits are the sampling of wide range of L-shells up to the polar cap, where the L-shell is not defined, and observing particles at the edge and inside the atmospheric loss cone. The satellite was designed for only three years lifetime but it survived over 20 years. The official mission ended in June 2004, but data were collected until the re-entry to the atmosphere in November 2012 (e.g., Fig. 1.6).

The observational basis of radiation belt studies was revolutionized by NASA's twin *Radiation Belt Storm Probes* (RBSP) mission, named after the launch as *Van Allen Probes* (Mauk et al. 2013).[4] The satellites were launched at the end of August 2012 and deactivated in July and October 2019. The initial perigees of the highly elliptical orbits were at 618 km and apogees at 30,414 km (geocentric altitude of 5.7 R_E) with inclination of 10.2° and orbital period of 9 hours. Thus the satellites travelled from the inner belt through the heart of the outer radiation belt several times each day. Numerous examples of *Van Allen Probes* observations can be found in Chaps. 5, 6, and 7. Detailed information on the satellites, their instrumentation, and an updated list of publications are available on the mission-specific web-site of the Applied Physics Laboratory of Johns Hopkins University.[5]

The orbits of the *Van Allen Probes* limited the observations inside of $L \approx 6$. However, as discussed in Sect. 6.6, a wider view of radiation belt evolution (e.g., Fig. 6.11) has been possible to obtain with joint analysis of observations with NASA's *Time History of Events and Macroscale Interactions during Substorms* (THEMIS, Angelopoulos 2008). The initial THEMIS constellation consisted of five small satellites launched to a HEO ($470 \times 87{,}300$ km) with an inclination of 16° in February 2007. During the first half year the spacecraft followed each other in a string-of-pearls configuration. During autumn 2007 the spacecraft were moved to a constellation optimal for substorm studies in the magnetotail. Furthermore, THEMIS has provided the most comprehensive coverage of Pc4–Pc5 ULF waves (Sect. 5.4). In 2010 two of the five units were redirected to the Moon and renamed *Acceleration, Reconnection, Turbulence and Electrodynamics of the Moon's Interaction with the Sun* (ARTEMIS). At the time of writing this book the remaining three THEMIS satellites continue to produce important magnetospheric

---

[4] https://www.nasa.gov/van-allen-probes.

[5] http://vanallenprobes.jhuapl.edu.

observations. Details of the mission can be found on the THEMIS web-site at the Space Sciences Laboratory of the University of California, Berkeley.[6]

The *Magnetospheric Multiscale mission* (MMS, Burch et al. 2016) of NASA was launched in March 2015 to HEO, initially $2550 \times 70,080$ km with the inclination $28.0°$. In the second phase of the mission the apogee was lifted to 152,900 km (about $25 R_E$ geocentric). It is a constellation of four identical satellites moving at variable distances from each other with the goal of understanding the reconnection process at the magnetopause and in the tail current sheet. As discussed in Sect. 5.4.1 the close constellation has provided unique opportunity to investigate the ULF waves up to large azimuthal mode numbers. Details of MMS and its instruments are on the NASA mission web-site.[7] The MMS science data center is at the Laboratory of Atmospheric and Space Physics of the University of Colorado, Boulder.[8]

The four-satellite constellation mission *Cluster*[9] has been ESA's (European Space Agency) workhorse in magnetospheric physics since the spacecraft were launched in 2000. The mission is sometimes referred to as *Cluster* 2, because the first set of four satellites were lost in a launch failure in 1996. The original orbit was an elliptical $(19,000 \times 119,000$ km) high-inclination $(135°)$ orbit with an orbital period of 57 h. The inter-spacecraft distances have been adjusted several times during the mission to optimize the configuration for different domains of the magnetosphere and its boundary layers. Most of the satellites' instruments are still returning valuable data at the time of writing this volume (examples in Chaps. 5 and 6).

The Japanese *Arase*[10] satellite, formerly known as *Exploration of Energization and Radiation in Geospace* (ERG), was launched in December 2016 to a $440 \times 32,000$-km orbit with inclination of $32°$ and orbital period of 9.4 h. The higher inclination complemented the orbital coverage of *Van Allen Probes* during the last two and half years of the latter, as exemplified in Chap. 6 (Fig. 6.9).

Numerous spacecraft with different main scientific objectives have also been very useful to radiation belt studies. Examples of these are:

* *Imager for Magnetopause-to-Aurora Global Exploration* (IMAGE)[11] (Chaps. 1 and 7)
* *Geotail*[12] (Chap. 5)
* *Solar Terrestrial Relations Observatory* (STEREO)[13] (Chap. 5)

---

[6] http://themis.ssl.berkeley.edu/index.shtml.

[7] https://www.nasa.gov/mission_pages/mms/index.html.

[8] https://lasp.colorado.edu/mms/sdc/public/.

[9] https://sci.esa.int/web/cluster.

[10] https://ergsc.isee.nagoya-u.ac.jp.

[11] https://image.gsfc.nasa.gov.

[12] https://www.isas.jaxa.jp/en/missions/spacecraft/current/geotail.html.

[13] https://stereo.gsfc.nasa.gov.

- *Wind*[14] (Chap. 5)
- *Dynamics Explorer*[15] and *Double Star*[16] (Chap. 5)

During the time of writing this book the also the very small satellites in the CubeSat class have gained importance in space physics, including the radiation belts. Two examples of these are found in Chap. 6 are *AeroCube 6B*[17] and FIREBIRD II.[18]

In Sect. 7.6 data from a number of Earth Observing spacecraft are cited: *Polar Orbiting Environmental Satellites* (POES),[19] *Envisat*,[20] *Thermosphere Ionosphere Mesosphere Energetics and Dynamics* (TIMED),[21] and *Earth Observing System— Aura* (EOS—Aura).[22]

---

[14] https://wind.nasa.gov.

[15] https://lasp.colorado.edu/timas/info/DE/DE_home.html.

[16] http://english.nssc.cas.cn/missions/PM/201306/t20130605_102885.html.

[17] https://www.nanosats.eu/sat/aerocube-6.

[18] https://ssel.montana.edu/firebird2.html.

[19] https://www.ngdc.noaa.gov/stp/satellite/poes/.

[20] https://earth.esa.int/web/guest/missions/esa-operational-eo-missions/envisat.

[21] https://www.nasa.gov/timed.

[22] https://aura.gsfc.nasa.gov.

References

Abel B, Thorne RM (1998) Electron scattering loss in Earth's inner magnetosphere 1. Dominant physical processes. J Geophys Res 103:2385–2396. https://doi.org/10.1029/97JA02919

Agapitov O, Mourenas D, Artemyev A, Mozer FS, Bonnell JW, Angelopoulos V, Shastun V, Krasnoselskikh V (2018) Spatial extent and temporal correlation of chorus and hiss: Statistical results from multipoint THEMIS observations. J Geophys Res (Space Phys) 123(10):8317–8330. https://doi.org/10.1029/2018JA025725

Albert JM, Bortnik J (2009) Nonlinear interaction of radiation belt electrons with electromagnetic ion cyclotron waves. Geophys Res Lett 36(12):L12110. https://doi.org/10.1029/2009GL038904

Alexeev II, Belenkaya KVV E S, Feldstein YI, Grafe A (1996) Magnetic storms and magnetotail currents. J Geophys Res 101:7737–7747. https://doi.org/10.1029/95JA03509

Alfvén H (1950) Cosmical electrodynamics. Clarendon Press, Oxford

Ali AF, Malaspina DM, Elkington SR, Jaynes AN, Chan AA, Wygant J, Kletzing CA (2016) Electric and magnetic radial diffusion coefficients using the Van Allen probes data. J Geophys Res (Space Phys) 121(10):9586–9607. https://doi.org/10.1002/2016JA023002

Anderson KA, Milton DW (1964) Balloon observations of X rays in the auroral zone, 3, high time resolution studies. J Geophys Res 69(21):4457–4479. https://doi.org/10.1029/JZ069i021p04457

Andersson ME, Verronen PT, Rodger CJ, Clilverd MA, Seppälä A (2014) Missing driver in the Sun-Earth connection from energetic electron precipitation impacts mesospheric ozone. Nature Communications 5:5197. https://doi.org/10.1038/ncomms6197

André M (1985) Dispersion surfaces. J Plasma Phys 33:1–19. https://doi.org/10.1017/S0022377800002270

Angelopoulos V (2008) The THEMIS mission. Space Sci Rev 141(1–4):5–34. https://doi.org/10.1007/s11214-008-9336-1

Angelopoulos V, Baumjohann W, Kennel CF, Coroniti FV, Kivelson MG, Pellat R, Walker RJ, Lühr H, Paschmann G (1992) Bursty bulk flows in the inner central plasma sheet. J Geophys Res 97:4027–4039. https://doi.org/10.1029/91JA02701

Angelopoulos V, Kennel CF, Coroniti FV, Pellat R, Kivelson MG, Walker RJ, Russell CT, Baumjohann W, Feldman WC, Gosling JT (1994) Statistical characteristics of bursty bulk flow events. J Geophys Res 99:21,257–21,280. https://doi.org/10.1029/94JA01263

Antonova AE, Shabansky VP (1968) Structure of the geomagnetic field at great distance from the earth. Geomagn Aeron 8:801–811

© The Author(s) 2022
H. E. J. Koskinen, E. K. J. Kilpua, *Physics of Earth's Radiation Belts*,
Astronomy and Astrophysics Library, https://doi.org/10.1007/978-3-030-82167-8

Archer MO, Horbury TS, Eastwood JP, Weygand JM, Yeoman TK (2013) Magnetospheric response to magnetosheath pressure pulses: A low-pass filter effect. J Geophys Res (Space Phys) 118(9):5454–5466. https://doi.org/10.1002/jgra.50519

Ashour-Abdalla M, Berchem JP, Büchner J, Zelenyi LM (1993) Shaping of the magnetotail from the mantle: Global and local structuring. J Geophys Res 98:5651–5676. https://doi.org/10.1029/92JA01662

Asikainen T, Ruopsa M (2016) Solar wind drivers of energetic electron precipitation. J Geophys Res (Space Phys) 121(3):2209–2225. https://doi.org/10.1002/2015JA022215

Axford WI, Hines CO (1961) A unifying theory of high-latitude geophysical phenomena and geomagnetic storms. Can J Phys 39:1433–1464. https://doi.org/10.1139/p61-172

Baker DN, Panasyuk MI (2017) Discovering Earth's radiation belts. Physics Today 70(12):46–51. https://doi.org/10.1063/PT.3.3791

Baker DN, Mason GM, Figueroa O, Colon G, Watzin JG, Aleman RM (1993) An overview of the Solar, Anomalous, and Magnetospheric Particle Explorer (SAMPEX) mission. IEEE Trans Geosci Remote Sens 31(3):531–541. https://doi.org/10.1109/36.225519

Baker DN, Pulkkinen TI, Angelopoulos V, Baumjohann W, McPherron RL (1996) Neutral line model of substorms: Past results and present view. J Geophys Res 101:12,975–13,010. https://doi.org/10.1029/95JA03753

Baker DN, Kanekal SG, Li X, Monk SP, Goldstein J, Burch JL (2004) An extreme distortion of the Van Allen belt arising from the 'Halloween' solar storm in 2003. Nature 432:878–881. https://doi.org/10.1038/nature03116

Baker DN, Kanekal SG, Hoxie VC, Henderson MG, Li X, Spence HE, Elkington SR, Friedel RHW, Goldstein J, Hudson MK, Reeves GD, Thorne RM, Kletzing CA, Claudepierre SG (2013) A long-lived relativistic electron storage ring embedded in Earth's outer Van Allen belt. Science 340:186–190. https://doi.org/10.1126/science.1233518

Baker DN, Jaynes AN, Hoxie VC, Thorne RM, Foster JC, Li X, Fennell JF, Wygant JR, Kanekal SG, Erickson PJ, Kurth W, Li W, Ma Q, Schiller Q, Blum L, Malaspina DM, Gerrard A, Lanzerotti LJ (2014) An impenetrable barrier to ultrarelativistic electrons in the Van Allen radiation belts. Nature 515:531–534. https://doi.org/10.1038/nature13956

Baker DN, Erickson PJ, Fennell JF, Foster JC, Jaynes AN, Verronen PT (2018) Space weather effects in the Earth's radiation belts. Space Sci Rev 214(1):17. https://doi.org/10.1007/s11214-017-0452-7

Balasis G, Daglis IA, Mann IR (eds) (2016) Waves, particles, and storms in geospace. Oxford University Pres, Oxford, UK. https://doi.org/10.1093/acprof:oso/9780198705246.001.0001

Balikhin MA, Shprits YY, Walker SN, Chen L, Cornilleau-Wehrlin N, Dand ouras I, Santolik O, Carr C, Yearby KH, Weiss B (2015) Observations of discrete harmonics emerging from equatorial noise. Nature Communications 6:7703. https://doi.org/10.1038/ncomms8703

Bell TF (1984) The nonlinear gyroresonance interaction between energetic electron and coherent VLF waves propagating at an arbitrary angle with respect to the earth's magnetic field. J Geophys Res 89(A2):905–918. https://doi.org/10.1029/JA089iA02p00905

Bellan PM (2006) Fundamentals of plasma physics. Gambridge University Press, Gambridge, UK. https://doi.org/10.1017/CBO9780511807183

Bentley SN, Watt CEJ, Owens MJ, Rae IJ (2018) ULF wave activity in the magnetosphere: Resolving solar wind interdependencies to identify driving mechanisms. J Geophys Res (Space Phys) 123(4):2745–2771. https://doi.org/10.1002/2017JA024740

Bernstein IB (1958) Waves in a plasma in a magnetic field. Phys Rev 109:10–21. https://doi.org/10.1103/PhysRev.109.10

Blake JB (2013) The shock injection of 24 March. 1991: Another look. In: Summers D, Mann IR, Baker DN, Schultz M (eds) Dynamics of the earth's radiation belts and inner magnetosphere, American Geophysical Union, Washington, D.C., Geophysical Monograph, vol 199, pp 189–193. https://doi.org/10.1029/2012GM001311

Blake JB, Kolasinski WA, Fillius RW, Mullen EG (1992) Injection of electrons and protons with energies of tens of MeV into L less than 3 on 24 March 1991. Geophys Res Lett 19:821–824. https://doi.org/10.1029/92GL00624

References

Blum LW, Artemyev A, Agapitov O, Mourenas D, Boardsen S, Schiller Q (2019) EMIC wave-driven bounce resonance scattering of energetic electrons in the inner magnetosphere. J Geophys Res (Space Phys) 124(4):2484–2496. https://doi.org/10.1029/2018JA026427

Bortnik J, Thorne RM, Meredith NP (2007) Modeling the propagation characteristics of chorus using CRRES suprathermal electron fluxes. J Geophys Res (Space Phys) 112(A8):A08204. https://doi.org/10.1029/2006JA012237

Bortnik J, Thorne RM, Inan US (2008a) Nonlinear interaction of energetic electrons with large amplitude chorus. Geophys Res Lett 35(21):L21102. https://doi.org/10.1029/2008GL035500

Bortnik J, Thorne RM, Meredith NP (2008b) The unexpected origin of plasmaspheric hiss from discrete chorus emissions. Nature 452(7183):62–66. https://doi.org/10.1038/nature06741

Bortnik J, Thorne RM, Li W, Tao X (2016) Chorus waves in geospace and their influence on radiation belt dynamics. In: Balasis G, Daglis IA, Mann IR (eds) Waves, particles and storms in geospace, Oxford University Press, Oxford, UK, pp 192–216. https://doi.org/10.1093/acprof:oso/9780198705246.001.0001

Bothmer V, Daglis IA (eds) (2007) Space weather, physics and effects. Springer, Praxis Publishing, Chichester, UK. https://doi.org/10.1007/978-3-540-34578-7

Boyd AJ, Turner DL, Reeves GD, Spence HE, Baker DN, Blake JB (2018) What causes radiation belt enhancements: A survey of the Van Allen probes era. Geophys Res Lett 45(11):5253–5259. https://doi.org/10.1029/2018GL077699

Boyd TJM, Sanderson JJ (2003) The physics of plasmas. Cambridge University Press, Cambridge, UK. https://doi.org/10.1017/CBO9780511755750

Brautigam DH, Albert JM (2000) Radial diffusion analysis of outer radiation belt electrons during the October 9, 1990, magnetic storm. J Geophys Res 105(A1):291–310. https://doi.org/10.1029/1999JA900344

Breneman AW, Halford A, Millan R, McCarthy M, Fennell J, Sample J, Woodger L, Hospodarsky G, Wygant JR, Cattell CA, Goldstein J, Malaspina D, Kletzing CA (2015) Global-scale coherence modulation of radiation-belt electron loss from plasmaspheric hiss. Nature 523(7559):193–195. https://doi.org/10.1038/nature14515

Breneman AW, Crew A, Sample J, Klumpar D, Johnson A, Agapitov O, Shumko M, Turner DL, Santolik O, Wygant JR, Cattell CA, Thaller S, Blake B, Spence H, Kletzing CA (2017) Observations directly linking relativistic electron microbursts to whistler mode chorus: Van Allen probes and FIREBIRD II. Geophys Res Lett 44(22):11,265–11,272. https://doi.org/10.1002/2017GL075001

Breuillard H, Zaliznyak Y, Krasnoselskikh V, Agapitov O, Artemyev A, Rolland G (2012) Chorus wave-normal statistics in the Earth's radiation belts from ray tracing technique. Annales Geophysicae 30(8):1223–1233. https://doi.org/10.5194/angeo-30-1223-2012

Brito T, Hudson MK, Kress B, Paral J, Halford A, Millan R, Usanova M (2015) Simulation of ULF wave-modulated radiation belt electron precipitation during the 17 March 2013 storm. J Geophys Res (Space Phys) 120(5):3444–3461. https://doi.org/10.1002/2014JA020838

Büchner J, Zelenyi LM (1989) Regular and chaotic charged particle motion in magnetotaillike field reversals. J Geophys Res 94:11,821–11,842. https://doi.org/10.1029/JA094iA09p11821

Burch JL, Moore TE, Torbert RB, Giles BL (2016) Magnetospheric multiscale overview and science objectives. Space Sci Rev 199(1–4):5–21. https://doi.org/10.1007/s11214-015-0164-9

Burtis WJ, Helliwell RA (1969) Banded chorus—A new type of VLF radiation observed in the magnetosphere by OGO 1 and OGO 3. J Geophys Res 74(11):3002. https://doi.org/10.1029/JA074i011p03002

Burtis WJ, Helliwell RA (1976) Magnetospheric chorus: Occurrence patterns and normalized frequency. Planet Space Sci 24(11):1007,IN1,1011–1010,IN4,1024. https://doi.org/10.1016/0032-0633(76)90119-7

Califf S, Li X, Zhao H, Kellerman A, Sarris TE, Jaynes A, Malaspina DM (2017) The role of the convection electric field in filling the slot region between the inner and outer radiation belts. J Geophys Res (Space Phys) 122:2051–2068. https://doi.org/10.1002/2016JA023657

Cao X, Ni B, Summers D, Bortnik J, Tao X, Shprits YY, Lou Y, Gu X, Fu S, Shi R, Xiang Z, Wang Q (2017a) Bounce resonance scattering of radiation belt electrons by H^+ band EMIC waves. J Geophys Res (Space Phys) 122(2):1702–1713. https://doi.org/10.1002/2016JA023607

Cao X, Ni B, Summers D, Zou Z, Fu S, Zhang W (2017b) Bounce resonance scattering of radiation belt electrons by low-frequency hiss: Comparison with cyclotron and landau resonances. Geophys Res Lett 44(19):9547–9554. https://doi.org/10.1002/2017GL075104

Cattell C, Wygant JR, Goetz K, Kersten K, Kellogg PJ, von Rosenvinge T, Bale SD, Roth I, Temerin M, Hudson MK, Mewaldt RA, Wiedenbeck M, Maksimovic M, Ergun R, Acuna M, Russell CT (2008) Discovery of very large amplitude whistler-mode waves in Earth's radiation belts. Geophys Res Lett 35(1):L01105. https://doi.org/10.1029/2007GL032009

Chapman S, Ferraro VCA (1931) A new theory of magnetic storms. Terr Magn Atmos Electr 36:77–97. https://doi.org/10.1029/TE036i002p00077

Chappell CR (1972) Recent satellite measurements of the morphology and dynamics of the plasmasphere. Rev Geophys Space Phys 10:951–972. https://doi.org/10.1029/RG010i004p00951

Chen H, Gao X, Lu Q, Wang S (2019) Analyzing EMIC Waves in the inner magnetosphere using long-term Van Allen probes observations. J Geophys Res (Space Phys) 124(9):7402–7412. https://doi.org/10.1029/2019JA026965

Chen J, Palmadesso PJ (1986) Chaos and nonlinear dynamics of single-particle orbits in a magnetotail-like magnetic field. J Geophys Res 91:1499–1508. https://doi.org/10.1029/JA091iA02p01499

Chen L, Hasegawa A (1974) A theory of long-period magnetic pulsations: 1. Steady state excitation of field line resonance. J Geophys Res 79(7):1024–1032. https://doi.org/10.1029/JA079i007p01024

Chen L, Hasegawa A (1991) Kinetic theory of geomagnetic pulsations 1. Internal excitations by energetic particles. J Geophys Res 96(A2):1503–1512. https://doi.org/10.1029/90JA02346

Chen L, Thorne RM, Jordanova VK, Horne RB (2010) Global simulation of magnetosonic wave instability in the storm time magnetosphere. J Geophys Res (Space Phys) 115(A11):A11222. https://doi.org/10.1029/2010JA015707

Chen Y, Reeves GD, Friedel RHW (2007) The energization of relativistic electrons in the outer Van Allen radiation belt. Nature Physics 3(9):614–617. https://doi.org/10.1038/nphys655

Chew GF, Goldberger ML, Low FE (1956) The Boltzmann equation and the one-fluid hydromagnetic equations in the absence of particle collisions. Proc Roy Soc Ser A 236(1024):112–118. https://doi.org/10.1098/rspa.1956.0116

Claudepierre SG, Hudson MK, Lotko W, Lyon JG, Denton RE (2010) Solar wind driving of magnetospheric ULF waves: Field line resonances driven by dynamic pressure fluctuations. J Geophys Res (Space Phys) 115:A11202. https://doi.org/10.1029/2010JA015399, 1010.3994

Claudepierre SG, O'Brien TP, Fennell JF, Blake JB, Clemmons JH, Looper MD, Mazur JE, Roeder JL, Turner DL, Reeves GD, Spence HE (2017) The hidden dynamics of relativistic electrons (0.7–1.5 MeV) in the inner zone and slot region. J Geophys Res (Space Phys) 122:3127–3144. https://doi.org/10.1002/2016JA023719

Colpitts C, Miyoshi Y, Kasahara Y, Delzanno GL, Wygant JR, Cattell CA, Breneman A, Kletzing C, Cunningham G, Hikishima M, Matsuda S, Katoh Y, Ripoll JF, Shinohara I, Matsuoka A (2020) First direct observations of propagation of discrete chorus elements from the equatorial source to higher latitudes, using the Van Allen probes and arase satellites. J Geophys Res (Space Phys) 125(10):e28315. https://doi.org/10.1029/2020JA028315

Coroniti FV, Kennel CF (1970) Electron precipitation pulsations. J Geophys Res 75:1279–1289. https://doi.org/10.1029/JA075i007p01279

Daglis IA, Thorne RM, Baumjohann W, Orsini S (1999) The terrestrial ring current: Origin, formation, and decay. Rev Geophys 37:407–438. https://doi.org/10.1029/1999RG900009

Delcourt DC, Sauvaud JA, Pedersen A (1990) Dynamics of single-particle orbits during substorm expansion phase. J Geophys Res 95:20,853–20,865. https://doi.org/10.1029/JA095iA12p20853

Dessler AJ, Parker EN (1959) Hydromagnetic theory of geomagnetic storms. J Geophys Res 64:2239–2252. https://doi.org/10.1029/JZ064i012p02239

References

Dungey JW (1961) Interplanetary magnetic field and auroral zones. Phys Rev Lett 6:47–48. https://doi.org/10.1103/PhysRevLett.6.47

Dungey JW (1965) Effects of the electromagnetic perturbations on particles trapped in the radiation belts. Space Sci Rev 4(2):199–222. https://doi.org/10.1007/BF00173882

Dysthe KB (1971) Some studies of triggered whistler emissions. J Geophys Res 76(28):6915–6931. https://doi.org/10.1029/JA076i028p06915

Elkington SR, Hudson MK, Chan AA (2003) Resonant acceleration and diffusion of outer zone electrons in an asymmetric geomagnetic field. J Geophys Res 108. https://doi.org/10.1029/2001JA009202

Fälthammar CG (1965) Effects of time-dependent electric fields on geomagnetically trapped radiation. J Geophys Res 70(11):2503–2516. https://doi.org/10.1029/JZ070i011p02503

Fei Y, Chan AA, Elkington SR, Wiltberger MJ (2006) Radial diffusion and MHD particle simulations of relativistic electron transport by ULF waves in the September 1998 storm. J Geophys Res (Space Phys) 111(A12):A12209. https://doi.org/10.1029/2005JA011211

Fok MC, Moore TE, Greenspan ME (1996) Ring current development during storm main phase. J Geophys Res 101:15,311–15,322. https://doi.org/10.1029/96JA01274

Foster JC, Wygant JR, Hudson MK, Boyd AJ, Baker DN, Erickson PJ, Spence HE (2015) Shock-induced prompt relativistic electron acceleration in the inner magnetosphere. J Geophys Res (Space Phys) 120:1661–1674. https://doi.org/10.1002/2014JA020642

Foster JC, Erickson PJ, Baker DN, Jaynes AN, Mishin EV, Fennel JF, Li X, Henderson MG, Kanekal SG (2016) Observations of the impenetrable barrier, the plasmapause, and the VLF bubble during the 17 March 2015 storm. J Geophys Res (Space Phys) 121:5537–5548. https://doi.org/10.1002/2016JA022509

Foster JC, Erickson PJ, Omura Y, Baker DN, Kletzing CA, Claudepierre SG (2017) Van Allen Probes observations of prompt MeV radiation belt electron acceleration in nonlinear interactions with VLF chorus. J Geophys Res (Space Phys) 122(1):324–339. https://doi.org/10.1002/2016JA023429

Friedel RWH, Korth A, Kremser G (1996) Substorm onsets observed by CRRES: Determination of energetic particle source regions. J Geophys Res 101:13,137–13,154. https://doi.org/10.1029/96JA00399

Fu S, He F, Gu X, Ni B, Xiang Z, Liu J (2018) Occurrence features of simultaneous H^+- and He^+-band EMIC emissions in the outer radiation belt. Adv Space Res 61(8):2091–2098. https://doi.org/10.1016/j.asr.2018.01.041

Ganushkina NY, Pulkkinen TI, Fritz T (2005) Role of substorm-associated impulsive electric fields in the ring current development during storms. Ann Geophys 23:579–591. https://doi.org/10.5194/angeo-23-579-2005

Glauert SA, Horne RB (2005) Calculation of pitch angle and energy diffusion coefficients with the PADIE code. J Geophys Res (Space Phys) 110(A4):A04206. https://doi.org/10.1029/2004JA010851

Gokani SA, Kosch M, Clilverd M, Rodger CJ, Sinha AK (2019) What fraction of the outer radiation belt relativistic electron flux at L $\approx$ 3–4.5 was lost to the atmosphere during the dropout event of the St. Patrick's day storm of 2015? J Geophys Res (Space Phys) 124(11):9537–9551. https://doi.org/10.1029/2018JA026278

Gold T (1959) Motions in the magnetosphere of the Earth. J Geophys Res 64:1219–1224. https://doi.org/10.1029/JZ064i009p01219

Goldstein J, Sandel BR, Thomsen MF, Spasojević M, Reiff PH (2004) Simultaneous remote sensing and in situ observations of plasmaspheric drainage plumes. J Geophys Res (Space Phys) 109(A3):A03202. https://doi.org/10.1029/2003JA010281

Grandin M, Aikio AT, Kozlovsky A, Ulich T, Raita T (2017) Cosmic radio noise absorption in the high-latitude ionosphere during solar wind high-speed streams. J Geophys Res (Space Phys) 122(5):5203–5223. https://doi.org/10.1002/2017JA023923

Greeley AD, Kanekal SG, Baker DN, Klecker B, Schiller Q (2019) Quantifying the contribution of microbursts to global electron loss in the radiation belts. J Geophys Res (Space Phys) 124(2):1111–1124. https://doi.org/10.1029/2018JA026368

Green JC, Kivelson MG (2004) Relativistic electrons in the outer radiation belt: Differentiating between acceleration mechanisms. J Geophys Res (Space Phys) 109(A3):A03213. https://doi.org/10.1029/2003JA010153

Häkkinen LVT, Pulkkinen TI, Nevanlinna H, Pirjola RJ, Tanskanen EI (2002) Effects of induced currents on Dst and on magnetic variations at midlatitude stations. J Geophys Res 107. https://doi.org/10.1029/2001JA900130

Hao YX, Zong QG, Wang YF, Zhou XZ, Zhang H, Fu SY, Pu ZY, Spence HE, Blake JB, Bonnell J, Wygant JR, Kletzing CA (2014) Interactions of energetic electrons with ULF waves triggered by interplanetary shock: Van Allen Probes observations in the magnetotail. J Geophys Res (Space Phys) 119:8262–8273. https://doi.org/10.1002/2014JA020023

Hartley DP, Kletzing CA, Chen L, Horne RB, Santolík O (2019) Van Allen probes observations of chorus wave vector orientations: Implications for the chorus-to-hiss mechanism. Geophys Res Lett 46(5):2337–2346. https://doi.org/10.1029/2019GL082111

Hietala H, Kilpua EKJ, Turner DL, Angelopoulos V (2014) Depleting effects of ICME-driven sheath regions on the outer electron radiation belt. Geophys Res Lett 41:2258–2265. https://doi.org/10.1002/2014GL059551

Horne RB (1989) Path-integrated growth of electrostatic waves: The generation of terrestrial myriametric radiation. J Geophys Res 94(A7):8895–8909. https://doi.org/10.1029/JA094iA07p08895

Horne RB, Wheeler GV, Alleyne HSCK (2000) Proton and electron heating by radially propagating fast magnetosonic waves. J Geophys Res 105(A12):27,597–27,610. https://doi.org/10.1029/2000JA000018

Horne RB, Thorne RM, Shprits YY, Meredith NP, Glauert SA, Smith AJ, Kanekal SG, Baker DN, Engebretson MJ, Posch JL, Spasojevic M, Inan US, Pickett JS, Decreau PME (2005) Wave acceleration of electrons in the van allen radiation belts. Nature 437:227–230. https://doi.org/10.1038/nature03939

Hudson M, Denton R, Lessard M, Miftakhova E, Anderson R (2004a) A study of Pc-5 ULF oscillations. Ann Geophys 22:289–302. https://doi.org/10.5194/angeo-22-289-2004

Hudson M, Jaynes A, Kress B, Li Z, Patel M, Shen XC, Thaller S, Wiltberger M, Wygant J (2017) Simulated prompt acceleration of multi-MeV electrons by the 17 March 2015 interplanetary shock. J Geophys Res (Space Phys) 122:10. https://doi.org/10.1002/2017JA024445

Hudson MK, Kress BT, Mazur JE, Perry KL, Slocum PL (2004b) 3D modeling of shock-induced trapping of solar energetic particles in the Earth's magnetosphere. J Atmospher Solar-Terrestrial Phys 66(15–16):1389–1397. https://doi.org/10.1016/j.jastp.2004.03.024

Hultqvist B, Øieroset M, Paschmann G, Treumann R (eds) (1999) Magnetospheric plasma sources and losses, space sciences series of ISSI, vol 6. Kluver Academic Publishers, Dordrecht, Holland. https://doi.org/10.1007/978-94-011-4477-3

Hwang KJ, Sibeck DG (2016) Role of low-frequency boundary waves in the dynamics of the dayside magnetopause and the inner magnetosphere. In: Low-frequency waves in space plasmas, Washington DC American Geophysical Union Geophysical Monograph Series, vol 216, pp 213–239. https://doi.org/10.1002/9781119055006.ch13

Jacobs JA, Kato Y, Matsushita S, Troitskaya VA (1964) Classification of geomagnetic micropulsations. J Geophys Res 69(1):180–181. https://doi.org/10.1029/JZ069i001p00180

James MK, Yeoman TK, Mager PN, Klimushkin DY (2016) Multiradar observations of substorm-driven ULF waves. J Geophys Res (Space Phys) 121(6):5213–5232. https://doi.org/10.1002/2015JA022102

Jaynes AN, Ali AF, Elkington SR, Malaspina DM, Baker DN, Li X, Kanekal SG, Henderson MG, Kletzing CA, Wygant JR (2018) Fast diffusion of ultrarelativistic electrons in the outer radiation belt: 17 March 2015 storm event. Geophys Res Lett 45(20):10,874–10,882. https://doi.org/10.1029/2018GL079786

Jordanova VK, Albert J, Miyoshi Y (2008) Relativistic electron precipitation by EMIC waves from self-consistent global simulations. J Geophys Res 113. https://doi.org/10.1029/2008JA013239

References

Jordanova VK, Zaharia S, Welling DT (2010) Comparative study of ring current development using empirical, dipolar, and self-consistent magnetic field simulations. J Geophys Res (Space Phys) 115(A12):A00J11. https://doi.org/10.1029/2010JA015671

Kalliokoski MMH, Kilpua EKJ, Osmane A, Turner DL, Jaynes AN, Turc L, George H, Palmroth M (2020) Outer radiation belt and inner magnetospheric response to sheath regions of coronal mass ejections: a statistical analysis. Annales Geophysicae 38(3):683–701. https://doi.org/10.5194/angeo-38-683-2020

Kanekal SG, Baker DN, Fennell JF, Jones A, Schiller Q, Richardson IG, Li X, Turner DL, Califf S, Claudepierre SG, Wilson LB III, Jaynes A, Blake JB, Reeves GD, Spence HE, Kletzing CA, Wygant JR (2016) Prompt acceleration of magnetospheric electrons to ultrarelativistic energies by the 17 March 2015 interplanetary shock. J Geophys Res (Space Phys) 121:7622–7635. https://doi.org/10.1002/2016JA022596

Kataoka R, Miyoshi Y (2006) Flux enhancement of radiation belt electrons during geomagnetic storms driven by coronal mass ejections and corotating interaction regions. Space Weather 4:09004. https://doi.org/10.1029/2005SW000211

Keika K, Takahashi K, Ukhorskiy AY, Miyoshi Y (2013) Global characteristics of electromagnetic ion cyclotron waves: Occurrence rate and its storm dependence. J Geophys Res (Space Phys) 118(7):4135–4150. https://doi.org/10.1002/jgra.50385

Kellogg PJ, Cattell CA, Goetz K, Monson SJ, Wilson I L B (2011) Large amplitude whistlers in the magnetosphere observed with wind-waves. J Geophys Res (Space Phys) 116(A9):A09224. https://doi.org/10.1029/2010JA015919

Kennel C (1966) Low-frequency whistler mode. Phys Fluids 9(11):2190–2202. https://doi.org/10.1063/1.1761588

Kennel CF (1995) Convection and substorms: Paradigms of magnetospheric phenomenology Oxford University Press, New York, NY. https://doi.org/10.1093/oso/9780195085297.001.0001

Kennel CF, Engelmann F (1966) Velocity space diffusion from weak plasma turbulence in a magnetic field. Phys Fluids 9:2377–2388. https://doi.org/10.1063/1.1761629

Kennel CF, Petschek HE (1966) Limit on stably trapped particle fluxes. J Geophys Res 71:1–28. https://doi.org/10.1029/JZ071i001p00001

Kepko L, Viall NM (2019) The source, significance, and magnetospheric impact of periodic density structures within stream interaction regions. J Geophys Res (Space Phys) 124(10):7722–7743. https://doi.org/10.1029/2019JA026962

Kilpua EKJ, Hietala H, Turner DL, Koskinen HEJ, Pulkkinen TI, Rodriguez JV, Reeves GD, Claudepierre SG, Spence HE (2015) Unraveling the drivers of the storm time radiation belt response. Geophys Res Lett 42:3076–3084. https://doi.org/10.1002/2015GL063542

Kilpua EKJ, Balogh A, von Steiger R, Liu YD (2017) Geoeffective properties of solar transients and stream interaction regions. Space Sci Rev 212(3–4):1271–1314. https://doi.org/10.1007/s11214-017-0411-3

Kilpua EKJ, Turner DL, Jaynes AN, Hietala H, Koskinen HEJ, Osmane A, Palmroth M, Pulkkinen TI, Vainio R, Baker D, Claudepierre SG (2019) Outer Van Allen radiation belt response to interacting interplanetary coronal mass ejections. J Geophys Res (Space Phys) 124(3):1927–1947. https://doi.org/10.1029/2018JA026238

Kim KC, Shprits Y (2018) Survey of the favorable conditions for magnetosonic wave excitation. J Geophys Res (Space Phys) 123(1):400–413. https://doi.org/10.1002/2017JA024865

Kim KC, Shprits Y (2019) Statistical analysis of hiss waves in plasmaspheric plumes using Van Allen probe observations. J Geophys Res (Space Phys) 124(3):1904–1915. https://doi.org/10.1029/2018JA026458

Kivelson MG, Southwood DJ (1986) Coupling of global magnetospheric MHD eigenmodes to field line resonances. J Geophys Res 91(A4):4345–4351. https://doi.org/10.1029/JA091iA04p04345

Koons HC, Edgar BC, Vampola AL (1981) Precipitation of inner zone electrons by whistler mode waves from the VLF transmitters UMS and NWC. J Geophys Res 86:640–648. https://doi.org/10.1029/JA086iA02p00640

Koskinen HEJ (2011) Physics of space storms, from solar surface to the earth. Springer-Praxis, Heidelberg, Germany. https://doi.org/10.1007/978-3-642-00319-6

Krall NA, Trivelpiece AW (1973) Principles of plasma physics. McGraw-Hill, New York, NY

Kurita S, Miyoshi Y, Shiokawa K, Higashio N, Mitani T, Takashima T, Matsuoka A, Shinohara I, Kletzing CA, Blake JB, Claudepierre SG, Connors M, Oyama S, Nagatsuma T, Sakaguchi K, Baishev D, Otsuka Y (2018) Rapid loss of relativistic electrons by EMIC waves in the outer radiation belt observed by arase, Van Allen probes, and the PWING ground stations. Geophys Res Lett 45(23):12,720–12,729. https://doi.org/10.1029/2018GL080262

Landau LD (1946) On the vibrations of the electronic plasma. J Phys (USSR) 10:25–34

Langel RA, Estes RH (1985) Large-scale, near-field magnetic fields from external sources and the corresponding induced internal field. J Geophys Res 90:2487–2494. https://doi.org/10.1029/JB090iB03p02487

Lee DY, Kim KC, Choi CR (2020) Nonlinear scattering of 90° pitch angle electrons in the outer radiation belt by large-amplitude EMIC waves. Geophys Res Lett 47(4):e86738. https://doi.org/10.1029/2019GL086738

Lejosne S (2019) Analytic expressions for radial diffusion. J Geophys Res (Space Phys) 124(6):4278–4294. https://doi.org/10.1029/2019JA026786

Lejosne S, Kollmann P (2020) Radiation belt radial diffusion at earth and beyond. Space Sci Rev 216(1):19. https://doi.org/10.1007/s11214-020-0642-6

Lerche I (1968) Quasilinear theory of resonant diffusion in a magneto-active, relativistic plasma. Phys Fluids 11(8):1720–1727. https://doi.org/10.1063/1.1692186

Li J, Bortnik J, An X, Li W, Angelopoulos V, Thorne RM, Russell CT, Ni B, Shen X, Kurth WS, Hospodarsky GB, Hartley DP, Funsten HO, Spence HE, Baker DN (2019) Origin of two-band chorus in the radiation belt of earth. Nature Communications 10:4672. https://doi.org/10.1038/s41467-019-12561-3

Li X, Roth I, Temerin M, Wygant JR, Hudson MK, Blake JB (1993) Simulation of the prompt enerigzation and transport of radiation belt particles during the March 24, 1991 SSC. Geophys Res Lett 20:2423–2426. https://doi.org/10.1029/93GL02701

Li X, Baker DN, Zhao H, Zhang K, Jaynes AN, Schiller Q, Kanekal SG, Blake JB, Temerin M (2017) Radiation belt electron dynamics at low l (< 4): Van allen probes era versus previous two solar cycles. J Geophys Res 122(5):5224–5234. https://doi.org/10.1002/2017JA023924

Liu W, Sarris TE, Li X, Elkington SR, Ergun R, Angelopoulos V, Bonnell J, Glassmeier KH (2009) Electric and magnetic field observations of Pc4 and Pc5 pulsations in the inner magnetosphere: A statistical study. J Geophys Res (Space Phys) 114(A12):A12206. https://doi.org/10.1029/2009JA014243

Longden N, Denton MH, Honary F (2008) Particle precipitation during ICME-driven and CIR-driven geomagnetic storms. J Geophys Res (Space Phys) 113:A06205. https://doi.org/10.1029/2007JA012752

Lugaz N, Farrugia CJ, Huang CL, Spence HE (2015) Extreme geomagnetic disturbances due to shocks within CMEs. Geophys Res Lett 42:4694–4701. https://doi.org/10.1002/2015GL064530

Lugaz N, Temmer M, Wang Y, Farrugia CJ (2017) The interaction of successive coronal mass ejections: A review. Solar Phys 292:64. https://doi.org/10.1007/s11207-017-1091-6, 1612.02398

Lyons LR, Speiser TW (1982) Evidence of current sheet acceleration in the geomagnetic tail. J Geophys Res 87:2276–2286. https://doi.org/10.1029/JA087iA04p02276

Lyons LR, Williams DJ (eds) (1984) Quantitative aspects of magnetospheric physics. D. Reidel, Dordrecht, NL

Lyons LR, Thorne RM, Kennel CF (1972) Pitch-angle diffusion of radiation belt electrons within the plasmasphere. J Geophys Res 77(19):3455. https://doi.org/10.1029/JA077i019p03455

Malmberg JH, Wharton CB (1964) Collisionless damping of electrostatic plasma waves. Phys Rev Lett 13:184–186. https://doi.org/10.1103/PhysRevLett.13.184

References

Manchester W, Kilpua EKJ, Liu YD, Lugaz N, Riley P, Török T, Vršnak B (2017) The physical processes of CME/ICME evolution. Space Sci Rev 212:1159–1219. https://doi.org/10.1007/s11214-017-0394-0

Mann IR, Ozeke LG, Murphy KR, Claudepierre SG, Turner DL, Baker DN, Rae IJ, Kale A, Milling DK, Boyd AJ, Spence HE, Reeves GD, Singer HJ, Dimitrakoudis S, Daglis IA, Honary F (2016) Explaining the dynamics of the ultra-relativistic third Van Allen radiation belt. Nature Physics 12:978–983. https://doi.org/10.1038/nphys3799

Mauk BH, Fox NJ, Kanekal SG, Kessel RL, Sibeck DG, Ukhorskiy A (2013) Science objectives and rationale for the radiation belt storm probes mission. Space Sci Rev 179(1–4):3–27. https://doi.org/10.1007/s11214-012-9908-y

Mayaud PN (1980) Derivation, meaning, and use of geomagnetic indices, geophysical monograph, vol 22. American Geophysical Union, Washington, DC. https://doi.org/10.1029/GM022

McCollough JP, Elkington SR, Baker DN (2012) The role of Shabansky orbits in compression-related electromagnetic ion cyclotron wave growth. J Geophys Res (Space Phys) 117(A1):A01208. https://doi.org/10.1029/2011JA016948

McCracken KG, Dreschhoff GAM, Zeller EJ, Smart DF, Shea MA (2001) Solar cosmic ray events for the period 1561–1994: 1. Identification in polar ice, 1561–1950. J Geophys Res 106(A10):21,585–21,598. https://doi.org/10.1029/2000JA000237

Mead GD (1964) Deformation of the geomagnetic field by the solar wind. J Geophys Res 69(7):1181–1195. https://doi.org/10.1029/JZ069i007p01181

Meredith NP, Thorne RM, Horne RB, Summers D, Fraser BJ, Anderson RR (2003) Statistical analysis of relativistic electron energies for cyclotron resonance with EMIC waves observed on CRRES. J Geophys Res 108. https://doi.org/10.1029/2002JA009700

Meredith NP, Horne RB, Clilverd MA, Horsfall D, Thorne RM, Anderson RR (2006) Origins of plasmaspheric hiss. J Geophys Res (Space Phys) 111(A9):A09217. https://doi.org/10.1029/2006JA011707

Meredith NP, Horne RB, Glauert SA, Baker DN, Kanekal SG, Albert JM (2009) Relativistic electron loss timescales in the slot region. J Geophys Res (Space Phys) 114(A3):A03222. https://doi.org/10.1029/2008JA013889

Meredith NP, Horne RB, Sicard-Piet A, Boscher D, Yearby KH, Li W, Thorne RM (2012) Global model of lower band and upper band chorus from multiple satellite observations. J Geophys Res (Space Phys) 117(A10):A10225. https://doi.org/10.1029/2012JA017978

Meredith NP, Horne RB, Shen XC, Li W, Bortnik J (2020) Global model of whistler mode chorus in the near-equatorial region ($|\lambda_m| < 18°$). Geophys Res Lett 47(11):e87311. https://doi.org/10.1029/2020GL087311

Miyoshi Y, Kataoka R, Kasahara Y, Kumamoto A, Nagai T, Thomsen MF (2013) High-speed solar wind with southward interplanetary magnetic field causes relativistic electron flux enhancement of the outer radiation belt via enhanced condition of whistler waves. Geophys Res Lett 40:4520–4525. https://doi.org/10.1002/grl.50916

Moldwin MB, Zou S, Heine T (2016) The story of plumes: the development of a new conceptual framework for understanding magnetosphere and ionosphere coupling. Annales Geophysicae 34(12):1243–1253. https://doi.org/10.5194/angeo-34-1243-2016

Morley SK, Henderson MG, Reeves GD, Friedel RHW, Baker DN (2013) Phase Space Density matching of relativistic electrons using the Van Allen Probes: REPT results. Geophys Res Lett 40(18):4798–4802. https://doi.org/10.1002/grl.50909

Mourenas D, Artemyev AV, Ma Q, Agapitov OV, Li W (2016) Fast dropouts of multi-MeV electrons due to combined effects of EMIC and whistler mode waves. Geophys Res Lett 43(9):4155–4163. https://doi.org/10.1002/2016GL068921

Mozer FS, Agapitov OV, Blake JB, Vasko IY (2018) Simultaneous observations of lower band chorus emissions at the equator and microburst precipitating electrons in the ionosphere. Geophys Res Lett 45(2):511–516. https://doi.org/10.1002/2017GL076120

264

References

Murphy KR, Inglis AR, Sibeck DG, Rae IJ, Watt CEJ, Silveira M, Plaschke F, Claudepierre SG, Nakamura R (2018) Determining the mode, frequency, and azimuthal wave number of ULF waves during a HSS and moderate geomagnetic storm. J Geophys Res (Space Phys) 123(8):6457–6477. https://doi.org/10.1029/2017JA024877

Nakamura S, Omura Y, Summers D, Kletzing CA (2016) Observational evidence of the nonlinear wave growth theory of plasmaspheric hiss. Geophys Res Lett 43(19):10,040–10,049. https://doi.org/10.1002/2016GL070333

Ni B, Bortnik J, Thorne RM, Ma Q, Chen L (2013) Resonant scattering and resultant pitch angle evolution of relativistic electrons by plasmaspheric hiss. J Geophys Res (Space Phys) 118(12):7740–7751. https://doi.org/10.1002/2013JA019260

Northrop TG (1963) The adiabatic motion of charged particles. Interscience Publishers, Wiley, New York

O'Brien TP, McPherron RL (2000) An empirical phase analysis of ring current dynamics: Solar wind control of injection and decay. J Geophys Res 105:7707–7719. https://doi.org/10.1029/1998JA000437

Omura Y, Nunn D, Summers D (2013) Generation processes of whistler mode chorus emissions: Current status of nonlinear wave growth theory. In: Summers D, mann IR, Baker DN, Schulz M (eds) Dynamics of the Earth's radiation belts and inner magntosphere. American Geophysical Union, Washington, D.C., Geophysical Monograph, vol 199, pp 243–254. https://doi.org/10.1029/2012GM001347

Omura Y, Nakamura S, Kletzing CA, Summers D, Hikishima M (2015) Nonlinear wave growth theory of coherent hiss emissions in the plasmasphere. J Geophys Res (Space Phys) 120(9):7642–7657. https://doi.org/10.1002/2015JA021520

Osmane A, Wilson I Lynn B, Blum L, Pulkkinen TI (2016) On the connection between microbursts and nonlinear electronic structures in planetary radiation belts. Astrophys J 816(2):51. https://doi.org/10.3847/0004-637X/816/2/51

Ozeke LG, Mann IR, Murphy KR, Degeling AW, Claudepierre SG, Spence HE (2018) Explaining the apparent impenetrable barrier to ultra-relativistic electrons in the outer Van Allen belt. Nature Communications 9:1844. https://doi.org/10.1038/s41467-018-04162-3

Ozeke LG, Mann IR, Olifer L, Dufresne KY, Morley SK, Claudepierre SG, Murphy KR, Spence HE, Baker DN, Degeling AW (2020) Rapid outer radiation belt flux dropouts and fast acceleration during the March 2015 and 2013 storms: The role of ultra-low frequency wave transport from a dynamic outer boundary. J Geophys Res (Space Phys) 125(2):e27179. https://doi.org/10.1029/2019JA027179

Parker EN (1960) Geomagnetic fluctuations and the form of the outer zone of the Van Allen radiation belt. J Geophys Res 65:3117. https://doi.org/10.1029/JZ065i010p03117

Paulikas GA, Blake JB (1976) Modulation of trapped energetic electrons at 6.6 earth radii by the direction of the interplanetary magnetic field. Geophys Res Lett 3:277–280. https://doi.org/10.1029/GL003i005p00277

Pellinen RJ, Heikkila WJ (1984) Inductive electric fields in the magnetotail and their relation to auroral and substorm phenomena. Space Sci Rev 37:1–61. https://doi.org/10.1007/BF00213957

Perry KL, Hudson MK, Elkington SR (2005) Incorporating spectral characteristics of Pc5 waves into three-dimensional radiation belt modeling and the diffusion of relativistic electrons. J Geophys Res (Space Phys) 110(A3):A03215. https://doi.org/10.1029/2004JA010760

Pinto VA, Bortnik J, Moya PS, Lyons LR, Sibeck DG, Kanekal SG, Spence HE, Baker DN (2018) Characteristics, occurrence, and decay rates of remnant belts associated with three-belt events in the Earth's radiation belts. Geophys Res Lett 45:12. https://doi.org/10.1029/2018GL080274

Rae IJ, Murphy KR, Watt CEJ, Halford AJ, Mann IR, Ozeke LG, Sibeck DG, Clilverd MA, Rodger CJ, Degeling AW, Forsyth C, Singer HJ (2018) The role of localized compressional ultra-low frequency waves in energetic electron precipitation. J Geophys Res (Space Phys) 123(3):1900–1914. https://doi.org/10.1002/2017JA024674

Ratcliffe H, Watt CEJ (2017) Self-consistent formation of a 0.5 cyclotron frequency gap in magnetospheric whistler mode waves. J Geophys Res (Space Phys) 122(8):8166–8180. https://doi.org/10.1002/2017JA024399

Reeves GD, McAdams KL, Friedel RHW (2003) Acceleration adn loss of relativistic electrons during geomagnetic storms. Geophys Res Lett 30. https://doi.org/10.1029/2002GL016513

Reeves GD, Spence HE, Henderson MG, Morley SK, Friedel RHW, Funsten HO, Baker DN, Kanekal SG, Blake JB, Fennell JF, Claudepierre SG, Thorne RM, Turner DL, Kletzing CA, Kurth WS, Larsen BA, Niehof JT (2013) Electron acceleration in the heart of the Van Allen radiation belts. Science 341(6149):991–994. https://doi.org/10.1126/science.1237743

Reeves GD, Friedel RHW, Larsen BA, Skoug RM, Funsten HO, Claudepierre SG, Fennell JF, Turner DL, Denton MH, Spence HE, Blake JB, Baker DN (2016) Energy-dependent dynamics of keV to MeV electrons in the inner zone, outer zone, and slot regions. J Geophys Res (Space Phys) 121:397–412. https://doi.org/10.1002/2015JA021569

Richardson IG (2018) Solar wind stream interaction regions throughout the heliosphere. Living Rev Solar Phys 15:1. https://doi.org/10.1007/s41116-017-0011-z

Roederer JG (1970) Dynamics of geomagnetically trapped radiation. Springer, Berlin, Germany. https://doi.org/10.1007/978-3-642-49300-3

Roederer JG, Zhang H (2014) Dynamics of magnetically trapped particles. Springer, Heidelberg, Germany. https://doi.org/10.1007/978-3-642-41530-2

Russell CT, Elphic RC (1979) ISEE observations of flux transfer events at the dayside magnetopause. Geophys Res Lett 6(1):33–36. https://doi.org/10.1029/GL006i001p00033

Russell CT, Holzer RE, Smith EJ (1970) OGO 3 observations of ELF noise in the magnetosphere: 2. The nature of the equatorial noise. J Geophys Res 75(4):755. https://doi.org/10.1029/JA075i004p00755

Santolík O, Chum J, Parrot M, Gurnett DA, Pickett JS, Cornilleau-Wehrlin N (2006) Propagation of whistler mode chorus to low altitudes: Spacecraft observations of structured ELF hiss. J Geophys Res (Space Phys) 111(A10):A10208. https://doi.org/10.1029/2005JA011462

Schulz M (1996) Canonical coordinates for radiation-belt modeling. In: Lemaire JF, Heynderickx D, Baker DN (eds) Radiation belts: Models and standards. American Geophysical Union, Washington, D.C., Geophysical Monograph, vol 97, pp 153–160. https://doi.org/10.1029/GM097p0153

Schulz M, Lanzerotti LJ (1974) Particle diffusion in the radiation belts, physics and chemistry in space, vol 7. Springer, New York, NY. https://doi.org/10.1007/978-3-642-65675-0

Selesnick RS, Looper MD, Mewaldt RA (2007) A theoretical model of the inner proton radiation belt. Space Weather 5(4):S04003. https://doi.org/10.1029/2006SW000275

Seppälä A, Matthes K, Randall CE, Mironova IA (2014) What is the solar influence on climate? Overview of activities during CAWSES-II. Progress Earth Planetary Sci 1:24. https://doi.org/10.1186/s40645-014-0024-3

Seppälä A, Douma E, Rodger CJ, Verronen PT, Clilverd MA, Bortnik J (2018) Relativistic electron microburst events: Modeling the atmospheric impact. Geophys Res Lett 45(2):1141–1147. https://doi.org/10.1002/2017GL075949

Shen XC, Zong QG, Shi QQ, Tian AM, Sun WJ, Wang YF, Zhou XZ, Fu SY, Hartinger MD, Angelopoulos V (2015) Magnetospheric ULF waves with increasing amplitude related to solar wind dynamic pressure changes: The Time History of Events and Macroscale Interactions during Substorms (THEMIS) observations. J Geophys Res (Space Phys) 120:7179–7190. https://doi.org/10.1002/2014JA020913

Shen XC, Hudson MK, Jaynes AN, Shi Q, Tian A, Claudepierre SG, Qin MR, Zong QG, Sun WJ (2017) Statistical study of the storm time radiation belt evolution during Van Allen Probes era: CME- versus CIR-driven storms. J Geophys Res (Space Phys) 122:8327–8339. https://doi.org/10.1002/2017JA024100

Shprits YY (2016) Estimation of bounce resonant scattering by fast magnetosonic waves. Geophys Res Lett 43(3):998–1006. https://doi.org/10.1002/2015GL066796

Shprits YY, Kellerman A, Aseev N, Drozdov AY, Michaelis I (2017) Multi-MeV electron loss in the heart of the radiation belts. Geophys Res Lett 44(3):1204–1209. https://doi.org/10.1002/2016GL072258

Shprits YY, Horne RB, Kellerman AC, Drozdov AY (2018) The dynamics of Van Allen belts revisited. Nature Physics 14:102–103. https://doi.org/10.1038/nphys4350

Shue JH, Song P, Russell CT, Steinberg JT, Chao JK, Zastenker G, Vaisberg OL, Kokubun S, Singer HJ, Detman TR, Kawano H (1998) Magnetopause location under extreme solar wind conditions. J Geophys Res 103(A8):17,691–17,700. https://doi.org/10.1029/98JA01103

Simms LE, Engebretson MJ, Clilverd MA, Rodger CJ, Reeves GD (2018) Nonlinear and synergistic effects of ULF Pc5, VLF chorus, and EMIC waves on relativistic electron flux at geosynchronous orbit. J Geophys Res (Space Phys) 123(6):4755–4766. https://doi.org/10.1029/2017JA025003

Singer SF (1958) Trapped albedo theory of the radiation belt. Phys Rev Lett 1(5):181–183. https://doi.org/10.1103/PhysRevLett.1.181

Southwood DJ, Hughes WJ (1983) Theory of hydromagnetic waves in the magnetosphere. Space Sci Rev 35:301–366. https://doi.org/10.1007/BF00169231

Southwood DJ, Dungey JW, Etherington RJ (1969) Bounce resonant interaction between pulsations and trapped particles. Planet Space Sci 17(3):349–361. https://doi.org/10.1016/0032-0633(69)90068-3

Spanswick E, Donovan E, Baker G (2005) Pc5 modulation of high energy electron precipitation: particle interaction regions and scattering efficiency. Annales Geophysicae 23(5):1533–1542. https://doi.org/10.5194/angeo-23-1533-2005

Storey LRO (1953) An investigation of whistling atmospherics. Phil Trans Roy Soc Lond A 246(908):113–141. https://doi.org/10.1098/rsta.1953.0011

Summers D (2005) Quasi-linear diffusion coefficients for field-aligned electromagnetic waves with applications to the magnetosphere. J Geophys Res (Space Phys) 110(A8):A08213. https://doi.org/10.1029/2005JA011159

Summers D, Thorne RM (2003) Relativistic pitch-angle scattering by electromagnetic ion cyclotron waves during geomagnetic storms. J Geophys Res 108. https://doi.org/10.1029/2002JA009489

Summers D, Thorne RM, Xiao F (1998) Relativistic theory of wave-particle resonant diffusion with application to electron acceleration in the magnetosphere. J Geophys Res 103(A9):20,487–20,500. https://doi.org/10.1029/98JA01740

Summers D, Omura Y, Nakamura S, Kletzing CA (2014) Fine structure of plasmaspheric hiss. J Geophys Res (Space Phys) 119(11):9134–9149. https://doi.org/10.1002/2014JA020437

Takahashi K, Lysak R, Vellante M, Kletzing CA, Hartinger MD, Smith CW (2018) Observation and numerical simulation of cavity mode oscillations excited by an interplanetary shock. J Geophys Res (Space Phys) 123:1969–1988. https://doi.org/10.1002/2017JA024639

Tao X, Li X (2016) Theoretical bounce resonance diffusion coefficient for waves generated near the equatorial plane. Geophys Res Lett 43(14):7389–7397. https://doi.org/10.1002/2016GL070139

Thomsen MF, Denton MH, Gary SP, Liu K, Min K (2017) Ring/shell ion distributions at geosynchronous orbit. J Geophys Res (Space Phys) 122(12):12,055–12,071. https://doi.org/10.1002/2017JA024612

Thorne RM, Smith EJ, Burton RK, Holzer RE (1973) Plasmaspheric hiss. J Geophys Res 78(10):1581–1596. https://doi.org/10.1029/JA078i010p01581

Thorne RM, O'Brien TP, Shprits YY, Summers D, Horne RB (2005) Timescale for MeV electron microburst loss during geomagnetic storms. J Geophys Res (Space Phys) 110(A9):A09202. https://doi.org/10.1029/2004JA010882

Thorne RM, Li W, Ni B, Ma Q, Bortnik J, Baker DN, Spence HE, Reeves GD, Henderson MG, Kletzing CA, Kurth WS, Hospodarsky GB, Turner D, Angelopoulos V (2013a) Evolution and slow decay of an unusual narrow ring of relativistic electrons near L ~ 3.2 following the September 2012 magnetic storm. Geophys Res Lett 40:3507–3511. https://doi.org/10.1002/grl.50627

References

Thorne RM, Li W, Ni B, Ma Q, Bortnik J, Chen L, Baker DN, Spence HE, Reeves GD, Henderson MG, Kletzing CA, Kurth WS, Hospodarsky GB, Blake JB, Fennell JF, Claudepierre SG, Kanekal SG (2013b) Rapid local acceleration of relativistic radiation-belt electrons by magnetospheric chorus. Nature 504:411–414. https://doi.org/10.1038/nature12889

Treumann RA, Baumjohann W (1997) Advanced space plasma physics. Imperial College Press, London, UK. https://doi.org/10.1142/p020

Tsurutani BT, Park SA, Falkowski BJ, Bortnik J, Lakhina GS, Sen A, Pickett JS, Hajra R, Parrot M, Henri P (2019) Low frequency (f < 200 Hz) polar plasmaspheric hiss: Coherent and intense. J Geophys Res (Space Phys) 124(12):10,063–10,084. https://doi.org/10.1029/2019JA027102

Tsyganenko NA (2013) Data-based modelling of the Earth's dynamic magnetosphere: A review. Annales Geophysicae 31(10):1745–1772. https://doi.org/10.5194/angeo-31-1745-2013

Tsyganenko NA, Sitnov MI (2005) Modeling the dynamics of the inner magnetosphere during strong geomagnetic storms. J Geophys Res (Space Phys) 110(A3):A03208. https://doi.org/10.1029/2004JA010798

Turner DL, Ukhorskiy AY (2020) Outer radiation belt losses by magnetopause incursions and outward radial transport: new insights and outstanding questions from the van allen probes era. In: Jaynes AN, Usanova ME (eds) The Dynamic loss of Earth's radiation belts. Elsevier, pp 1–28. https://doi.org/10.1016/B987-0-12-813371-2-00001-9

Turner DL, Angelopoulos V, Li W, Hartinger MD, Usanova M, Mann IR, Bortnik J, Shprits Y (2013) On the storm-time evolution of relativistic electron phase space density in Earth's outer radiation belt. J Geophys Res (Space Phys) 118:2196–2212. https://doi.org/10.1002/jgra.50151

Turner DL, Claudepierre SG, Fennell JF, O'Brien TP, Blake JB, Lemon C, Gkioulidou M, Takahashi K, Reeves GD, Thaller S, Breneman A, Wygant JR, Li W, Runov A, Angelopoulos V (2015) Energetic electron injections deep into the inner magnetosphere associated with substorm activity. Geophys Res Lett 42:2079–2087. https://doi.org/10.1002/2015GL063225

Turner DL, Kilpua EKJ, Hietala H, Claudepierre SG, O'Brien TP, Fennell JF, Blake JB, Jaynes AN, Kanekal S, Baker DN, Spence HE, Ripoll JF, Reeves GD (2019) The response of Earth's electron radiation belts to geomagnetic storms: Statistics from the Van Allen probes era including effects from different storm drivers. J Geophys Res (Space Phys) 124(2):1013–1034. https://doi.org/10.1029/2018JA026066

Turner NE, Baker DN, Pulkkinen TI, McPherron RL (2000) Evaluation of the tail current contribution to Dst. J Geophys Res 105:5431–5439. https://doi.org/10.1029/1999JA000248

Tyler E, Breneman A, Cattell C, Wygant J, Thaller S, Malaspina D (2019) Statistical occurrence and distribution of high-amplitude whistler mode waves in the outer radiation belt. Geophys Res Lett 46(5):2328–2336. https://doi.org/10.1029/2019GL082292

Ukhorskiy AY, Sitnov MI (2013) Dynamics of radiation belt particles. Space Sci Rev 179(1–4):545–578. https://doi.org/10.1007/s11214-012-9938-5

Usanova ME, Mann IR (2016) Understanding the role of EMIC waves in radiation belt and ring current dynamics: Recent advances. In: Balasis G, Daglis IA, Mann IR (eds) Waves, particles and storms in geospace. Oxford University Press, Oxford, UK, pp 244–276. https://doi.org/10.1093/acprof:oso/9780198705246.001.0001

Van Allen JA (1983) Origins of magnetospheric physics. Smithsonian Institution Press, Washington, D.C., USA

Van Allen JA, Ludwig GH, Ray EC, McIlwain CE (1958) Observation of high intensity radiation by satellites 1958 alpha and gamma. Jet Propulsion 28:588–592. https://doi.org/10.2514/8.7396

Verronen PT, Rodger CJ, Clilverd MA, Wang S (2011) First evidence of mesospheric hydroxyl response to electron precipitation from the radiation belts. J Geophys Res (Atmospheres) 116(D7):D07307. https://doi.org/10.1029/2010JD014965

Wang C, Rankin R, Zong Q (2015) Fast damping of ultralow frequency waves excited by interplanetary shocks in the magnetosphere. J Geophys Res (Space Phys) 120:2438–2451. https://doi.org/10.1002/2014JA020761

Wygant J, Mozer F, Temerin M, Blake J, Maynard N, Singer H, Smiddy M (1994) Large amplitude electric and magnetic field signatures in the inner magnetosphere during injection of 15 MeV electron drift echoes. Geophys Res Lett 21:1739–1742. https://doi.org/10.1029/94GL00375

Xia Z, Chen L, Dai L, Claudepierre SG, Chan AA, Soto-Chavez AR, Reeves GD (2016) Modulation of chorus intensity by ULF waves deep in the inner magnetosphere. Geophys Res Lett 43(18):9444–9452. https://doi.org/10.1002/2016GL070280

Xiao F, Yang C, Su Z, Zhou Q, He Z, He Y, Baker DN, Spence HE, Funsten HO, Blake JB (2015) Wave-driven butterfly distribution of Van Allen belt relativistic electrons. Nature Communications 6:8590. https://doi.org/10.1038/ncomms9590

Young DT, Perraut S, Roux A, de Villedary C, Gendrin R, Korth A, Kremser G, Jones D (1981) Wave-particle interactions near Ω_{He+} observed in GEOS 1 and 2 1. Propagation of ion cyclotron waves in He$^+$ -rich plasma. J Geophys Res 86(A8):6755–6772. https://doi.org/10.1029/JA086iA08p06755

Yu J, Li LY, Cao JB, Chen L, Wang J, Yang J (2017) Propagation characteristics of plasmaspheric hiss: Van Allen Probe observations and global empirical models. J Geophys Res (Space Phys) 122(4):4156–4167. https://doi.org/10.1002/2016JA023372

Zhang D, Liu W, Li X, Sarris T, Xiao C, Wygant JR (2018) Observations of Impulsive Electric Fields Induced by Interplanetary Shock. Geophys Res Lett 45:7287–7296. https://doi.org/10.1029/2018GL078809

Zhang XJ, Mourenas D, Artemyev AV, Angelopoulos V, Sauvaud JA (2019) Precipitation of MeV and Sub-MeV Electrons Due to Combined Effects of EMIC and ULF Waves. J Geophys Res (Space Phys) 124(10):7923–7935. https://doi.org/10.1029/2019JA026566

Zhao H, Baker DN, Li X, Malaspina DM, Jaynes AN, Kanekal SG (2019a) On the acceleration mechanism of ultrarelativistic electrons in the center of the outer radiation belt: A statistical study. J Geophys Res (Space Phys) 124(11):8590–8599. https://doi.org/10.1029/2019JA027111

Zhao H, Ni B, Li X, Baker DN, Johnston WR, Zhang W, Xiang Z, Gu X, Jaynes AN, Kanekal SG, Blake JB, Claudepierre SG, Temerin MA, Funsten HO, Reeves GD, Boyd AJ (2019b) Plasmaspheric hiss waves generate a reversed energy spectrum of radiation belt electrons. Nature Physics 15(4):367–372. https://doi.org/10.1038/s41567-018-0391-6

Index

© The Author(s) 2022
H. E. J. Koskinen, E. K. J. Kilpua, *Physics of Earth's Radiation Belts*,
Astronomy and Astrophysics Library, https://doi.org/10.1007/978-3-030-82167-8

Printed in the United States
by Baker & Taylor Publisher Services